Life Chemistry & Molecular Biology

Life Chemistry & Molecular Biology

E.J. Wood
C.A. Smith
&
W.R. Pickering

Published by Portland Press Ltd, 59 Portland Place, London W1N 3AJ, U.K.

In North America orders should be sent to Princeton University Press, 41 William Street, Princeton, NJ 08540, U.S.A.

First edition 1997
Reprinted 1999

© 1997 Portland Press Ltd, London

ISBN 1 85578 064 X

British Library Cataloguing-in-Publication Data
A catalogue record for this book is available from the British Library

All rights reserved. Apart from any fair dealing for the purposes of research or private study, or criticism or review, as permitted under the Copyright, Designs and Patents Act, 1988, this publication may be reproduced, stored or transmitted, in any forms or by any means, only with the prior permission in writing of the publishers, or in the case of reprographic reproduction in accordance with the terms of licences issued by the Copyright Licensing Agency. Inquiries concerning reproduction outside those terms should be sent to the publishers at the above-mentioned address.

Although, at the time of going to press, the information contained in this publication is believed to be correct, neither the authors nor the publisher assume any responsibility for any errors or omissions herein contained. Opinions expressed in this book are those of the authors and are not necessarily held by the publishers.

All profits made by the sale of this publication are returned to The Biochemical Society for the promotion of the molecular life sciences.

Design by Angie Moyes
Typeset by Portland Press Ltd
Printed in Great Britain by Information Press Ltd

Contents

Preface	ix

Chapter 1 • Life Chemistry and the Ecosystem

1.1	Introduction	1
1.2	**Chemistry and life**	3
	■ Bonding in biological systems	4–5
	■ Energy flow through an ecosystem	8–9
1.3	**Nutrition**	10
1.4	**The biological importance of water**	14
	■ Properties of water	16–17
1.5	**Cells**	18
1.6	**Biological membranes**	19
1.7	**Intracellular organization**	21
1.8	**Prokaryotes**	21
	■ Structure and function of membranes	22–23
1.9	**Eukaryotes**	26
	■ A prokaryotic cell has no true organelles	28–29
	■ Comparison of plant and animal cell structure	30–31
1.10	**Further reading**	32
1.11	**Examination questions**	33

■ Denotes double-page Spread

Chapter 2 • Biological Molecules

2.1	Introduction	35
2.2	**The monomers and how they are joined together**	35
2.3	**The proteins — very versatile macromolecules**	36
	■ Proteins: solubility, structure, shapes	44–45
2.4	**The carbohydrates — storage and structure**	47
	■ Functions of carbohydrates	48–49
	■ Carbohydrates: monosaccharides, disaccharides and polysaccharides	52–53
2.5	**The nucleic acids**	55
	■ The nucleic acids: deoxyribonucleic acid (DNA)	56–57
	■ The nucleic acids: ribonucleic acid (RNA)	60–61
2.6	**The lipids: fats and oils**	62
	■ Structure and function of lipids	64–65
2.7	**Other biomolecules**	67
2.8	**Further reading**	67
2.9	**Examination questions**	68

Chapter 3 • Enzymes

3.1	Introduction	71
3.2	**Enzyme-catalysed reactions**	72
	■ Naming and classifying enzymes	74–75
3.3	**Catalysis**	76
3.4	**Enzymes and rate enhancement**	77
	■ Measuring enzyme activity	78–79
3.5	**The active site**	80
	■ Binding of substrate to enzyme	82–83
3.6	**Specificity of enzyme action**	84
3.7	**Cofactors, coenzymes and prosthetic groups**	85

3.8	**Effects of temperature and pH**	**86**
3.9	**Inhibition of enzyme activity**	**87**
■	Factors that affect enzyme activity: temperature and pH	88–89
3.10	**Enzymes in metabolism**	**90**
3.11	**Allosteric enzymes**	**91**
■	Effect of inhibitors on enzyme activity	92–93
3.12	**The role of enzymes in metabolism**	**94**
3.13	**Further reading**	**94**
3.14	**Examination questions**	**95**

Chapter 4 • Obtaining Energy

4.1	**Introduction**	**99**
■	Adenosine triphosphate: ATP	100–101
4.2	**Photosynthesis**	**102**
■	Chloroplast structure and photosynthesis	106–107
4.3	**ATP production in heterotrophs**	**110**
4.4	**The Krebs tricarboxylic acid cycle**	**113**
■	Mitochondria and ATP production	114–115
■	Mitchell's chemiosmotic theory	118–119
■	The Krebs tricarboxylic acid cycle	120–121
4.5	**Energy from carbohydrates**	**122**
■	Glycolysis	124–125
4.6	**Energy from fats**	**126**
4.7	**Energy from proteins**	**127**
■	The oxidation of fatty acids	128–129
4.8	**Conclusion**	**131**
4.9	**Further reading**	**131**
4.10	**Examination questions**	**132**

Chapter 5 • Using Metabolic Energy

5.1	**Introduction**	**136**
5.2	**Biosynthesis**	**137**
■	Metabolic relationships in plant cells	138–139
5.3	**The Calvin cycle**	**140**
■	The Calvin cycle	142–143
5.4	**Nitrogen fixation**	**144**
■	The nitrogen cycle	146–147
5.5	**Biosynthesis of polymers**	**148**
5.6	**Locomotion**	**148**
■	Biosynthesis of polysaccharides and fats	150–151
5.7	**Transport across biological membranes**	**154**
5.8	**Further reading**	**157**
5.9	**Examination questions**	**158**

Chapter 6 • DNA: Dealing with Information

6.1	**Introduction**	**161**
6.2	**How is DNA replicated at cell division?**	**161**
■	Replication of DNA	164–165
6.3	**The genetic code**	**166**

	■ Packaging of DNA	168–169
	■ Regulation of gene activity	172–173
	■ Mutations produce sickle cell disease	174–175
6.4	**Protein biosynthesis**	**177**
	■ Protein synthesis	180–181
6.5	**Are prokaryotic cells different?**	**182**
6.6	**Further reading**	**183**
6.7	**Examination questions**	**184**

Chapter 7 • Molecular Biology and Applied Biochemistry

7.1	**Introduction**	**189**
7.2	**The technology**	**190**
	■ DNA sequencing	194–195
7.3	**Transgenic organisms**	**198**
	■ Transgenic plants	200–201
7.4	**Genetic diseases**	**202**
	■ The Human Genome Project	204–205
	■ Gene therapy	208–209
	■ Viruses, HIV and AIDS	214–215
7.5	**Further reading**	**217**
7.6	**Examination questions**	**218**
	Index	**219**

Preface

It is possible to describe the complicated processes that go on in living cells and organisms in terms of the laws of chemistry and physics and, indeed, the science of molecular biology attempts to describe biological phenomena in terms of what is happening at the molecular level.

This book is about life chemistry and molecular biology and will help those of you in your last years at school, or first year at college, to understand some of the 'mysteries of life'. Actually, they are not really mysteries at all. As we gain more knowledge about how cell chemistry and information processing work in living systems, the better we can understand how living organisms function. Although the activities that constitute life are extremely complex, what we try to demonstrate in this book is that, because all cells operate by the same type of chemistry, if we understand a process in one cell, then we will usually be able to recognize an identical, or closely similar, process in all other cells. The more that is learned as a result of the vast amount of research going on in biology laboratories all round the world, the more this unity of life is brought home to us. For example, in practically all forms of life the genetic message is carried in long chains of DNA, proteins are always synthesized on ribosomes, and enzymes work in the same ways on a range of organic molecules in metabolism. All of biology is being affected by the explosion of knowledge which is currently taking place. Scientists working in biochemistry, genetics, microbiology and biology laboratories are all to a large extent using the techniques of biochemistry and molecular biology to find out what makes life tick. The term 'molecular and cellular life sciences' is used to embrace all of this knowledge about living organisms.

As you read this book you will begin to realize that while biochemical processes are complex in detail, they are based on very simple themes that recur again and again in life. We summarize the important features of life chemistry in an easy-to-remember '4Ms' — Macromolecules, Membranes, Metabolism and Memory. Macromolecules are the giant molecules that make life possible and they mostly exist inside cells and organelles surrounded by Membranes. The cellular chemistry that provides the energy and building blocks for carrying out life chemistry we call Metabolism, and finally the nucleic acids and especially DNAs carry the genetic information — or Memory — from generation to generation. There is a lot to learn, but it is not as complicated as you might think! At intervals throughout each chapter, you will find annotated, highly illustrated, double-page Spreads that summarize background information, give up-to-date details and important pieces of biochemical knowledge, or provide explanations of diverse and important phenomena and techniques, ranging from the AIDS virus to DNA sequencing. We hope we have included sufficient examples to show that life chemistry affects a great deal in our daily lives — from providing the energy of all life on Earth (photosynthesis) to offering the prospect of gene therapy to attempt to cure people born with serious genetic diseases.

We hope that you will enjoy finding out about life chemistry and molecular biology. The more that you learn, the more connections you will make, for example, recognizing the molecular structures and biochemical phenomena that are common to all life forms. When you come to choose your future career, we hope also that some of you will be inspired to continue the work of today's molecular life scientists, not only to add to our fund of knowledge, but to exploit these discoveries for the common good: for example, to design better drugs and vaccines, to develop new therapies, to design proteins with novel properties and to develop more productive microbes, yeasts and plants, using the power that this knowledge will give you.

Acknowledgements

We are grateful to all those who gave us permission to use their microphotographs and images and to those with whom we have discussed and debated about how best to put things. We acknowledge the permission of the various Examination Boards for allowing us to reproduce questions taken from recent Advanced level papers: UODLE material (Oxford Local Biology and O&C questions) is reproduced by permission of the University of Cambridge Local Examinations Syndicate; JMB material is reproduced by permission of the Northern Examinations and Assessment Board; AEB material is reproduced by permission of the Associated Examining Board; ULEAC material is reproduced by permission of the Edexcel Foundation, London Examinations.

We also acknowledge the following companies for allowing us to include photographs of their products: McNeil Consumer Products Company (Lactaid); Wm Morrison Supermarkets plc (Sunflower Margarine); Procter and Gamble Limited (Ariel Automatic); Repsol Petroleum Limited (Repsol Antifreeze), and Perkin Elmer Applied Biosystems Division (ABI PRISM™ 377 DNA Sequencer).

We are also very happy to record our thanks to Sarah Bell and the other staff at Portland Press for being both encouraging and extremely patient with us while this volume was being written.

1.1 Introduction

The Earth is believed to be about 4500 million years old. When it was first formed conditions would have been too harsh for any life to have existed. However, the presence of microfossils in rocks suggests that some sort of life did exist at least 3000 million years ago. These primitive organisms were cellular, were probably built up of the same sorts of compounds as we are, and could undoubtedly do many of the things that modern-day organisms can do. They must have been able to take in chemicals and energy from their surroundings and use the energy to transform these chemicals into their own sorts of compounds in order to maintain themselves and build new cells.

There are some important ideas here. The concept of cells implies that there were biological *membranes* acting as controlling barriers. The concept of transforming one chemical into another using energy is summarized by the idea of *metabolism* — building up new compounds (anabolism) and breaking down compounds (catabolism). Almost certainly these processes were controlled and catalysed by very large molecules called *macromolecules*. Biochemistry is the study of these 3Ms — membranes, metabolism and macromolecules. In addition there is another M — for *memory*, because information on how to make macromolecules is passed on from generation to generation, usually as the double helix of DNA.

Chapter 1
Life Chemistry and the Ecosystem

The origin of life

Life appeared on Earth a long time ago when the prevailing conditions were favourable. We can only guess at how this happened and whether it happened just once by chance or if it inevitably happened many times. Many experiments have been carried out to try to recreate the conditions of the Earth's early atmosphere in the laboratory. These experiments assume that the atmosphere was probably oxygen-free and consisted of a mixture of hydrogen, methane, ammonia and water vapour. There would have been energy in the form of heat, intense ultra-violet radiation and lightning discharges. In the laboratory, passing an electrical discharge through such an atmosphere results in the generation of many of the smaller types of molecule that are found in present-day living cells (Figure 1.1).

It is highly likely that, as time went on, the organisms that were evolving on the primitive Earth gradually changed the conditions in the biosphere. The most important of these changes was from a reducing atmosphere (hydrogen, methane, etc.) to one containing about 20% oxygen. This allowed cells to evolve to obtain energy in different and much more efficient ways and almost certainly paved the way for the evolution of so-called higher forms of life.

Apart from this, however, organisms have constantly been changing the Earth's geology. The sedimentary deposits of chalk are there as a result of living processes, as are sulphur deposits (from so-called sulphur bacteria). The decaying residues of millions of organisms over millions of years have produced the deposits of peat, coal and oil on which we

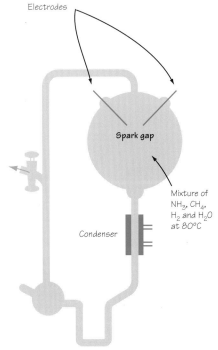

Figure 1.1 Electrical discharge apparatus.

depend for the majority of our energy supply. Living things still exert major influences on the cycles of the elements such as carbon, nitrogen and phosphorus. One particular organism, *Homo sapiens*, has the power to modify the ecosystem very dramatically — for better or for worse.

The origin of biochemistry

The study of the chemistry of living things is called *biochemistry*, or life chemistry, but sometimes instead of approaching the subject through chemistry, we approach it via biology. The scientific disciplines of *cell biology* and *molecular biology* aim to describe biology at the cellular, and then the molecular, level. Describing biological events at the molecular level is essentially the same as talking about 'the chemistry of life'.

The idea that it is possible to describe life processes in molecular terms is an important one. When knowledge of chemistry increased in the latter part of the eighteenth century, people began to ask what characteristics distinguished living matter from non-living matter. Both were made up of chemical compounds, and the Swedish chemist, Berzelius, classified all the known chemicals into two groups, *organic* and *inorganic*. Substances such as salt and water, coming from the inanimate world of air, soil and ocean, were classified as inorganic, while compounds derived directly or indirectly from living organisms were classified as organic. It was believed that only living organisms could manufacture organic chemicals. Although it was recognized that inorganic compounds obeyed a set of 'laws' governing their behaviour, it was believed that when the very same elements were present in organic compounds they obeyed quite different laws. At that time most scientists agreed that the chemistry of living organisms was distinct from that of the inanimate world, and the concept of a 'vital force' was invoked to try to explain phenomena that were encountered only in living animals or plants.

In 1828, Wöhler, one of Berzelius' students, caused a revolution in chemistry that was to have far-reaching effects. He heated an inorganic compound, ammonium cyanate, and found that urea was produced. Urea is a compound found in urine, and is clearly organic because it is produced by living organisms.

Organic chemistry

'Organic chemistry' refers to the branch of chemistry dealing with the compounds of the element carbon [except for carbon dioxide (CO_2) and the carbonates, which are regarded as inorganic]. These organic chemical compounds are the same whether they are made in living cells or manufactured in the chemical laboratory. Millions of different organic compounds are known today: the majority of them have been synthesized in laboratories rather than in cells.

$$NH_4\,O\,CN \longrightarrow O=C\begin{smallmatrix}NH_2\\NH_2\end{smallmatrix}$$

Ammonium cyanate → Urea

Although there was controversy about Wöhler's discovery, it soon became clear that there could be no doubting what had happened in his experiment. An inorganic substance had been converted into an organic one without the intervention of a living cell: the vital force theory was dead! It gradually became accepted that the activities and properties of living organisms were potentially explainable in chemical terms. This was a very fundamental realization: the physical and chemical laws applied equally to living and non-living matter.

1.2 Chemistry and life

Life on Earth is based on the element carbon. Because of its unique atomic structure, carbon can combine with other elements, especially hydrogen, oxygen, nitrogen and sulphur, but it can also combine with itself (Figure 1.2). It is this second property that gives us the enormous range of organic chemicals, and some of these compounds are themselves enormous. The proteins, the nucleic acids and the polysaccharides are compounds found in all life forms, and they are all giant molecules called *macromolecules*.

A good deal of biochemistry is concerned with these macromolecules: what their structures are, how they are built up, how they interact with each other and with smaller molecules. The word 'interact' is important here. Chemists typically talk about molecules *reacting* with one another: strong covalent bonds are broken and reformed in different ways. In life chemistry, although molecules are indeed taken apart and built up with corresponding changes in covalent bonds during metabolism, many vital life processes occur because weak bonds can break and reform in different ways. This idea is important because, as we shall see, it provides the basis for forming cells and membranes, for 'matching up' structures so that they 'recognize' each other, and for a multitude of other processes that are absolutely vital to life chemistry.

The molecules of life

There are about 100 elements in the Earth's crust, in the waters that cover much of the earth's surface, and in the atmosphere, and yet only about 16 of these are essential for life, and only four of them — carbon, hydrogen, oxygen and nitrogen — make up 95% of all living matter.

We may ask why these few elements were 'chosen' to form the molecules of life. Certainly the proportions of these elements in organisms does not match their abundance in the biosphere. The elements silicon and aluminium, for example, are highly abundant but play little part in life chemistry. These, and other elements, are not easily available; for example they do not easily leach out of the rocks. But probably the most important factor is carbon's ability to form bonds with itself and with the small group of elements mentioned above, namely H, O and N. Carbon is unique in this property (see Spread 1.1).

When we look at the chemical composition of living things we find that not only are they composed of the same elements — the four mentioned above, plus phosphorus, sulphur and a small range of metallic elements — but also that they contain the same types of compounds. Comparing a bacterium with a plant cell and with a human cell, we find that all are made up of about 70% water, and that all contain some sugars, some fats, some amino acids, and the three types of macromolecule mentioned above: proteins, nucleic acids and polysaccharides (Table 1.1). True, humans have human-type proteins and bacteria have bacteria-type proteins, but the basic molecular design of these is identical. Proteins are proteins, and if we understand their structures and how they function in cells, we can understand life processes in all organisms (see section 2.3).

Figure 1.2 Organic molecules.

(a) Carbon atoms can form chains of various lengths:

Alanine, an amino acid

Part of a fatty acid chain

(b) Carbon atoms may form covalent bonds with other elements:

Methionine, an amino acid

(c) Carbon atoms can form branched chains:

Valine, an amino acid

(d) Carbon and other atoms can join to form ring structures:

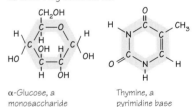

α-Glucose, a monosaccharide

Thymine, a pyrimidine base

(e) Rings and chains may be joined, and multiple ring structures may be formed:

Histidine, an amino acid

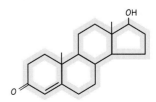

Testosterone, a steroid hormone

Bonding in biological systems

objectives
- To describe the various types of bonds that occur within biological molecules
- To appreciate the significance of these bonds in different biological systems

COVALENT BONDS are the **strongest** bonds occurring in biological molecules, in that they require the greatest energy input if they are to be broken. Covalent bonds provide great stability, and the ability of carbon to form up to four of such bonds, with other carbon atoms or with atoms of other elements, contributes greatly to the enormous range of organic molecules.

VAN DER WAALS INTERACTIONS may be positive (attractions) or negative (repulsions). The interactions occur because of random, short-lived inequalities in the charge distribution between the two atoms which share a covalent bond. Van der Waals interactions are responsible for:

- the optimum packing of molecules around one another, or of atoms within molecules
- some interactions between enzyme and substrate, and between enzyme and product.

HYDROGEN BONDS are electrostatic attractions between oppositely charged regions of neighbouring polar molecules. These small 'charges' result from the differing electronegativities of two atoms linked by a covalent bond. Hydrogen bonds are particularly important in giving water its unique properties (see page 14).

IONIC or ELECTROSTATIC BONDS involve the transfer of electrons between ions. Such bonds in solids can be very strong, but they are weakened by the presence of other charged particles. These other particles may form a neutralizing 'coat' over the surface of the ions, reducing the attraction between them. Water molecules (which have both positive and negative regions) are very effective in this respect so that ions can easily separate in aqueous solution.

Some biological systems take advantage of this variation in strength; for example, many enzymes bind substrate and release product molecules by forming and breaking relatively weak ionic bonds. Some ionic bonds form in locations which exclude water, setting up strong and stable ionic interactions; for example, metallic ions are tightly bound to the active sites of some enzymes (e.g. catalase) and in haemoglobin.

HYDROPHOBIC INTERACTIONS occur when hydrophobic (i.e. 'water-hating') molecules are mixed with water. Examples of these interactions that are important in biology include:

- those that stabilize the phospholipid bilayers in biological membranes
- those that help to maintain 'water-free' regions in large molecules, thus maintaining ionic bond strength
- those that limit the solubility of fibrous proteins, such as collagen and keratin, in water.

The significance of weak interactions is that they can be broken and reassembled very easily. This is most important in many biochemical reactions in which molecular flexibility plays a part. For example, the unwinding and reassembly of the DNA molecule during replication and transcription depends on the individual weakness but collective strength of hydrogen bonds.

A single COVALENT BOND is formed between two atoms when each atom contributes one electron to a SHARED ELECTRON PAIR, e.g.

H:H or H—H represent a covalent bond between two hydrogen atoms in a hydrogen molecule.

Covalent bonds may also be formed by the sharing of two or three electron pairs, thus

O::O or O=O   and   N⋮⋮N or N≡N

Some important covalent bonds in biological systems are:

the GLYCOSIDIC BOND between two glucose units in a starch molecule,

the PEPTIDE BOND between adjacent amino acids in a polypeptide chain,

the 3', 5' PHOSPHODIESTER BRIDGE between adjacent pentose molecules in the nucleotide units of nucleic acids,

the DISULPHIDE LINK joins the side chains of cysteine residues from different regions of a polypeptide chain and plays a part in the maintenance of protein conformation.

Cys—S—S—Cys

HYDROPHOBIC INTERACTIONS occur because hydrophobic molecules suspended in water disrupt the hydrogen-bonded structure of water. In order to get to this state, the hydrophobic molecules must be dispersed in the water. This requires breaking and rearranging of hydrogen bonds, a process which consumes energy.

Hydrophobic (non-polar) molecules

Water molecule

Hydrophobic molecules therefore tend to come together since in this way they minimize disruption to water's hydrogen-bonded structure; this is energetically more favourable.

Life chemistry and molecular biology

1.1 Bonding in biological systems

HYDROGEN BONDS are electrostatic attractions between oppositely charged regions of neighbouring polar molecules:

Oxygen is more electronegative than carbon and so the shared electrons spend more of their time located close to the oxygen atom.

Hydrogen is less electronegative than nitrogen and so the shared electrons spend less of their time located close to the hydrogen atom.

Important examples of hydrogen bonds in biological systems include:
- those joining the peptide links between amino acids on adjacent turns of an α-helix.

- those between purine and pyrimidine bases which stabilize the double helical form of some nucleic acids.

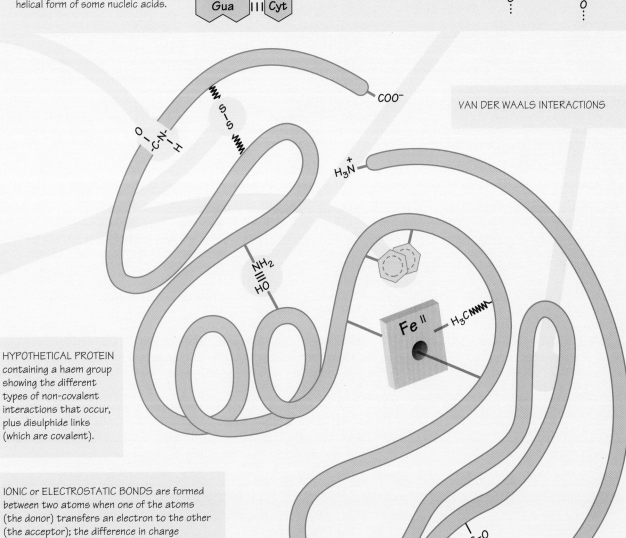

VAN DER WAALS INTERACTIONS

HYPOTHETICAL PROTEIN containing a haem group showing the different types of non-covalent interactions that occur, plus disulphide links (which are covalent).

IONIC or ELECTROSTATIC BONDS are formed between two atoms when one of the atoms (the donor) transfers an electron to the other (the acceptor); the difference in charge between the two atoms now sets up a force of attraction between them. Thus:

$$Na + Cl \xrightarrow{\text{Electron transfer}} Na^+ Cl^-$$

Chapter 1 • **Life Chemistry and the Ecosystem**

Table 1.1
Substances found in cells and the building blocks of which these are composed

Substance	Percentage of total weight	Role	Constituents
Water	70		
MACROMOLECULES			
Proteins	15	Enzymes, structural, antibodies, etc.	20 types of amino acid
Polysaccharides	3	Storage of energy, structural	10–20 types of sugar
Nucleic acids	7	Storage and transmission of information	Four nitrogen-containing bases and sugar–phosphate backbone
SMALL MOLECULES			
Lipids	2	Storage of energy, insulation, structural	Fats, fatty acids, phospholipids
Metabolic intermediates	1–2	Stages in the breakdown or build-up of other molecules	Thousands of different compounds
Cofactors	<1	Participants in enzyme reactions	Vitamins, and many other substances
INORGANIC IONS AND TRACE ELEMENTS			
	1	Many roles, including cofactors	Na^+, K^+, Ca^{2+}, Cl^-, PO_4^{3-}, Fe, Cu, Zn, Mg, Co, Mn.

The unity of life

When we investigate how the molecules of life are built up (*anabolism*, or more commonly *biosynthesis*) and broken down (*catabolism*, or sometimes *degradative metabolism*), we find that the processes and mechanisms used in different types of organism are more or less identical. The process by which proteins are biosynthesized in bacterial cells is essentially the same as that by which proteins are manufactured in human cells. There are minor differences, of course, but then bacteria and humans diverged a long way back in the course of evolution.

What we can see, therefore, is that there is a *unity of life*: the chemical composition of cells and the metabolic processes that go on in them are all broadly similar despite the enormous range of life forms on the Earth (Figure 1.3). Although the zoologist, the botanist and the microbiologist see vastly different forms, at the biochemical level all organisms operate by the same type of chemistry. Biochemistry provides a unifying basis to the study of Biology. Although organisms evolved to produce the multitude of different life forms that presently exist on the Earth, their basic chemistry did not change very much at all. This unity of life has a number of very significant consequences. Three of them will be mentioned here to illustrate the unity, although there are many more.

The first is that the underlying unity of life makes biochemistry much easier to understand than would otherwise be the case. For example, a molecule found in one type of cell is likely to perform the same functions and be involved in the same interactions in any other cell in which it might be present. Proteins are always formed from the same 20 amino acids, nucleic acids are always made up of the same four/five nucleotide units, and membranes are always made up of lipids and proteins. Organisms do not operate by mutually alien chemistries on the Earth (Figure 1.3). (If we ever encounter living organisms from outer space, this may not be the case.)

The second is that organisms can eat each other. If humans eat plant or animal tissues ('vegetables' or 'meat' to put it in culinary terms) they obtain proteins, fats, nucleic acids, sugars, etc. They can very easily rearrange the component parts of these to make more of their own type of cellular material. The plant proteins are broken down (metabolized) to amino acids which are then reassembled into human-type proteins. (This would not be possible with alien, silicon-based creatures from the planet Zorg.)

The third is really the other side of the same coin, and this is that practically all organisms can be parasitized. When a virus invades your cells it finds a range of compounds and processes that can be manipulated for its own ends. When a bacterium or a tapeworm resides in your intestine, it can absorb nutrients from your body and your food. These nutrients are the chemicals that both you and the parasite need and can use in metabolism.

All organisms are related by chemistry, and many biological phenomena — such as nutrition and parasitism — have a very clear biochemical basis or explanation. The next section deals with nutrition.

Figure 1.3 The elephant (a) and the small weed Arabidopsis thaliana (b) both operate by the same type of chemistry.

Energy flow through an ecosystem

objectives
- To explain how solar energy drives virtually all the ecosystems on Earth
- To describe how photosynthesis converts light energy into chemical bond energy in a form in which it is available to living organisms

ENERGY TRANSFER FROM PRODUCER TO PRIMARY CONSUMER is typically of the order of 5–10% of net primary production (NPP). This is because:
1. much of plant biomass (NPP) is indigestible to herbivores – there are no animal enzymes to digest lignin and cellulose;
2. much of the plant biomass may not be consumed by any individual herbivore species – roots may be inaccessible, or trampled grass may be considered inedible.

In many Indian and Pakistani villages, cow and buffalo dung is used for fuel. The material is dried on the sides of the houses before being burnt to supply heat for cooking. In this case the organic material in the dung is immediately re-oxidized to produce atmospheric CO_2. However, valuable nitrogenous materials that might act as fertilizer are lost when this is done.

At the Equator, the SOLAR FLUX (sunlight energy which reaches the Earth's upper atmosphere) is almost constant at 1.4 $kJm^{-2}s^{-1}$. Most of this incoming sunlight energy is reflected by the atmosphere, used to heat the atmosphere and Earth's surface or causes the evaporation of water. Less than 0.1% actually falls on leaves and is thus available for photosynthesis.

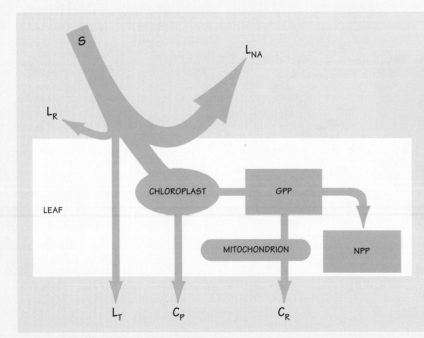

ENERGY BUDGET OF A PRIMARY PRODUCER

- S = solar energy available at leaf surface
- L_R = energy lost by reflection from leaf
- L_T = energy lost by transmission through leaf
- L_{NA} = energy lost since not of correct wavelength for absorption by photosynthetic pigments
- C_P = energy consumed by the reactions of photosynthesis
- C_R = energy consumed by respiration
- GPP = gross primary production
   = total energy fixed by photosynthesis
- NPP = net primary production
   = total energy fixed as biomass and available for heterotrophs

Life chemistry and molecular biology

1.2 Energy flow through an ecosystem

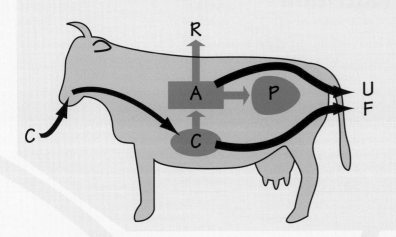

ENERGY BUDGET OF A PRIMARY CONSUMER

C = energy consumed; N.B. this does NOT equal NPP of producer since feeding is inefficient
A = energy assimilated
R = energy consumed by respiration and eventually lost as heat
U = energy lost in urine
F = energy lost in faeces
P = PRODUCTION OF BIOMASS, i.e. energy available for secondary consumer

thus $C = P + R + U + F$

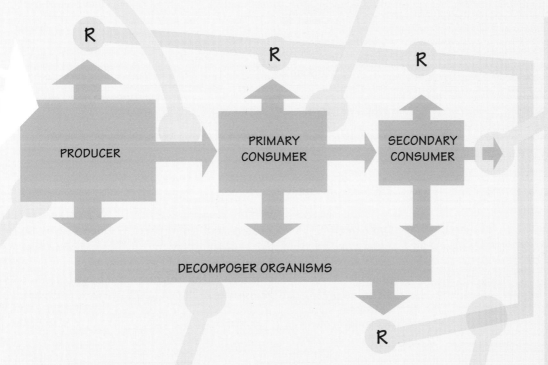

THE LIMIT TO THE NUMBER OF TROPHIC LEVELS is determined by:
1. the total producer biomass;
2. the efficiency of energy transfer between trophic levels.

In practice, the energy losses limit the number of levels to three or four, very rarely five or six. The longest food chains can only be supported by an enormous producer biomass, e.g. a six-level chain will only have about 10% x 10% x 10% x 10% of NPP available to the top carnivores. Only the enormous volume of the oceans can provide sufficient biomass to support the longest food chains.

The DECOMPOSERS are fungi and bacteria which obtain energy and raw materials from animal and plant remains. In some situations 80% or more of the productivity at any trophic level may go through a decomposer pathway (e.g. forest floors of tropical forest). In some ecosystems — peat bogs for example — the cold, wet, acidic and anaerobic conditions inhibit decomposition to such an extent that only about 10% of the material entering the decomposer food chain is broken down and the remainder accumulates as peat.

RESPIRATORY ENERGY LOSSES: a proportion of the energy consumed by organisms at all trophic levels is used in respiration, i.e. the organic compounds are oxidized to release energy, which is then used to drive metabolic reactions, e.g.

- active transport of ions
- synthesis of other compounds such as proteins and fats
- cell division
- muscle contraction

None of these processes is 100% efficient so that all respiratory energy is EVENTUALLY LOST AS HEAT.

Chapter 1 • **Life Chemistry and the Ecosystem**

1.3 Nutrition

Living organisms show a number of characteristic life activities, summarized in Table 1.2. At least some of these, such as irritability and locomotion, clearly require energy, and obviously growth, reproduction and the formation of new individuals also need an input of both raw materials and energy. Perhaps less obviously, simply maintaining an ordered state amongst the chaos of the surroundings incurs an energy expenditure. If you do not maintain and look after your property it gradually decays or returns to chaos. Mammals and birds also spend energy on maintaining their body temperature above that of their surroundings.

Table 1.2
The characteristic properties of life

Property	Comments
Nutrition	Ingestion of raw materials and energy
Respiration	Concerned with oxidative processes to produce energy
Irritability (sensitivity)	Response to changes in the environment
Locomotion	Movement of organism towards e.g. food, energy sources
Excretion	Elimination of waste generated by metabolism
Reproduction	Replication of genes and formation of new individuals

These maintenance, growth and energy requirements may be summarized by saying that in order to stay alive an organism must take raw materials from its environment and reorganize these into the molecules it requires for maintaining its structure or for growth. It must obtain energy to drive these processes. It must also have ways of dealing with waste materials it does not want. These may potentially be toxic and it will probably cost energy to get rid of them.

Not all organisms have the same nutritional requirements

The whole process of getting hold of raw materials and energy is called *nutrition*, but not all organisms have the same nutritional requirements (Figure 1.4). At one extreme some organisms — those containing the green pigment, chlorophyll — use very simple raw materials (CO_2, H_2O, NH_3) and the energy in sunlight. At the other extreme, many organisms including humans, have very complicated nutritional requirements, and they obtain the energy they need by breaking down food materials (Spread 1.2).

During the history of living organisms, different ways of obtaining energy and raw material have evolved. The very first organisms would not have had a very wide range of substances to choose from — some of them at least must have been able to synthesize all of their requirements

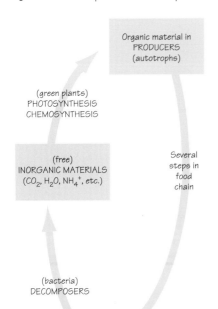

Figure 1.4 Autotrophs and heterotrophs.

from simple inorganic molecules such as carbon dioxide, ammonium salts, water and a few mineral ions. Organisms able to manufacture all of their organic requirements from simple inorganic precursors still exist and are called *autotrophs*; the energy required to drive this synthesis may come from sunlight (the organisms are then called *photoautotrophs*) or from the oxidation of certain chemicals (these organisms are *chemo-autotrophs*); see Figure 1.4.

Other organisms have evolved which are not able to use simple precursors in this way. Such organisms must obtain their complex organic molecules in a ready-made form (although they must be able to reorganize some of these molecules chemically, since the molecules may be too large to be absorbed unaltered or may not be in a form ideal for their function). These organisms are *heterotrophs* ('other feeders'). Although humans and all other animals are heterotrophic, heterotrophs are not necessarily complicated multicellular organisms. Fungi and many bacteria, for example those which live as parasites and cause disease, are also heterotrophic.

The relationships between the different types of organisms and their methods of nutrition can be expressed as a food chain which represents the flow of energy through an ecosystem. Figures 1.4 and 1.5 illustrate the facts that:

- the base of every food chain is occupied by autotrophs; these are the *producers* of the ecosystem, and
- the heterotrophs include the *decomposers,* which break down animal and plant remains and make the materials available once more to other organisms.

The pathway of nutrients between autotrophs and heterotrophs and back again constitutes a *biogeochemical cycle*. Elements move through such cycles, and although the pathway for each element is different, the basic principle is common to all of them: the element moves between simple and complex forms.

Figure. 1.6 illustrates these principles for carbon. Atmospheric carbon dioxide (CO_2) represents the most accessible source of carbon: although only about 0.3% of the atmosphere is CO_2, this represents about 700×10^{12} kg. Respiration, most commonly using glucose, e.g.:

$$\text{Glucose} + 6O_2 \longrightarrow 6CO_2 + 6H_2O$$

returns about 210×10^{12} kg of carbon to the atmosphere. This respiratory effort is split approximately equally between terrestrial and aquatic ecosystems and, within the former, nearly equally between plants and animals on the one hand and fungi and bacteria on the other.

Carbon fixation involves the reduction of carbon dioxide to produce organic molecules. Simply:

$$CO_2 + 4[H] \longrightarrow CH_2O + H_2O$$

It is carried out by photosynthesis (99%) and by chemosynthesis (1%). The amount of carbon fixed into organic molecules is considerable:

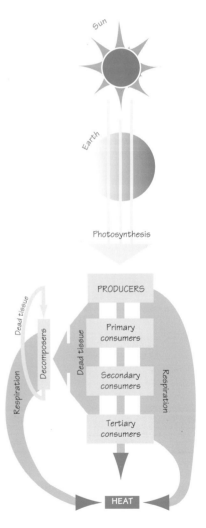

Figure 1.5 Energy flows from sunlight via producers, consumers and decomposers to heat.

Figure 1.6 *The carbon cycle depends upon both biochemical and physical processes.*

animals and plants (150); long-lived plants such as trees (700); soil organisms (1500); litter, i.e. animal and plant remains (60); peat (~150) [all figures $\times 10^{12}$ kg]. Note that less than 1% of organic carbon in living organisms is in the marine ecosystem.

Fossil fuels are formed under conditions which do not permit oxidation/decomposition of organic matter. For example, deep oceans and waterlogged soils may be anaerobic, and increasing acidity (fall in pH) inhibits decomposers. As the pressure increases, plant debris turns into peat which may turn into coal and eventually anthracite. Alternatively, peat may give rise to methane and gases, which subsequently form oil. Fossil fuels may contain $3,000,000 \times 10^{12}$ kg of carbon, which can only be returned to the atmosphere by combustion.

Calcareous rocks such as limestone and chalk are formed from the skeletons of microscopic organisms which combined calcium with CO_2. Limestone and chalk contain the greatest carbon reservoir — $100,000,000 \times 10^{12}$ kg. This is released extremely slowly.

Weathering and precipitation account for a very small proportion of the turnover of carbon (considerably less than 0.1% of the respiration/photosynthesis turnover). The average turnover time for a carbon atom in rock is about 100 million years!

The *nitrogen cycle* is illustrated in Figure 1.7. All living organisms require nitrogen because it is a component of amino acids, proteins, nucleic acids and many other compounds of life. While there is an abundance of nitrogen in the atmosphere (approximately 78% of air is nitrogen), it is first necessary to turn this into usable compounds by *nitrogen fixation*, a process which involves the reduction of nitrogen to ammonium ions. Some nitrogen fixation occurs in the atmosphere as the result of electrical energy such as lightning. In this case, high-energy discharge may combine nitrogen and oxygen,

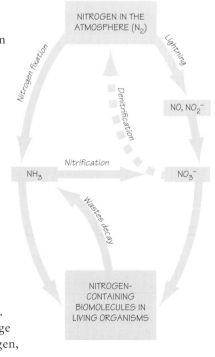

Figure 1.7 *The nitrogen cycle depends on micro-organisms.*

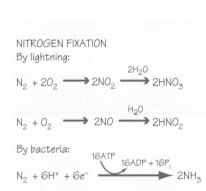

NITROGEN FIXATION
By lightning:

$$N_2 + 2O_2 \longrightarrow 2NO_2 \xrightarrow{2H_2O} 2HNO_3$$

$$N_2 + O_2 \longrightarrow 2NO \xrightarrow{H_2O} 2HNO_2$$

By bacteria:

$$N_2 + 6H^+ + 6e^- \xrightarrow[16ADP + 16P_i]{16ATP} 2NH_3$$

and the products may dissolve in rainwater and fall to the ground as weakly acidic solutions. However, the majority of nitrogen fixation takes place in the rhizosphere (soil atmosphere) in the presence of nitrogen-fixing bacteria. The basic reaction is a multi-stage process catalysed by the enzyme complex called *nitrogenase*, and requires both iron and molybdenum for enzyme function. There are some free-living nitrogen-fixing bacteria, such as *Azotobacter*, but by far the greatest proportion of nitrogen fixation is carried out by symbiotic bacteria such as *Rhizobium* and cyanobacteria such as *Nostoc* (see Spread 5.3, page 146).

Virtually all ammonia reaching the soil is oxidized to *nitrate* by soil bacteria. This is the form of nitrogen that is most easily absorbed by plant roots. The series of reactions is known as *nitrification*, and it releases energy for the bacteria involved.

Plants and many bacteria can readily reduce nitrate to ammonia by a reaction that takes place in two stages, each catalysed by a different enzyme and occurring in different parts of the cell. *Nitrate reductase* requires molybdenum for activity and is abundant in fast-growing crops but much less abundant in acidic, nitrate-deficient soils. *Nitrite reductase* is about 10 times more abundant than nitrate reductase so that nitrite never accumulates (luckily for human consumers as nitrite may be converted into carcinogens in the gut).

The ammonia so formed is incorporated in plants by the synthesis of the amino acid glutamate from the oxo acid α-oxoglutarate. Glutamate may then take part in many transamination reactions to produce any of the other 19 amino acids. ('Oxo acids' were formerly referred to as 'α-keto acids', i.e. they have a ketone group next to a carboxyl group. You will find both nomenclature used in different books.)

When organisms die, the microbial degradation of their proteins returns ammonia to the soil, where nitrifying bacteria convert it into nitrite and nitrate again. To maintain a balance between fixed nitrogen and atmospheric nitrogen, bacteria will convert nitrate to nitrogen, by a process known as *denitrification*, under anaerobic conditions, such as might occur in a waterlogged soil.

Putrefaction releases nitrogen from combined organic forms. For example:

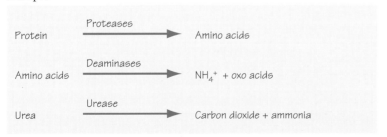

These processes are carried out by saprophytic bacteria and fungi (decomposers), which are sometimes called ammonifiers since NH_4^+ is the end product.

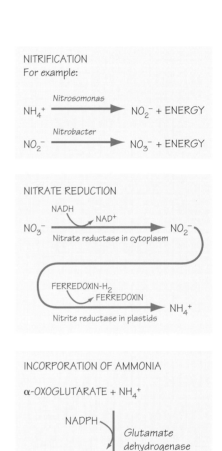

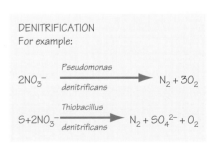

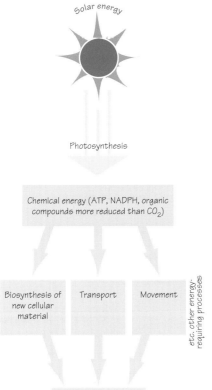

Figure 1.8 Flow of energy in the biosphere.

Each of the cycles described above also emphasizes the significance of micro-organisms, both heterotrophic and autotrophic, in the cycling of elements. No such cycle exists for energy — photosynthesis traps light energy and converts it into a chemical form (organic molecule) from which it can be released later by oxidation. Eventually all of this energy will be lost as heat so that there must be a continuous replenishment from the sun. The flow of energy through the ecosystem is illustrated in Figure 1.8.

Before we go on to how life processes operate using macromolecules and membranes, we need to understand the importance of an apparently very simple compound, water. This will provide the basis for an understanding of many life processes.

1.4 The biological importance of water

Life on earth began in water, and today, almost wherever water is found, life is present. Conversely, life is impossible in the absence of water: practically all cellular reactions and interactions go on in a watery (aqueous) environment. Where water is scarce, organisms have evolved elaborate ways of retaining it. If they dry out, they must either die or go into suspended animation. Three-quarters of the Earth's surface is covered by water, and in fact if all the Earth's surface were smooth, it would be 1–2 miles under water. Thus water is abundant on Earth and most living organisms have water making up about 70% of their mass.

Although water is an apparently rather simple molecule, in fact, compared with other liquids, it is an exceptional and extraordinary molecule. Most of these special properties of water have their origin in the fact that water molecules have a tendency to form *intermolecular hydrogen bonds*. This gives water a number of very important physical properties and makes life as we know it possible.

Hydrogen bonds

Hydrogen bonds form between water molecules because the oxygen atom in H_2O tends to be a little more attractive to electrons than the two hydrogen atoms. Consequently the oxygen atom tends to be slightly negative in charge (because electrons carry a negative charge), and the hydrogen atoms tend to be slightly positive in charge. Since unlike charges attract one another, hydrogen atoms in a group of water molecules tend to be attracted to oxygen atoms to form so-called hydrogen bonds. Although hydrogen bonds are individually weak, in a body of water there will be millions of them. This makes water much more cohesive than apparently similar molecules such as H_2S and NH_3, and gives rise to the physical properties of water explained in Spread 1.3.

Hydrophilic and hydrophobic

One other thing is important here. Individual hydrogen bonds are weak and they tend to be breaking and re-forming all the time. In a body of water, a large proportion of the water molecules will be connected by hydrogen bonds at any one instant. An instant later, these will have changed to another configuration. Thus the 'structure' of water is both very complicated and very fluid. The higher the temperature, the more this will happen. A consequence of this is that other molecules capable of forming hydrogen bonds — and there are many such molecules — can interact with water by forming their own hydrogen bonds with the water molecules. We call this *dissolving*.

In contrast, it is a familiar experience that many substances will not dissolve in water. Examples are the oils and fats, substances that biochemists call lipids, as well as a vast range of organic molecules such as benzene and toluene, and hydrocarbons such as propane and butane. These substances are incapable of forming hydrogen bonds and therefore to get them dispersed into a body of water takes some effort; it does not happen spontaneously. You can test this by shaking together olive oil and vinegar as you would to make French dressing (Figure 1.9). If you put in some energy by shaking you can achieve dispersal, but the emulsion formed soon separates out on standing. Physical chemists say that in order to create gaps in the hydrogen-bonded water structure, energy has to be supplied.

These simple observations are summed up in a rather fundamental classification. Some compounds are said to be water-loving or *hydrophilic*; such compounds would include sugar, ethanol and common salt. They form hydrogen bonds with water and tend to dissolve. Other compounds are said to be water-hating or *hydrophobic*; such compounds would include olive oil and benzene. This simple classification is self-evident when we observe a fat droplet inside a living cell. However, it has a much more fundamental importance than this simple division of properties. When we deal with biological molecules we find that it is perfectly possible for some parts of the molecule to be hydrophilic and other parts to be hydrophobic. This makes molecules orientate themselves with respect to each other: the hydrophilic regions attract each other and the hydrophobic regions tend to come together. When molecules come together in certain orientations like this, complicated biological structures such as membranes can be formed and biological macromolecules can 'recognize' each other. These interactions are vital to life processes.

Figure 1.9 Oil and water do not mix. **(a)** A bottle containing water (lower layer) and oil (upper layer, stained with an oil-soluble dye). If the bottle is shaken vigorously the two fluids mix **(b)**, but on standing for a few minutes they separate again **(c)**.

Properties of water

objectives
- To explain that water (H_2O) tends to form hydrogen bonds with neighbouring molecules
- To list the properties of water that are vital for life
- To describe the ionization of water and to be able to explain the pH scale

Because hydrogen and oxygen atoms are different in SIZE and ELECTRONEGATIVITY, the water molecule (H_2O) is NON-LINEAR and POLAR.

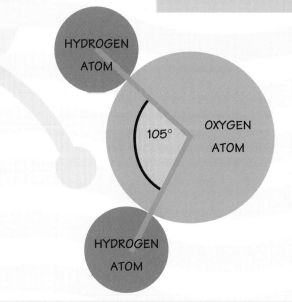

SOLVENT PROPERTIES

The polarity of water makes it an excellent solvent for other polar molecules …

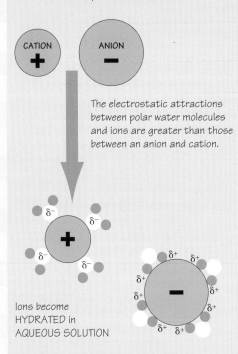

The electrostatic attractions between polar water molecules and ions are greater than those between an anion and cation.

Ions become HYDRATED in AQUEOUS SOLUTION

Such polar substances which dissolve in water are said to be HYDROPHILIC ('water-loving')

… but means that non-polar (HYDROPHOBIC or 'water-hating') substances do not readily dissolve in water.

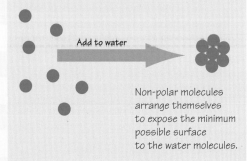

Non-polar molecules arrange themselves to expose the minimum possible surface to the water molecules.

DISSOCIATION

Water dissociates to a very slight extent:

$$H_2O \rightleftharpoons H^+ + OH^-$$

In pure water the hydrogen ion concentration [H^+] is 0.0000001 mol dm^{-3} (that is, 1 in every 10^7 water molecules is dissociated into H^+ and OH^-). This is an awkward number to handle and is more conventionally expressed in a logarithmic form, i.e. in pure water:

$$[H^+] = 10^{-7} \text{ mol} \cdot dm^{-3}$$

This is more commonly written as:

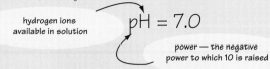

pH = 7.0 — hydrogen ions available in solution; power — the negative power to which 10 is raised

The pH range is from 0 to 14, because in 'neutral' water at pH 7.0, [H^+] = [OH^-] = 10^{-7} mol·dm^{-3}. The product of the concentrations of these two ions is thus 10^{-14} and, by the Law of Chemical Equilibrium, as the concentration of one of them increases the other must decrease to maintain this product at 10^{-14}.

Life chemistry and molecular biology

1.3 Properties of water

HYDROGEN ION CONCENTRATIONS IN AQUEOUS SOLUTIONS ARE MEASURED ON THE pH SCALE

	$[H^+]$	pH	Examples
Log ↑	10	0	1 mol·dm^{-3} HCL
	10^{-1}	1	gastric juice (pH 1.0–3.0), lemon juice
	10^{-2}	2	
Increasingly acidic	10^{-3}	3	vinegar, orange juice
	10^{-4}	4	tomato juice
	10^{-5}	5	rain water
	10^{-6}	6	urine (5–7), milk (6.5), saliva (6.5–7.4)
Neutral	10^{-7}	7	pure water, blood (7.3–7.5)
	10^{-8}	8	duodenal contents (7.4–8.0), seawater (7.8–8.3)
	10^{-9}	9	phosphate detergents
	10^{-10}	10	soap solutions, Milk of Magnesia
Increasingly basic	10^{-11}	11	non-phosphate detergents
	10^{-12}	12	washing soda (Na_2CO_3)
	10^{-13}	13	
↓	10^{-14}	14	

Remember that the pH scale is logarithmic, so a change in pH of 1 corresponds to a change in hydrogen ion concentration of 10 mol·dm^{-3}. Thus, gastric juice is 10x more concentrated in $[H^+]$ than lemon juice and 100x more concentrated than a soft drink.

HIGH SPECIFIC HEAT CAPACITY

The specific heat capacity of water (the amount of heat, measured in joules, required to raise 1 kg of water through 1°C) is very high; much of the heat absorbed is used to break the hydrogen bonds which hold the water molecules together.

HYDROGEN BOND

One water molecule may form hydrogen bonds with up to four other water molecules.

HIGH LATENT HEAT OF VAPOURIZATION

Hydrogen bonds attract molecules of liquid water to one another and make it difficult for the molecules to escape as vapour; thus a relatively high energy input is necessary to vapourize water, and water has a much higher boiling point than other molecules of the same type, e.g. H_2S.

MOLECULAR MOBILITY

The weakness of individual hydrogen bonds means that individual water molecules continually jostle one another when in the liquid phase.

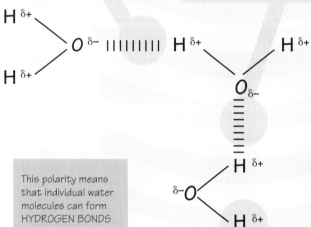

This polarity means that individual water molecules can form HYDROGEN BONDS with other water molecules ... although these individual hydrogen bonds are weak, collectively they make WATER A MUCH MORE STABLE SUBSTANCE THAN WOULD OTHERWISE BE THE CASE.

COHESION AND SURFACE TENSION

Hydrogen bonding causes water molecules to 'stick together', and also to other molecules – the phenomenon of cohesion. At the surface of a liquid the inwardly acting cohesive forces produce a 'surface tension' as the molecules are particularly attracted to one another.

COLLOID FORMATION

Some molecules have strong intramolecular forces which prevent the solution forming in water, but have charged surfaces which attract a covering of water molecules. This covering ensures that the molecules remain dispersed throughout the water, rather than forming large aggregates which could settle out. The dispersed particles and the liquid around them collectively form a COLLOID.

DENSITY AND FREEZING PROPERTIES

As water cools towards its freezing point the individual molecules slow down sufficiently for each one to form its maximum number of hydrogen bonds. To do this the water molecules in liquid water must move further apart to give enough space for all four hydrogen bonds to fit into. As a result water expands as it freezes, so that ice is less dense than liquid water and therefore floats upon its surface.

Chapter 1 • **Life Chemistry and the Ecosystem**

A definition of a cell might be as follows:

A unit of biological activity, delimited by a differentially permeable membrane, and capable of self-reproduction in a medium free of other living systems.

1.5 Cells

The fundamental unit of life is the cell: only cells can carry out the various life activities mentioned above (Table 1.2). The existence of cells, and of organisms as collections of cells, has been known since the microscope was invented, and the *cell theory* is a most significant concept in Biology. Schleiden and Schwann, in about 1837, proposed that all plants and all animals are made of one or more similar units called cells (Figure 1.10). (Schleiden was a botanist and Schwann a physiologist, bringing the plant and animal kingdoms together.) Soon after, in 1855, the pathologist Virchow was able to state: *omnis cellula e cellula* — "every cell arises from a cell". Table 1.3 summarizes our current interpretation of cells and their importance.

Table 1.3
Cells and their importance

All organisms are composed of cells which are distinctly organized and are separated from the environment or other cells by a selectively permeable membrane.

Cells arise only from other cells: growth of an organism is achieved by an increase in cell number by cell division.

Within an organism cells may be differentiated to form distinct cell types having characteristic features specific to a tissue, e.g. nerve cells, muscle cells, skin cells.

The majority of the metabolic processes that go on in living organisms take place within cells.

Cells contain the hereditary material of the organisms, usually in a nucleus, although in some cell types (e.g. red blood cell) this may be lost during development. Prokaryotic cells have no nucleus, although they do have hereditary material (see Spread 1.5).

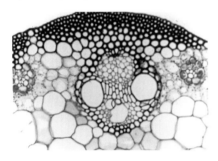

Figure 1.10 Cells in a plant seen in the light microscope. This cross-section of a leaf (*Zea mays*) shows cells of a variety of sizes. Notice that most of the larger cells have thick cell walls as do those near the epidermis (top). The large cells typically are mostly occupied by very large vacuoles (see Spread 1.6). Magnification × 30.

This is all Biology.... but what we want to get on to is a chemical explanation for life processes. How can the structure of cells be understood in chemical terms? One way, already mentioned, is to study the molecules involved; and as explained above all cells contain more or less the same collection of chemicals. But a living cell is more than a collection of chemicals — macromolecules, lipids, sugars, ions, water, etc. — there is also *organization*.

Although we recognize many many different cell types, one feature is common and fundamental to them all. All cells are surrounded by a membrane: cells are compartments, separated from their environment. This is a very important concept. To remain alive and functioning, a cell must maintain its internal composition within narrow limits despite wide and often unpredictable fluctuations in the external environment. Moreover, a cell cannot create a suitable internal environment and then seal itself off from its surroundings, because it needs raw materials from the outside, and wastes need to be expelled. Therefore a selectively permeable membrane is required, controlling what enters and leaves the cell. This role is played by the *plasma membrane* (also called the cell membrane or *plasmalemma*). This membrane has other functions, too. Because it is the outer boundary of the cell, one of its important roles is

Figure 1.11 Electron micrograph showing cell membrane between cells of rat seminiferous tubule. Magnification × 1800.

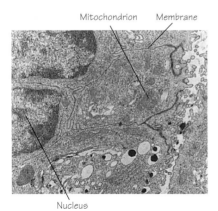

in cell–cell recognition, and this, too, is explainable by its chemical nature. The reason why liver cells all stay together to form a liver and do not get mixed up with blood cells flowing through it is because the liver cells recognize each other.

In addition to a cell membrane, most animal and plant cells also have internal compartments. Certain regions of the cell have different functions, and these regions may be separated from the rest of the cell by membranes. Such subdivisions of the cell are called *organelles* or 'little organs'. Examples include the mitochondria and chloroplasts, as well as the nucleus, lysosomes, endoplasmic reticulum and Golgi apparatus (see section 1.9). Here again the key property of the biological membrane is its selective permeability, controlling what gets into and out of the subcellular compartment. It is important, therefore, to be able to understand how biological membranes are formed and how they work — in chemical terms.

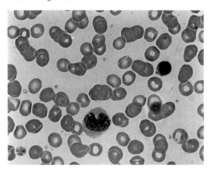

Figure 1.12 Red blood cells seen in the light microscope. Mature red cells are simply bags of haemoglobin with no nucleus and few other organelles. The cell membrane may easily be isolated after rupturing the cells by immersing them in hypotonic solution ('osmotic shock'). One white cell can be seen. Magnification × 425.

1.6 Biological membranes

Biological membranes are extremely thin, and perhaps somewhat surprisingly are rather fragile structures. Under the electron microscope they are observed as thin double lines about 7–8 nm thick (see Figure 1.11) and staining can show them to be made of mostly lipid and protein with some carbohydrate.

Mature mammalian red cells are little more than membranous bags of concentrated haemoglobin solution (Figure 1.12), and it is possible to isolate almost intact plasma membranes from red blood cells by osmotically shocking them. This allows membranes to be analysed chemically, and the analyses confirm that they are composed of lipid and protein with some carbohydrate.

How do these chemical components fit together to form a membrane? The key to the structure is the lipid molecules (Figure 1.13). The major proportion of these lipid molecules belong to the class called *phospholipids*. The structure of these is dealt with in more detail later on; the important feature to note here is that the molecules are *amphipathic*.

Amphipathic: a molecule which has some hydrophobic regions and some hydrophilic regions. If these regions are at opposite ends of a long molecule, such molecules will tend to orientate themselves at water:lipid interfaces.

Figure 1.13 Molecular model of lipid bilayer. (Note the smaller cholesterol molecules.)

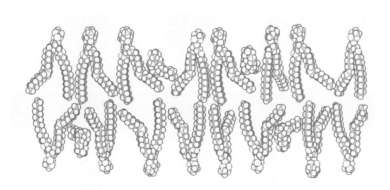

Chapter 1 • **Life Chemistry and the Ecosystem**

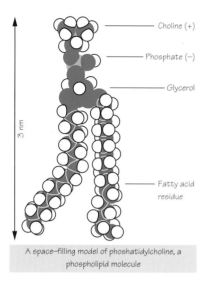

Figure 1.14 Membrane components.

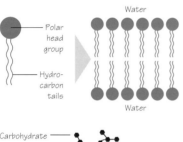

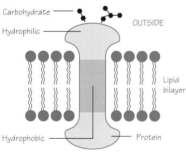

This means simply that the molecules are water-soluble (hydrophilic) at one end and water-insoluble (hydrophobic) at the other. The consequence of this is that when phospholipids and water are mixed together, they tend to form a *lipid bilayer*. This is shown in Figure 1.13. This is the basis of the membrane: the structure is fluid and flexible because the phospholipid molecules are not covalently joined but are interacting in a dynamic state. The dimensions of the molecules are in accord with the measured thickness of biological membranes, too (a typical phospholipid molecule is about 3 nm long; see Figure 1.14).

However, this is not the whole story, for we have not yet accounted for the protein component of membranes. A membrane solely composed of phospholipid molecules does not perform the biological functions of a true membrane because it is essentially impermeable. Hydrophilic molecules will tend to stay near one face and cannot penetrate the hydrophobic core, while hydrophobic molecules find difficulty in approaching the hydrophilic exterior of such a membrane.

So where do the protein molecules fit in? Surprisingly, that is exactly what they do do — they fit in or pass through the membrane. Proteins are macromolecules (see section 2.3), and membrane proteins typically have several distinct regions on their surface. Some of these regions are hydrophobic and some are hydrophilic. We can imagine a membrane protein to have a 'dumb-bell' shape with a hydrophobic central region (see Figure 1.14) that can 'sit' in the membrane in a stable configuration.

This will be a dynamic structure because it is not covalently bound to the lipid molecules and therefore can move around laterally in the membrane. If such a protein had a hydrophilic pore or cavity through it, this channel could conduct hydrophilic molecules through the membrane. This structure is now well established as the *fluid-mosaic model* of membrane structure, and its significant features are described in Spread 1.4.

Two other features of membranes are worth pointing out here (see also Spread 1.4), because they help us to understand their biological functions. The first is that membranes may be *asymmetric*; in other words, one surface may be chemically different from the other. Although the lipid and protein molecules of the membrane may move around laterally, there is very little 'flipping'. It would require a great deal of energy to move hydrophilic regions through hydrophobic ones, and consequently it rarely happens. Typically it is found that the outside of a membrane has carbohydrate attached to it (usually to the proteins) and this is called the *glycocalyx* (see Figure 1.14). This type of structure is at least partly responsible for the cell–cell recognition mentioned earlier.

The second is the question of how relatively large molecules get into and out of cells. Many cells can take up sizeable molecules and other chunks of matter, such as when an amoeba ingests food material; and they can also send out material too, such as proteins that they have synthesized for export. This process depends on the ability of membranes to form 'sacs' or invaginations, and this is in turn dependent on the dynamic nature of the structure. It is probably controlled by the membrane proteins, and the whole process may be referred to as *endocytosis* (uptake) or *exocytosis* (export) (Figure 1.15).

Cell membranes should not be confused with cell walls, which are quite different in both structure and function. Cell walls (Figure 1.16) are found only in plants, fungi and micro-organisms. They are relatively porous structures whose role is mostly to support cells. They show little selective permeability.

1.7 Intracellular organization

In the study of cells and membranes, two techniques have been of enormous importance in helping us to understand these biological structures. These techniques are (a) electron microscopy (Figure 1.17) and (b) subcellular fractionation (Figure 1.18), in which cells are 'taken apart' in order to study their component parts both chemically and biologically.

The cell is subdivided into 'working units' by membrane systems giving rise to the subcellular organelles. Most of these can be observed under the electron microscope and some can be isolated by subcellular fractionation. In this way we can attempt to put together their observed structures with their biological functions. This is how we know that the mitochondria are the energy-supplying powerhouses of the cell, for example. The compartmentalization is vital to the proper functioning of the cell, and the fine detail of cell structure as revealed by electron microscopy is referred to as the *ultrastructure*.

When many different types of cell had been investigated using the electron microscope, it became clear that all cells could be classified into two major groups: the *prokaryotes* and *eukaryotes*. Prokaryotic cells lack separate internal membrane-bounded compartments, and are generally much smaller than eukaryotic cells. Bacteria, of which there are very many species, and a few other groups, are prokaryotes. They do not exist in complex, multicellular, arrangements. In contrast, the characteristic of the larger, eukaryotic cells is that they do have internal compartments — nuclei, mitochondria, chloroplasts — separated from the rest of the cell and from each other by membranes. Eukaryotic organisms may be apparently simple and unicellular (e.g. *Amoeba*) or highly complex and multicellular (e.g. mammals and higher plants).

1.8 Prokaryotes

The earliest fossil remains of living cells seem to be those of primitive organisms with structures similar to those of present-day cyanobacteria (formerly blue–green algae). Modern prokaryotes have presumably evolved from these, and although 'primitive' they are actually very successful forms of life, having colonized practically every habitat on Earth. There is some evidence that the organelles in eukaryotic cells, such as mitochondria and chloroplasts, are the living remnants of prokaryotic cells that colonized other cells and remained in a symbiotic relationship and lost their independence.

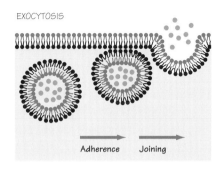

Figure 1.15 The processes of exocytosis and endocytosis: how particles get out of and into cells 'through' a biological membrane. [Redrawn with permission from B.M. Alberts et al. (1989), Molecular Biology of the Cell, 2nd edn., Garland, New York.]

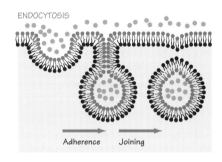

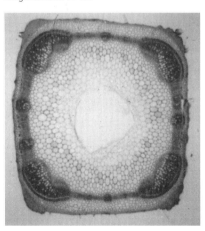

Figure 1.16 Cross-section of a stem of dead nettle (*Lamium*) which has a square plan. The cells at the corners have thick cellulose cell walls for supporting this structure. Magnification × 10.

Structure and function of membranes

objectives
- To describe the structural features of a cell membrane that underlie its 'dynamic' nature
- To explain the role of protein components in the biological functions of a membrane

The PHOSPHOLIPID BILAYER is impermeable to water-soluble charged solutes and is a barrier between adjacent aqueous environments. Diffusion across the lipid bilayer is responsible for the movement of small, uncharged molecules.

The hydrophobic unsaturated fatty acid tails may be 'kinked', limiting close packing and so increasing fluidity.

The various types of MEMBRANE PROTEINS each have distinct properties that are essential to the overall biological function of the membrane. In particular, they regulate the transport of molecules in and out of cells.

The polypeptide chains of TRANSMEMBRANE PROTEINS are folded into a conformation which maximizes both hydrophobic interactions with phospholipid tails and hydrophilic interactions with water. *Within* the membrane the protein adopts a largely helical conformation. Transmembrane proteins may act as:

- Transmembrane transport systems by recognizing and binding to small molecules and ions which otherwise would not cross the phospholipid bilayer. The system may be either active or passive.
- Hormone receptors by recognizing and binding to circulating chemical messengers.
- Transmembrane pumps to set up electrochemical gradients between the cytoplasm and the environment.
- Ligand receptors to initiate receptor-mediated endocytosis, i.e. the selective uptake into the cell of molecules or particles too large to cross the membrane by a more conventional transport system.

Attached to the outside of the membrane, the GLYCOCALYX — composed of oligosaccharides joined to membrane proteins and lipids — is important in recognition: cell–cell contacts of this type may regulate cell growth, division and development.

Some proteins span the membrane as a single α-helix with their ends lying folded across the outer and inner surfaces. Other proteins have polypeptides that go backwards and forwards across the membrane (again usually as α-helices) up to 12 times. These are shown as 'blobs' in the picture, but in fact, the tight cluster of polypeptides traversing the membrane can form highly specific channels. In either case the outside portion of the membrane proteins may be glycosylated (i.e. have carbohydrate residues attached).

MEMBRANE-BOUND ENZYMES may be involved in the biosynthesis or assembly of extracellular materials such as CELL WALLS which surround plant, fungal and bacterial cells.

CHOLESTEROL acts as a fluidity buffer — at low temperatures it prevents crystallization of hydrocarbon tails and at high temperatures it prevents excessive fatty acid mobility which might otherwise affect membrane permeability (see Figure 1.13).

PERIPHERAL PROTEINS may play a part in intracellular signalling systems in response to protein movements caused by binding to extracellular messengers such as hormones.

PORE PROTEINS are transmembrane proteins with aqueous channels through which charged molecules may pass and thus avoid the hydrophobic tails of the phospholipid molecules. Some channels are open all the time, but others are GATED (they open and close only in response to a stimulus, such as a change in the membrane's electrical potential).

ACTIVE TRANSPORT uses a CARRIER PROTEIN to transport a solute across a membrane. Energy, in the form of ATP, is required since transport is against a concentration gradient. There are three types of active transport:

A UNIPORT SYSTEM transports a single solute molecule or ion in one direction.

A SYMPORT SYSTEM transports two solutes in the same direction, e.g. the uptake of both glucose and amino acids by gut epithelial cells requires coupled Na^+ uptake: this is CO-TRANSPORT.

AN ANTIPORT SYSTEM transports two solutes in opposite directions, e.g. for each three Na^+ pumped out of a cell, two K^+ are pumped in: this is the ATP-DEPENDENT Na^+/K^+ PUMP.

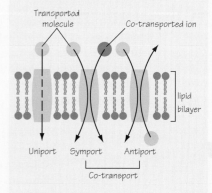

RECEPTOR-MEDIATED ENDOCYTOSIS The molecule or LIGAND binds to a specific receptor protein and is then enclosed by molecules called CLATHRIN to form a COATED PIT. This moves into the cell where the coat is removed and the molecules used as appropriate (see Figure 1.15).

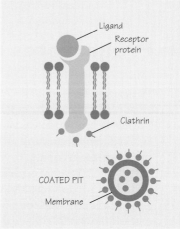

1.4 Structure and function of membranes

CHOLESTEROL molecules orient themselves in the membrane so that their polar hydroxy groups are close to the polar heads of the phospholipid molecules and their plate-like steroid rings partly immobilize the regions of the hydrophobic tails closest to the polar heads.

CELL ADHESION PROTEINS firmly attach adjacent cells to one another — this is particularly important in epithelia. These proteins also serve as internal anchorage points for the proteins of the cytoskeleton.

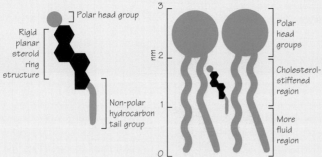

The GLYCOCALYX is composed of oligosaccharides and is important in recognition.

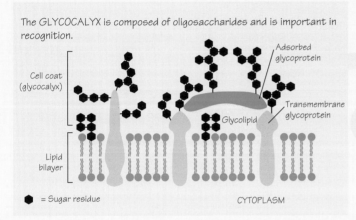

TRANSMEMBRANE PROTEINS may act as:
(a) Transmembrane transport systems: either passive or active.
(b) Hormone receptors.
(c) Transmembrane pumps.
(d) Ligand receptors to initiate receptor-mediated endocytosis.

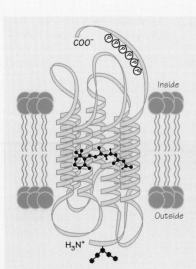

Model of the membrane protein, rhodopsin, from retina. It has seven transmembrane α-helices and contains one molecule of vitamin A aldehyde (retinal). The part inside the cell (C-terminus) carrries phosphate residues and that outside the cell carbohydrate residues (black hexagons). [Redrawn with permission from L. Stryer et al. (1995), *Biochemistry*, 4th edn., W.H. Freeman & Sons, New York.]

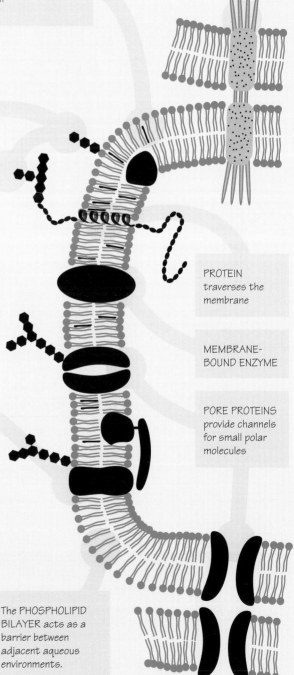

PROTEIN traverses the membrane

MEMBRANE-BOUND ENZYME

PORE PROTEINS provide channels for small polar molecules

The PHOSPHOLIPID BILAYER acts as a barrier between adjacent aqueous environments.

Chapter 1 • **Life Chemistry and the Ecosystem**

Figure 1.17 Transmission electron microscope.

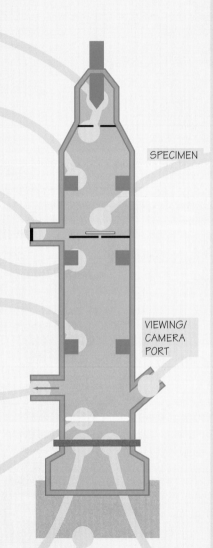

CATHODE: metal electrode (commonly platinum) which emits high-voltage electron beam. Electrons are negatively charged particles and are attracted towards the positively charged ANODE (+50 kV with respect to cathode). This accelerates the electron beam.

CONDENSER: Electromagnetic lens which focuses the electron beam on to the specimen.

AIR LOCK/SPECIMEN PORT: allows the introduction of the specimen into the microscope without the loss of vacuum.

OBJECTIVE: Electromagnetic lens which focuses and magnifies (depending on applied voltage) the first image.

PROJECTOR: further magnification by selection of region of image to be viewed.

TO VACUUM PUMP: creation of vacuum to minimize electron scattering and any heating caused by electron–air molecule collision.

FLUORESCENT SWING-OUT SCREEN: coated with electron-sensitive compounds — necessary since deflected electron beam (the image) cannot be viewed directly.

PHOTOGRAPHIC PLATE: allows a black and white image to be made. Printing may offer further magnification.

CONCRETE BASE: stable support which minimizes vibration and thus eliminates unwanted deflection of electron beam.

SPECIMEN PREPARATION

Sample is:

FIXED — to avoid deformation of cell components. Use small sample (for rapid penetration) and immerse in GLUTARALDEHYDE or GLUTARALDEHYDE/OSMIC ACID.

DEHYDRATED — to prepare material for infiltration by embedding or in infiltration medium which is not miscible with water. Dehydration should be gradual to preserve fine detail, using a series of progressively increasing concentrations of ETHANOL or PROPANONE.

CLEARED — alcohol or propanone may be immiscible with embedding agents, so is replaced with a clearing agent (commonly XYLOL) which is miscible and also makes the material transparent.

EMBEDDED — PLASTIC or RESIN is used to support the material so that it is not distorted during sectioning.

SECTIONED — the material must be cut into ULTRATHIN SECTIONS (20–100 nm thick) since the electron beam has very low penetration.

Ribbon of sections collects on water. Embedded sample. Ultramicrotome moves embedded sample forward in 20 nm steps. Diamond or glass knife cuts ultrathin sections.

STAINED — biological structures are virtually transparent to electrons. To increase electron beam deflection (i.e. contrast between different structures) sections are treated with solutions of heavy metal salts such as URANYL or LEAD ACETATE.

MOUNTED — sections are supported on a small copper grid (approx. 3 mm diameter); the electron beam may pass through the gaps in the grid (a standard glass slide would not permit penetration by the electron beam).

IMAGE INTERPRETATION

A number of ultrathin sections e.g. A–A', B–B' must be examined to provide a true three-dimensional representation of the sample.

High magnification means that several photographs may be necessary to give a composite image of the specimen.

Figure 1.18 Subcellular fractionation.

ISOTONIC SOLUTION prevents osmotic damage to cells and organelles

MOTOR-DRIVEN HOMOGENIZER forces cells between wall of tube and rotating pestle. Shearing forces developed are just sufficient to rupture the cells but not damage the organelles

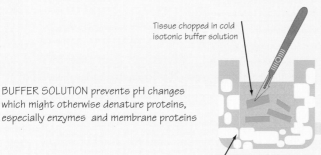

Tissue chopped in cold isotonic buffer solution

BUFFER SOLUTION prevents pH changes which might otherwise denature proteins, especially enzymes and membrane proteins

Ice: low temperature limits thermal damage to tissues (keeps cool but not frozen). Also stops growth of micro-organisms.

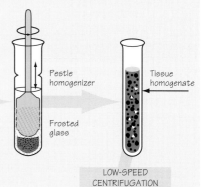

Pestle homogenizer

Frosted glass

Tissue homogenate

LOW-SPEED CENTRIFUGATION

Supernatant contains cell components too small to sediment at this speed

Pellet contains whole cells, nuclei, cytoskeletons, chloroplasts

Supernatant subjected to MEDIUM-SPEED CENTRIFUGATION

Cell extracts (tissue homogenate) are centrifuged at progressively greater speeds in order to separate (fractionate) their components. The main factors governing sedimentation are:
- magnitude of centrifugal force, which depends on the speed of rotation
- size of suspended particles
- density of organelles relative to that of suspension medium.

Exact times and speed of centrifugation vary from one tissue to another, and are determined by trial and error.
Typical values:
LOW SPEED –   1 000 × *g* for 10 min.
MEDIUM SPEED – 20,000 × *g* for 20 min.
HIGH SPEED –   100,000 × *g* for 60 min.

Supernatant subjected to HIGH-SPEED CENTRIFUGATION

Pellet contains mitochondria, lysosomes, peroxisomes

Pellet contains fragments of endoplasmic reticulum called microsomes, small vesicles, membranes

The microsomal suspension can be further purified by EQUILIBRIUM DENSITY CENTRIFUGATION, which involves layering the suspension on top of a SUCROSE DENSITY GRADIENT. When the gradient is centrifuged, organelles and membranes migrate and form bands in the region of the gradient equal to their density. The purified cell fractions are then collected by puncturing the tube bottom and collecting samples of the solution.

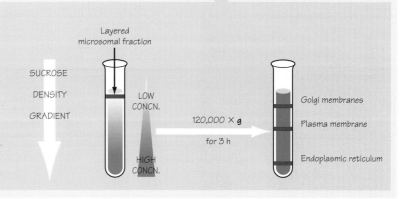

Layered microsomal fraction

SUCROSE DENSITY GRADIENT

LOW CONCN.

HIGH CONCN.

120,000 × *g* for 3 h

Golgi membranes
Plasma membrane
Endoplasmic reticulum

Chapter 1 • **Life Chemistry and the Ecosystem**

Present-day prokaryotes are divided into four groups called *bacteria*, *cyanobacteria* (formerly the blue–green algae), *archaebacteria* or archaea (most of which live in 'extreme' environments such as the Dead Sea or hot sulphur springs), and *mollicutes* (formerly mycoplasmas, which unusually do not have cell walls and some of which are responsible for pneumonia). Virtually all prokaryotes have a rigid cell wall made of polysaccharide linked by short peptides, and sometimes there is also a layer of lipopolysaccharide present. However, mycoplasmas do not have a cell wall ('mollicute' means soft skin) and tend to show all sorts of shapes.

Prokaryotes have very few cellular inclusions or organelles compared with eukaryotes (see below). This is illustrated in Spread 1.5, which shows the structure of a typical prokaryotic cell.

1.9 Eukaryotes

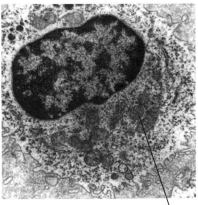

Figure 1.19 Electron micrograph of a section of human foetal kidney showing the nucleus. Compare the size of the mitochondria. Magnification × 5000.

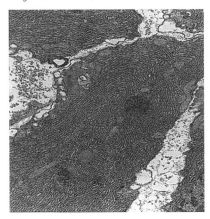

Figure 1.20 Electron micrograph to show endoplasmic reticulum. Magnification × 14,000.

An enormous number of types of eukaryotic cell is known and part of the reason for this is that many eukaryotes are multicellular organisms. Division of labour results in certain cells becoming highly specialized (differentiated) for certain functions. We shall concern ourselves here with the common features and with some typical organelles (see Spread 1.6).

Flagella and cilia may be present, but their detailed structures are very different from those found in prokaryotes. For example, they are much larger [cilia (2–10 μm) × 0.5 μm; flagella (100–200 μm) × 0.5 μm] and are constructed from sets of comparatively rigid elements surrounded by an extension of the plasma membrane.

Eukaryotic cells typically contain many organelles, a number of which are larger in size than the average bacterial cell. We know a great deal about eukaryotic organelles, not only because their structures can be seen under the electron microscope, but also because these organelles can be isolated from cells by special techniques. The chief method that has been used is that of cell fractionation in which broken-up cells are spun in a centrifuge at different speeds. Because of their different sizes and densities the various organelles sediment towards the bottom of the centrifuge tube at different rates. By means of this technique the various organelles have been separated from one another (see Figure 1.18) and this has made it possible for biochemists to study their biological function in isolation. Thus the reactions of photosynthesis may be studied with a preparation of chloroplasts.

The following are the major organelles of eukaryotic cells, but a complete list would be very much longer. Plant and animal cells are compared in Spread 1.6.

Nucleus

The nucleus is the most prominent body in the cell and is surrounded by a double membrane (Figure 1.19). These membranes appear to form circular nuclear pores in certain places of 80–100 nm diameter. These may occupy up to one-third of the total surface area of the nucleus and are thought to be responsible for the selective passage of materials into and out of it. The chief component of the nucleus is deoxyribonucleic acid (DNA), but ribonucleic acid (RNA) and protein are also present. Another structure often clearly defined within the nucleus is the nucleolus, which contains a large number of granules rich in RNA. These granules are the precursors of the ribosomes. The nucleolus has no membrane of its own.

Endoplasmic reticulum

In the cytoplasm of most eukaryotic cells there is a network of membranes called the endoplasmic reticulum. When the cell is disrupted fragments of this may be separated (see Figure 1.20) and are referred to as the microsomal fraction. The cavities within the endoplasmic reticulum function as channels and reservoirs in internal transport and storage. Ribosomes are often found attached to the endoplasmic reticulum, giving it a granular appearance. Such regions of endoplasmic reticulum are called rough endoplasmic reticulum to distinguish them from smooth endoplasmic reticulum which has no attached ribosomes. As in the case of prokaryotic ribosomes, eukaryotic ribosomes are responsible for the synthesis of protein. It is believed that proteins destined for export from the cell are synthesized by the ribosomes attached to the endoplasmic reticulum.

Golgi apparatus

Another system of complex vesicles and membranes found in most eukaryotic cells is referred to as the Golgi apparatus (Figure 1.21). Material produced within the cell for export is processed within the Golgi apparatus and is packaged in pinched-off vesicles derived from it. Eventually the vesicles fuse with the plasma membrane and their contents are released to the exterior. The digestive enzymes of the pancreas are produced and released in this way.

Mitochondria

The oxidative processes of cell metabolism which are concerned with energy production from foodstuffs take place in the mitochondria. These may be spherical or elongated or even ramifying filamentous structures, but are always enclosed by a double membrane (Figure 1.22). The inner membrane is infolded to give characteristic structures called cristae. Mitochondria behave almost as if they are autonomous organisms within the cell. They have ribosomes, a protein-synthesizing apparatus and their own DNA, and 'reproduce' by binary fission.

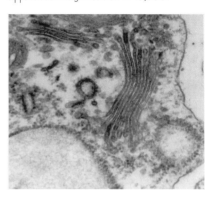

Figure 1.21 Electron micrograph of Golgi apparatus. Magnification × 19,000.

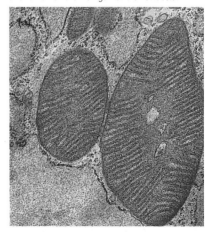

Figure 1.22 (a) Electron micrograph of a mitochondrion (section of rat pancreas). Magnification × 20,000. (b) Cutaway diagram to show structure of mitochondrion and arrangement of cristae.

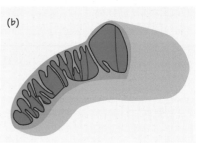

A prokaryotic cell has no true organelles

objectives
- To outline the structure and function of the various components of a prokaryotic cell
- To list the differences between prokaryotic and eukaryotic cells

Prokaryotic cells are typically very small, ranging in size from 1 to 10 μm (0.001–0.01mm) in diameter. They frequently have appendages such as flagella for locomotion, and pili which are shorter projections involved in cell–cell attachment. The flagellum is much simpler than that of a eukaryotic cell, being composed of a single cylinder of protein subunits (flagellin). It does not 'beat', but instead rotates about a 'bearing' anchored in the cell wall to produce a corkscrew motion which drives the cell along.

The cell is often coated with a gelatinous CAPSULE. This gummy layer of mucilage may unite bacteria into colonies (e.g. *Bacillus anthracis*) or confer protection (e.g. smooth strain of *Streptococcus pneumoniae*) to the cell. These outer structures are secreted by the cell, although under certain conditions the cell can live perfectly well without them.

Virtually all prokaryotic cells are surrounded by a cell wall, consisting of a rigid framework of murein, a polysaccharide cross-linked by peptide chains. In GRAM-POSITIVE bacteria the wall is thickened with further polysaccharide and protein deposits, whilst in GRAM-NEGATIVE bacteria the wall is thinner but coated with a lipid layer which provides protection against LYSOZYME and PENICILLIN. The rigidity of the cell wall prevents osmotic damage (penicillin interferes with this in susceptible Gram-positive bacteria) and confers shape on the cell.

Prokaryotic cells also have a plasma membrane, but have very few inclusions or organelles compared with eukaryotes. There may be granules of various sorts which appear and disappear depending upon the metabolic condition of the cell, such as granules of β-hydroxybutyric acid which some bacterial cells accumulate when carbon and energy sources are plentiful. Many bacteria form tiny resistant vegetative bodies called spores.

Photosynthetic prokaryotes (the cyanobacteria and the photosynthetic bacteria) have complex arrangements of membranes within the cell that contain the photosynthetic apparatus. These membranes are surfaces for light-absorbing pigments, principally BACTERIOCHLOROPHYLL, in green (e.g. *Chlorobium*) and purple (e.g. *Chromatium*) bacteria. The photosynthetic apparatus in photosynthetic bacteria has a different structure from eukaryotic chloroplasts, and does not evolve oxygen.

The genetic apparatus of prokaryotes, usually a single piece of DNA, is to some extent folded, but it is not contained within a membrane as is the case in eukaryotes. The region of the cell containing the DNA is sometimes referred to as the nuclear body. Small circles of DNA (plasmids) may also be present.

Apart from the photosynthetic apparatus and the nuclear body the only true organelles of prokaryotes are the ribosomes. These are small (20 nm) approximately spherical particles made of nucleic acid and protein, and are essential for the synthesis of proteins. Some ribosomes are free in the cytoplasm while others are attached to the cell membrane.

CAPSULE

PLASMA MEMBRANE is similar to phospholipid bilayers.

The rigid CELL WALL has a protective function and also confers shape on the cell. The three most common shapes are:

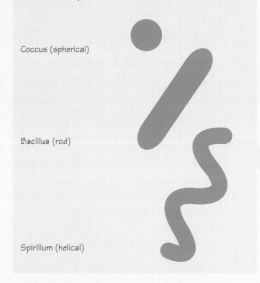

Coccus (spherical)

Bacillus (rod)

Spirillum (helical)

PLASMIDS are short pieces of circular DNA which replicate independently of the cell genome. They have been widely exploited in recombinant DNA technology (see Figure 7.10).

PILI (or FIMBRIAE) are protein rods concerned with cell–cell attachment. The SEX PILUS is involved in DNA transfer during sexual reproduction.

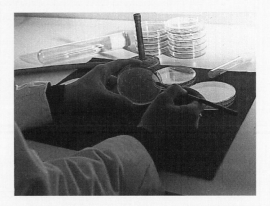

Growing cultures of bacteria on agar plates under sterile conditions.

1.5 A prokaryotic cell has no true organelles

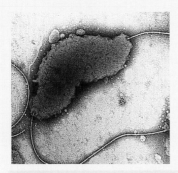

Scanning electron micrograph of *Campylobacter jejuni*, which has a single polar flagellum. The organism is a major cause of food poisoning and diarrhoea. It is spread by the faecal–oral route, originating in farm animals, but may also occur in unpasteurised milk. (Courtesey of A. Curry, Public Health Laboratory, Withington Hospital, Manchester.) Magnification × 14,000.

The GENETIC MATERIAL is composed of a circle of double-stranded DNA which is not enclosed within a nuclear membrane. There are typically about 2000 genes — about 0.2% of the number found in a eukaryotic cell.

PHOTOSYNTHETIC MEMBRANES

RIBOSOMES are smaller than those found in eukaryotes. They are scattered throughout the cytoplasm, rather than being supported on the endoplasmic reticulum.

FOOD STORES are typically lipid globules or glycogen granules.

The FLAGELLUM is responsible for motility of many bacteria. It is driven by a bacterial motor.

Rotor — Bearing — Plasma membrane — Cell wall — Flagellar filament

Bacterial motors run on gradients of protons (see Figure 5.3).

MESOSOMES are infoldings of the plasma membrane on which the enzymes associated with respiration are located. A proton gradient generated across these membranes is used to drive the synthesis of ATP. Their structure is uncertain and they may even be an artifact of the preparation for the electron microscope.

Chapter 1 • **Life Chemistry and the Ecosystem**

Comparison of plant and animal cell structure

objectives
- To list the differences and similarities between animal and plant cells
- To describe the structure and functions of the various cell organelles

Cells are surrounded by a selectively permeable membrane called the PLASMA MEMBRANE (or PLASMALEMMA). Plant cells are also surrounded by a cellulose CELL WALL. The function of the cell wall is a mechanical one — pressure from the cell protoplast maintains cell turgidity. It is freely permeable to water and most solutes so that it represents an important transport route — the APOPLAST SYSTEM — throughout the plant body. The plasmalemma is flexible enough to move close to or away from the cell wall as the water content of the cytoplasm changes. It is also responsible for the synthesis and assembly of cell wall components.

Plant and animal cells are subdivided by membrane systems, which give rise to the subcellular ORGANELLES. The NUCLEUS is the centre of the regulation of cell activities since it contains the hereditary material, DNA. There is usually only one nucleus per cell, although there may be many in large animal cells (syncitia) such as those of striated muscle.

The nucleus is continuous with the ENDOPLASMIC RETICULUM, the network of membranes responsible for the internal transport and storage of cellular products. Another organelle is the GOLGI APPARATUS, which, in animal cells, modifies products such as trypsinogen, insulin and mucin. The Golgi is also involved in lipid modification in cells of the ileum, and in the formation of lysosomes. In plants, the Golgi (or DICTYOSOME) synthesizes polysaccharides and packages them in vesicles which migrate to the plasma membrane for eventual incorporation in the cell wall.

Other organelles which can be identified within eukaryotic cells are illustrated. An organelle which is found only in plant cells is the chlorophyll-containing CHLOROPLAST. This is one of a number of PLASTIDS, all of which develop from PROPLASTIDS. Other typical plastids are CHROMOPLASTS, which may develop from chloroplasts by internal rearrangements. Chromoplasts are coloured due to the presence of carotenoid pigments and are found in cells of flower petals or fruit skins. LEUCOPLASTS are a third type of plastid common in cells of higher plants — they include AMYLOPLASTS which synthesize and store starches and ELAIOPLASTS which synthesize oils.

Cellular organelles are suspended within the CYTOSOL, which consists principally of water, with many solutes including glucose, proteins and ions. It is permeated by the CYTOSKELETON, which is the main architectural support of the cell. In mature plant cells, ~90% of the volume may be occupied by the VACUOLE, which is filled with cell sap (a solution of salts, sugars and organic acids) and helps to maintain turgor pressure inside the cell. The vacuole also contains anthocyanins — pigments responsible for many of the colours of flowers — and enzymes involved in recycling of cell components. The vacuolar membrane is called the TONOPLAST.

PLASMODESMATA are minute strands of cytoplasm which pass through pores in the cell wall and connect the protoplasts of adjacent cells. This represents the SYMPLAST pathway for the movement of water and solutes throughout the plant body. These cell–cell cytoplasmic connections are important in cell survival during periods of drought.

The CHLOROPLAST is the site of photosynthesis.

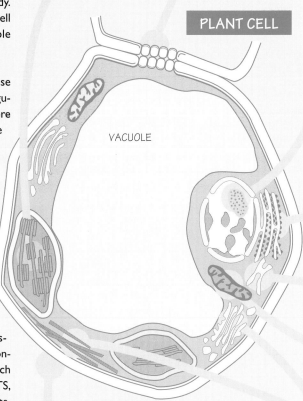

PLANT CELL

VACUOLE

The CELL WALL is composed of long cellulose molecules grouped in bundles called MICROFIBRILS which, in turn, are twisted into rope-like MACROFIBRILS. The macrofibrils are embedded in a matrix of PECTINS (which are very adhesive) and HEMICELLULOSES (which are quite fluid). There may be a SECONDARY CELL WALL, in which case the outer covering of the cell is arranged as:

- **Plasma membrane**
- **Secondary cell wall:** laid down on inside of primary wall. Often impregnated with LIGNIN (gives mechanical strength to xylem) or SUBERIN (waterproofs endodermis).
- **Primary cell wall:** laid down first, by plasma membrane.
- **Middle lamella:** contains gums and calcium pectate to cement cells together.

1.6 Comparison of plant and animal cell structure

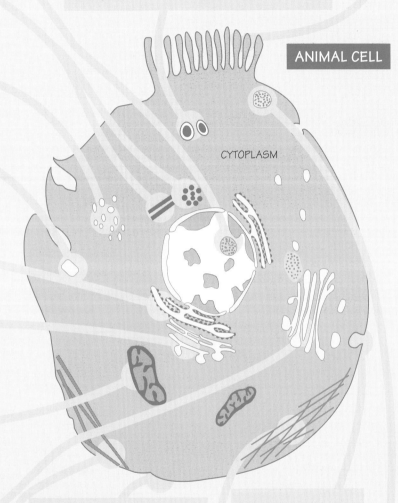

ANIMAL CELL

The NUCLEUS is the centre of the regulation of cell activities. It contains one or more nucleoli in which ribosome subunits and transfer RNA are manufactured. The nucleus is surrounded by a double nuclear membrane, crossed by a number of nuclear pores, and is continuous with the endoplasmic reticulum.

The CENTRIOLES are a pair of structures, held at right angles to one another, which act as organizers of the nuclear spindle in preparation for the separation of chromosomes or chromatids during nuclear division.

FREE RIBOSOMES are the sites of protein synthesis, principally for proteins destined for intracellular use. There may be 50,000 or more in a typical eukaryote cell.

The ENDOCYTIC VESICLE may contain molecules or structures too large to cross the membrane by active transport or diffusion.

The ROUGH ENDOPLASMIC RETICULUM is so-called because of the many ribosomes attached to its surface. This intracellular membrane system aids cell compartmentalization and transports proteins synthesized at the ribosomes towards the Golgi bodies for secretory packaging.

The SMOOTH ENDOPLASMIC RETICULUM is a series of flattened sacs and sheets and is the site of lipid synthesis and secretion.

MITOCHONDRIA are the sites of aerobic respiration. In animals, they are abundant in cells which are physically (skeletal muscle) or metabolically (hepatocytes) active. In plants they may be abundant in sieve tube companion cells, root endodermal cells and dividing meristematic cells.

The GOLGI APPARATUS consists of a stack of sacs called CISTERNAE. It modifies a number of cell products delivered to it, often enclosing them in vesicles to be secreted.

MICROTUBULES are hollow tubes of the protein tubulin, about 25 nm in diameter. They are involved in intracellular transport of organelles, and they are components of specialized structures such as the centrioles, spindles and cell plates of dividing cells, and the basal bodies of cilia and flagella. They also have a structural role as part of the cytoskeleton.

MICROVILLI are extensions of the plasma membrane which increase the cell surface area. They are found in cells with a high absorptive capacity (e.g. liver or kidney cells). Collectively the microvilli represent a BRUSH BORDER to the cell.

LYSOSOMES are sacs that contain high concentrations of hydrolytic (digestive) enzymes. These enzymes are kept apart from the cell contents, and are kept inactive by an alkaline environment within the lysosome. They are especially abundant in cells with a high phagocytic activity, such as some LEUCOCYTES.

CYTOPLASM

The PLASMA MEMBRANE (PLASMALEMMA) is the differentially permeable cell surface, responsible for the control of solute movements between the cell and its environment.

MICROFILAMENTS are threads of the protein ACTIN. They are usually situated in bundles just beneath the cell surface and play a role in endo- and exo-cytosis, and possibly in cell motility.

SECRETORY VESICLE undergoing exocytosis. May be carrying a synthetic product of the cell or the products of degradation by lysosomes.

The PEROXISOME is one of a group of vesicles known as MICROBODIES. Each of them contains oxidative enzymes such as CATALASE, and they are particularly important in delaying cell aging.

Chapter 1 • **Life Chemistry and the Ecosystem**

Chloroplasts

Only photosynthetic eukaryotic cells possess the chlorophyll-containing structures called chloroplasts. These have a highly characteristic structure of membranous vesicles contained within a plasma membrane. Like mitochondria they are semi-autonomous and have their own DNA and ribosomes (Figure 1.23).

Many other organelles can be identified within eukaryotic cells, including liposomes, peroxisomes, cytoplasmic filaments, microtubules, centrioles and vacuoles, each having a specific function within the cell.

Figure 1.23 Electron micrograph of a chloroplast. Magnification × 10,000.

Cytosol

Traditionally the whole of the cell contents less the nucleus was called the cytoplasm. It is now more usual to say that the cellular organelles, including the cytoskeleton, are suspended in a fluid called the cytosol.

1.10 Further reading

Baum, H. (1992) Mitochondria. Biological Sciences Review **4(3)**, 38–40

Berner, R.A. and Lasage, A.C. (1989) Modelling the geochemical carbon cycle. Scientific American **260 (Mar)**, 54–60

Brown, B.S. (1990) The wonderful water molecule. Biological Sciences Review **2(4)**, 18–20

Butler, R. (1989) The bees knees. Biological Sciences Review **2(1)**, 21–24

Lichtman, J.W. (1994) Confocal microscopy. Scientific American **269 (Aug)**, 30–35

Milsom, D. (1994) Biological membranes. Biological Sciences Review **6(4)**, 16–20

Orgel, L.E. (1994) The origin of life on earth. Scientific American **263 (Oct)**, 52–61

Postgate, J. (1989) Fixing nitrogen. Biological Sciences Review **1(4)**, 2–6

Vidal, G. (1984) The oldest eukaryotic cells. Scientific American **250 (Feb)**, 32–41

Weinberg, R.A. (1985) The molecules of life. Scientific American **254 (Oct)**, 34–43 [and other papers in the same issue]

1.11 Examination questions

1. (a) Grass assimilates approximately 1% of the sun's radiant energy that falls on it. Give two reasons why plants assimilate such a relatively small amount of the radiant energy. (2 marks)

   (b) In a grassland the energy flow through part of the food chain has been estimated as shown in the diagram. Figures are in $kJ \cdot m^{-2} \cdot year^{-1}$.
   - (i) In what form was the energy transferred between the trophic levels shown? (1 mark)
   - (ii) Give the gross production, in $kJ \cdot m^{-2} \cdot year^{-1}$, of the grass. (1 mark)
   - (iii) Give the total energy, in $kJ \cdot m^{-2} \cdot year^{-1}$, lost by the rabbits in respiration, excretion and egestion. (1 mark)
   - (iv) Which organisms in the food chain are heterotrophs? (2 marks)
   - (v) Explain why the number of links in a food chain does not normally exceed five. (2 marks)

   [ULEAC (AS Biology) June 1991, Paper 1, No. 3]

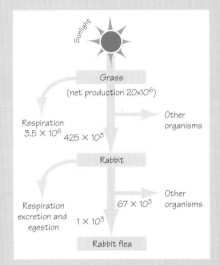

2. In the diagram, the numbers on the arrows show the energy flow through grassland in kilojoules per square metre per day. The numbers in the boxes show the energy stored in biomass as $kJ \cdot m^2$.
   (a) What percentage of gross productivity is used by:
   - (i) cattle?
   - (ii) humans? (2 marks)

   (b) (i) Give two examples of possible 'other herbivores' in this ecosystem.
   - (ii) State their relationship with the cattle. (2 marks)

   (c) Identify the losses of energy from the system shown at points A and B. (2 marks)

   [Northern Examinations and Assessment Board (formerly JMB) (Social and Environmental Biology) June 1991, Paper II, No. 5]

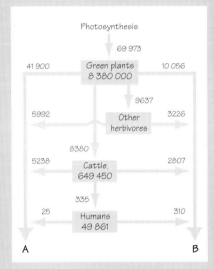

3. The diagram shows a model of the plasma membrane (cell surface membrane).
   (a) State two functions of proteins in the plasma membrane. (2 marks)

   (b) Explain the importance of phospholipids in the structure of the plasma membrane. (3 marks)

   (c) (i) Suggest one function of the part labelled 'X'. (1 mark)
   - (ii) On the diagram write the word 'OUTSIDE' on the surface of the membrane which faces outward. State the reason for your choice. (2 marks)

   (d) The membrane contains both hydrophilic and hydrophobic molecules. Explain briefly how both water and hydrophobic molecules can pass through the membrane. (3 marks)

   (e) Indicate how ions may accumulate inside a cell against the prevailing concentration gradient. (4 marks)

   (f) (i) A pH gradient may often be detected across a membrane. Explain how such a gradient could be developed. (3 marks)

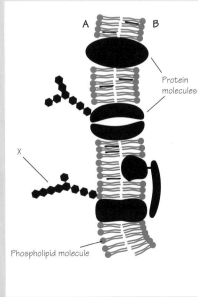

Chapter 1 • **Life Chemistry and the Ecosystem** 33

(ii) What is the significance of this gradient in relation to ATP synthesis in a cell? (2 marks)

(g) What part do cell membranes play in nerve conduction? (4 marks)

[Composite]

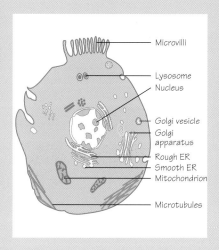

4. The figure shows a generalized diagram of an animal cell.
   (a) Which one of the following represents the approximate width of the cell type shown in the diagram?
   1 μm, 20 μm, 500 μm, 1 mm, 5 mm (1 mark)
   (b) (i) Name the organelle associated with:
       protein synthesis;
       respiration;
       transport of cell material in vesicles;
       lipid and steroid synthesis. (4 marks)
   (ii) On the diagram, label the site of glycolysis. (1 mark)
   (c) Name two structures shown which would **not** be clearly visible when viewed under the light microscope. (2 marks)
   (d) Indicate the function of lysosomes. (2 marks)
   (e) What is the significance of the microvilli? (2 marks)

[O & C (Biology) June 1991, Paper 1, No. 10]

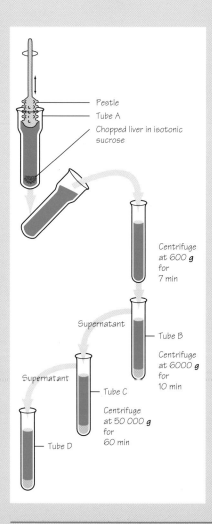

5. The diagram shows the stages involved in separating liver-cell components.
   (a) Why is the tissue placed in an isotonic solution? (2 marks)
   (b) The rotating pestle is lowered into Tube A. What is the purpose of this? (1 mark)
   (c) The sediment in tube C would be rich in mitochondria. Give one way in which you could determine that this mitochondrial fraction was:
       (i) pure;
       (ii) metabolically active. (2 marks)
   (d) Name one organelle that would be found in the sediment in Tube D. (1 mark)

[AEB (Biology) June 1992, Paper 1, No. 10]

2.1 Introduction

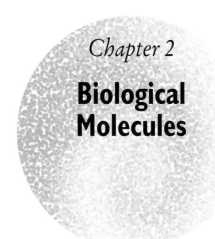

Chapter 2
Biological Molecules

All organisms are made up of the same types of chemical compounds: proteins, carbohydrates, nucleic acids and lipids, along with a range of smaller molecules and inorganic ions, all having their own biological functions. Many of the important molecules in cells are very large — giant molecules, usually called *macromolecules*. Some examples of macromolecules are: haemoglobin, a protein; starch, a polysaccharide; and deoxyribonucleic acid (DNA), a nucleic acid.

Even the simplest of cells contains several thousand different types of macromolecule, and the relative molecular masses (M_r) of these range from a few thousand to several million. To take an example, haemoglobin, the iron-containing protein in red blood cells, has an M_r of about 68,000, and this is only a relatively small protein. This also gives us an example of biological function: haemoglobin carries oxygen. In contrast, another type of protein, immunoglobulin, functions as an antibody in the blood. It is still a protein, but one quite different from haemoglobin and is about twice its size. Proteins can have many different biological functions and, as we might expect, their structures will reflect this.

Like many of the common materials upon which modern life depends (e.g. plastics), biological macromolecules are polymers, albeit rather special and complicated ones. This complexity has made macromolecules very difficult to study, and one of the triumphs of modern biochemistry has been the unravelling of the *structures* of these complex molecules, providing a firm basis from which to understand biological *function*. The key to understanding all macromolecules, biological and other, is to realize that they are all built up from relatively few similar building blocks, or *monomers*, which are joined together covalently to produce the polymers.

2.2 The monomers and how they are joined together

Biological macromolecules are made by linking monomer units into chains, usually but not always unbranched ones. The type of macromolecule (protein, polysaccharide or nucleic acid) depends on the type of monomer unit employed. Thus proteins are made up of amino acid monomer units, polysaccharides of monosaccharide sugar units, and nucleic acids from nucleotide units. The order in which the monomer units are linked together is vitally important to the function of the macromolecule because it is this order that determines the three-dimensional shape the macromolecule will fold into and consequently decides its biological properties. The order is also something that is 'remembered' and passed on from generation to generation, so that bacteria always make bacteria-type proteins and humans make human-type proteins. The memory for this order is itself stored in a macromolecule, usually DNA.

Figure 2.1 Linking monomers together.

The basic reaction is a condensation. Two compounds are linked and water is removed:

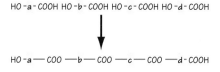

$$\text{RCOOH} + \text{HOR}' \longrightarrow \text{RCOOR}' + H_2O$$
Acid   Alcohol   Ester   Water

The reaction shown is the production of an ester, in other words, an esterification, but the general name for the reaction in which water is removed is a condensation. The reverse reaction is a hydrolysis.

Condensation may be used to join monomer units to form a polymer:

HO–a–COOH HO–b–COOH HO–c–COOH HO–d–COOH

↓

HO–a—COO—b—COO—c—COO—d–COOH

Note that the units condensing together here are all different (a, b, c, d). If they were joined in a different order a different polymer would be produced.

Note also that the monomers are directional, so the chain is directional:

Finally, note that groups other than –CO$_2$H and HO– may be involved but with the same general result.

- Links in proteins are called peptide bonds.
- Links in polysaccharides are called glycosidic bonds.
- Links in nucleic acids are called phosphodiester bonds.

An analogy is helpful here. Buildings are constructed of brick or stone, or sometimes wood or concrete, but very few types of building material are actually needed in the construction of thousands of buildings with different functions. This design for different function — house, office, theatre, shop — is referred to as 'architecture'. It is the design that is fundamental to the function of the building rather than the nature of the building blocks themselves. There will also be different sorts of houses, all composed of brick, but having different functions. Some houses are built of wood. Houses, of course, are for living in, but some are for big families and some for small, and different countries have houses of different design. The same is true of biological macromolecules. Amylase, the protein enzyme from humans that digests starch, has the same function as that from yeast, but the structures of the two proteins are different. Both are constructed of amino acid monomer units. Sometimes the three-dimensional structure of macromolecules is referred to as their 'architecture'.

The covalent bonds between monomer units in biological macromolecules are almost always formed by a reaction in which the elements of water are removed. This sort of reaction is a *condensation* (Figure 2.1). The monomer units in the polymer therefore lack the elements of water and are called *residues*. The reverse process, breaking the link, is called a *hydrolysis*. The links between the monomers in proteins, polysaccharides and nucleic acids are quite distinct from one another, although all are formed by condensation, and they have different names, too. We shall see what they are in the different sections below.

Links have a *polarity*: each 'end' of the monomer unit is different, meaning that the resultant polymer or macromolecule has a 'direction' to its chain (see Figure 2.1). The 'bricks' are more like jigsaw pieces than the bricks used to build a house.

In the next sections we will look at the three different types of biological macromolecule — proteins, polysaccharides and nucleic acids — in turn, bearing these principles in mind. The main concern will be to understand how different biological functions are achieved by macromolecules with different architectures. In addition, substances that do not form macromolecules, the lipids, are also dealt with in this chapter. This will bring together the four fundamental compounds found in all living organisms — proteins, polysaccharides, nucleic acids and lipids — and upon which all forms of life on Earth depend.

2.3 The proteins — very versatile macromolecules

Proteins perform a very wide variety of functions in organisms and take on a variety of physical forms. Many proteins, such as antibodies, enzymes and haemoglobin, are water-soluble molecules, whereas others, such as the keratins of hair, hoof and feather, are insoluble in water and are extremely resistant and tough. The leather of your shoes is made of another protein, collagen, somewhat modified chemically in leather, but which forms the main structural material in the vertebrate body, in bone, tendon and skin. There are tens of thousands of different types of proteins all with different biological functions. The functions of some proteins are listed in Table 2.1. However, when we analyse practically any protein, we find that it is made up of a selection chosen from 20 basic building block monomer units.

Table 2.1
Some proteins and their functions

Protein	Function
Pepsin	Digestive enzyme in stomach (extracellular)
Hexokinase	Enzyme in cytosol of cells
Immunoglobulin	Antibody in blood serum
Haemoglobin	Oxygen carrier in red cells
Insulin	Hormone that controls blood glucose level
Cholera toxin	Toxin produced by bacteria
Actin	Protein of muscle
Collagen	Fibrous protein of bone and tendon (extracellular)
Keratin	Fibrous protein of hair and feather (intracellular)
Silk (fibroin)	Fibrous protein of insects

Proteins are typically unbranched chains, or *polypeptides*, made by linking together several hundred, or even several thousand, *amino acid* monomer units by *peptide bonds*. Peptide bonds (see below) are strong covalent links. A protein is formed when one or more polypeptides coil up in certain precise and repeatable ways to form a three-dimensional structure with certain biological properties (Figure 2.2). This three-dimensional structure is typically rather unstable, and raising the temperature a few degrees above the natural temperature of the cell may cause it to unravel. This will normally lead to a loss of biological function which may be irreversible; the protein is then said to be *denatured*. Other things, such as strong acids and alkalis, organic solvents and heavy metal ions, can also denature proteins, and soluble proteins when denatured often precipitate. Usually it is the three-dimensional structure of the protein that is destroyed; the peptide bonds remain intact.

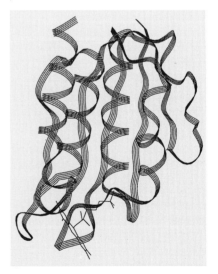

Figure 2.2 Computer-generated representation of a protein molecule to show some of the typical features of protein architecture. The model is highly simplified. No amino acid residues are shown. Regions of α-helix are shown as spirals, and regions of β-sheet are shown as ribbons lying side by side. Other parts are loops and turns. This molecule also has a small molecule or prosthetic group. If a space-filling model were constructed it would show that there are very few open spaces in the middle of protein molecules.

Amino acids

Twenty amino acid monomers are found in proteins. All amino acids have the same fundamental chemical structure but all are rather different in character. This is why stringing them together in different combinations produces proteins with thousands of different biological functions.

Each amino acid has, not surprisingly, an amino group and a carboxylic acid group. These are the groups that participate in the condensation

Figure 2.3 Amino acids and peptide bonds.

All the naturally occurring amino acids in proteins have the structure:

$$H_2N - \underset{R}{\underset{|}{C}} - H \quad (\text{with } COOH \text{ up})$$

Although at neutral pH they exist as 'zwitterions':

$$H_3\overset{+}{N} - \underset{R}{\underset{|}{C}} - H \quad (\text{with } COO^- \text{ up})$$

(a) Amino acids link to give peptides:

$$H_3\overset{+}{N} - \underset{R_1}{\underset{|}{CH}} - COO^- \; + \; H_3\overset{+}{N} - \underset{R_2}{\underset{|}{CH}} - COO^-$$

↓

$$H_3\overset{+}{N} - \underset{R_1}{\underset{|}{CH}} - \underset{\|}{\overset{O}{C}} - \underset{H}{\underset{|}{N}} - \underset{R_2}{\underset{|}{CH}} - COO^-$$

Peptide bond

Because the –CO–NH– bond has some double-bond character, free rotation is not allowed.

$$\underset{H}{\overset{O}{\underset{|}{C-N}}} \longleftrightarrow \underset{H}{\overset{O^-}{\underset{|}{C=\overset{+}{N}}}}$$

This, along with the three-dimensional structures of the amino acids themselves, restricts the shapes that polypeptides can adopt, so that proteins tend to form into certain shapes.

(b) Ball and stick model of the peptide bond showing bond angles and bond lengths (in nm). The C–N bond is midway between a double and a single bond and this causes the structure to be almost planar. There is no rotation about the C–N bond.

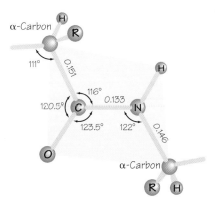

reaction to form a peptide bond (Figure 2.3). The other feature of amino acids is the group designated as 'R', sometimes called the 'side chain'. R can be any one of 20 different chemical structures (see Table 2.2), giving a very wide range of properties. It can be large or small, water-soluble or insoluble (i.e. hydrophilic or hydrophobic), acidic or basic, or aromatic (benzene-ring-containing), for example. Therefore linking together amino acids by the hundred produces an almost limitless range of proteins with different characters and biological functions.

It is important to realize that organic chemicals have a three-dimensional structure. This arises because the atoms themselves have bonds that point out at certain fixed angles. Carbon's four bonds, for example, point out towards the apices of a regular tetrahedron. Furthermore, because a peptide bond formed by the linking of any two amino acids has some 'double bond' character to it, the groups on either side are not free to rotate about the C–N bond (Figure 2.3). Thus there are restrictions on the shape of the polypeptide and hence on the three-dimensional arrangement a protein can adopt. Proteins have biological functions both because of their particular sequences of amino acids and because of their particular three-dimensional shapes.

A glance at Table 2.2 shows that amino acids with all sorts of different R side chains are available for the construction of proteins. It is as if you had 20 different types of bead with which to make a necklace, some large, some small, some smooth, some prickly, and so on. Clearly the type of necklace you made would depend on which beads you chose and on the order in which they were strung together. Order is very important, and in polypeptides we call it the *sequence* or *primary structure*. In the past much effort has gone into determining the sequence — or 'sequencing' — polypeptides. Today, over 100,000 sequences are known. It used to be a very tedious and painstaking job. The first person to sequence a rather small protein, insulin, was Fred Sanger, who received a Nobel Prize for the method he devised for doing this. Sequencing is mostly routine these days. It is normally carried out automatically on a machine called a *sequenator* which requires only tiny amounts of polypeptide.

Interestingly, although the sequences of many thousands of proteins are now known, and are stored in international computer databases, we are still unable to predict, from the sequence, what three-dimensional shape a protein will adopt. Good guesses can be made, based on knowledge of known proteins, but the success rate is little better than 60%. This is known as 'the protein folding problem', the problem being that the rules for how a polypeptide will fold up to give a biologically active protein are still largely unknown.

The only way to find out the three-dimensional shape of a large protein is to carry out X-ray crystallographic analysis (Figure 2.4), although the shapes of some smaller proteins (up to M_r 20,000) are now being determined by nuclear magnetic resonance (NMR) methods. X-ray crystallography of a protein is a major undertaking in effort, although it is becoming more accessible with high-speed computing to take the labour out of the millions of calculations required. However, it all depends on being able to crystallize the protein of interest, and for some proteins this is not possible.

Table 2.2

Neutral (hydrophobic)

Glycine (Gly, G)

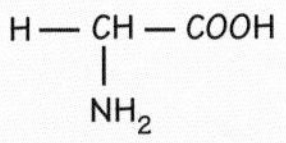

$$H-CH-COOH$$
$$|$$
$$NH_2$$

Alanine (Ala, A)

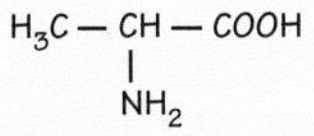

$$H_3C-CH-COOH$$
$$|$$
$$NH_2$$

Valine (Val, V)

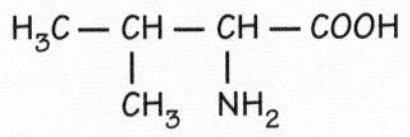

$$H_3C-CH-CH-COOH$$
$$|\quad\quad|$$
$$CH_3\quad NH_2$$

Leucine (Leu, L)

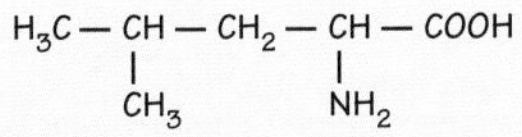

$$H_3C-CH-CH_2-CH-COOH$$
$$|\quad\quad\quad\quad|$$
$$CH_3\quad\quad\quad NH_2$$

Isoleucine (Ile, I)

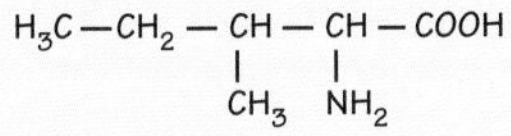

$$H_3C-CH_2-CH-CH-COOH$$
$$|\quad\quad|$$
$$CH_3\quad NH_2$$

Sulphur-containing

Cysteine (Cys, C)

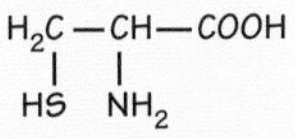

$$H_2C-CH-COOH$$
$$|\quad\quad|$$
$$HS\quad NH_2$$

Methionine (Met, M)

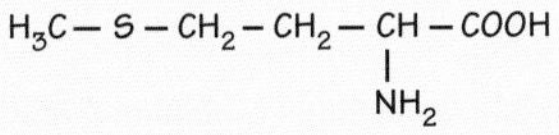

$$H_3C-S-CH_2-CH_2-CH-COOH$$
$$|$$
$$NH_2$$

Neutral (hydrophilic)

Serine (Ser, S)

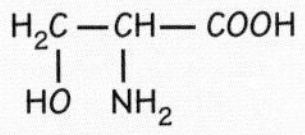

$$H_2C-CH-COOH$$
$$|\quad\quad|$$
$$HO\quad NH_2$$

Threonine (Thr, T)

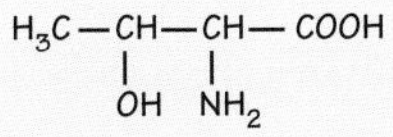

$$H_3C-CH-CH-COOH$$
$$|\quad\quad|$$
$$OH\quad NH_2$$

Imino acid

Proline (Pro, P)

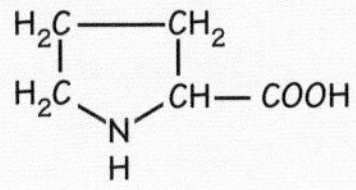

$$\begin{array}{c}H_2C-CH_2\\ |\quad\quad\quad|\\ H_2C\quad CH-COOH\\ \backslash\;/\\ N\\ |\\ H\end{array}$$

Aromatic

Phenylanine (Phe, F)

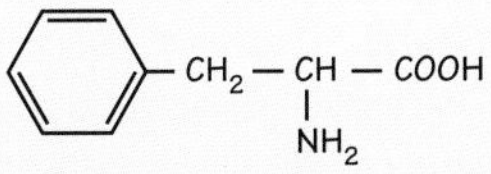

(benzene ring)$-CH_2-CH-COOH$
$\quad\quad\quad\quad\quad\quad\quad\;|$
$\quad\quad\quad\quad\quad\quad\quad NH_2$

Tyrosine (Tyr, Y)

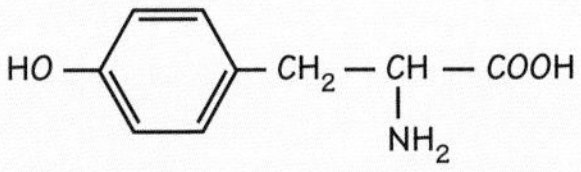

$HO-$(benzene ring)$-CH_2-CH-COOH$
$\quad\quad\quad\quad\quad\quad\quad\quad\quad\quad|$
$\quad\quad\quad\quad\quad\quad\quad\quad\quad\;NH_2$

Tryptophan (Trp, W)

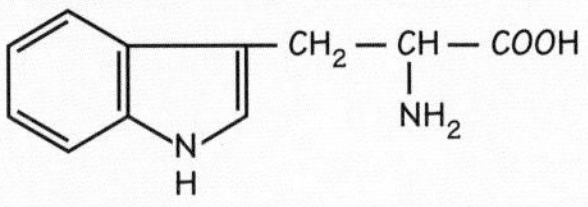

(indole ring)$-CH_2-CH-COOH$
$\quad\quad\quad\quad\quad\quad\quad|$
$\quad\quad\quad\quad\quad\;NH_2$

Histidine (His, H)

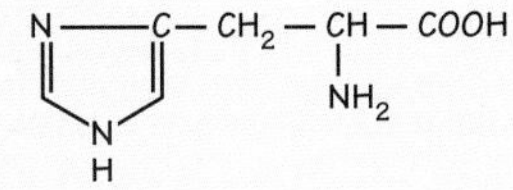

(imidazole ring)$-CH_2-CH-COOH$
$\quad\quad\quad\quad\quad\quad\quad\quad|$
$\quad\quad\quad\quad\quad\quad\;NH_2$

Basic

Lysine (Lys, K)

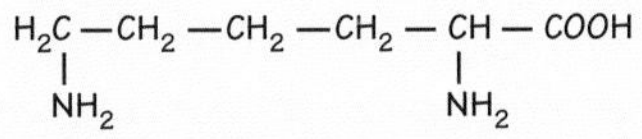

$$H_2C-CH_2-CH_2-CH_2-CH-COOH$$
$$|\quad\quad\quad\quad\quad\quad\quad\quad\quad|$$
$$NH_2\quad\quad\quad\quad\quad\quad\quad\quad NH_2$$

Arginine (Arg, R)

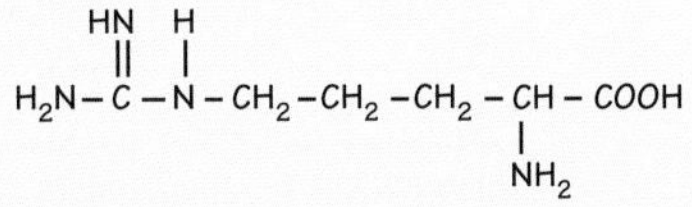

$$\begin{array}{c}HN\quad H\\ \|\quad\;|\\ H_2N-C-N-CH_2-CH_2-CH_2-CH-COOH\\ |\\ NH_2\end{array}$$

Acidic

Aspartic acid (Asp, D)

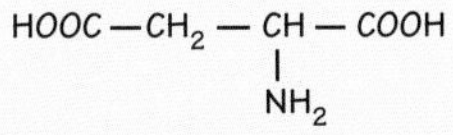

$$HOOC-CH_2-CH-COOH$$
$$|$$
$$NH_2$$

Glutamic acid (Glu, E)

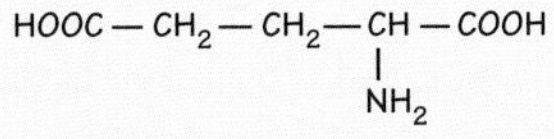

$$HOOC-CH_2-CH_2-CH-COOH$$
$$|$$
$$NH_2$$

Amides

Asparagine (Asn, N)

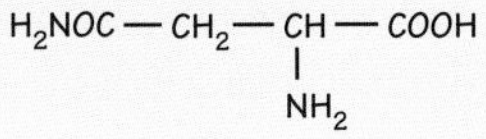

$$H_2NOC-CH_2-CH-COOH$$
$$|$$
$$NH_2$$

Glutamine (Gln, Q)

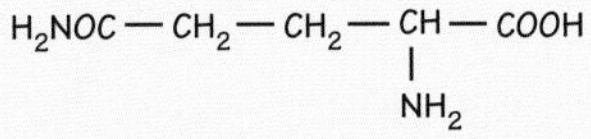

$$H_2NOC-CH_2-CH_2-CH-COOH$$
$$|$$
$$NH_2$$

Table 2.2 The 20 amino acids found in proteins are divided into groups according to the character of their side chain (R–) groups. Although all the amino acids are shown un-ionized, they will tend to be ionized at the pH values found in cells. Thus they will exist as zwitterions, with glycine being represented as H_3N^+–CH_2–CO_2^-. The side chain R groups may also be ionized in the cell and in proteins. The three-letter and one-letter abbreviations for the amino acids are given.

Figure 2.4 If proteins will crystallize, then analysis of their X-ray diffraction patterns can reveal the structure of the protein. The picture shows the diffraction pattern from a crystal of turkey lysozyme. A pattern of 'spots' can be recognized, but an enormous amount of calculation, from the positions and intensities of the spots, is required before the three-dimensional structure of the protein may be constructed.

Figure 2.5 (a) Highly simplified representation of the molecule of haemoglobin. The molecule has four polypeptides (two α, two β). Regions of α-helix are shown as fat sausages and the haem groups (one per subunit) are shown as simple discs. In individuals with sickle cell anaemia just one amino acid in each of the β-chains is 'wrong' because of a mutation. This causes the haemoglobin molecules to stack up in the red blood cells at low oxygen concentrations and deform the shape of the corpuscles. **(b)** Some red cells undergoing 'sickling'. The deformed corpuscles get trapped in the tiny capillaries and are destroyed, leading to anaemia. Magnification × 1000.

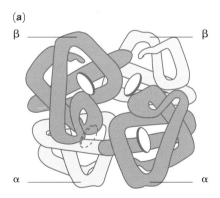

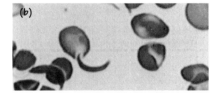

The consensus view at present seems to be that most, but perhaps not all, of the information that decides how a polypeptide should fold up to form an active protein resides in the sequence, i.e. in the primary structure (even though we do not know what the rules are). What is important is that the information for making proteins of a certain fixed sequence is carried from generation to generation. Human haemoglobin always has the same defined sequence of amino acids, and is different from, say, horse haemoglobin. This information is stored in the DNA of the nucleus of each cell of the body, and is carried on from generation to generation. Furthermore, if the information is corrupted, then faulty proteins may be manufactured which cannot carry out their biological function properly. They may have one or more 'wrong' amino acids in the sequence and we say that a *mutation* has occurred. One such mutation is the one that causes sickle cell anaemia (see Figure 2.5 and Spread 6.4, page 174).

The shapes of protein molecules

Proteins play many different roles in living organisms and, not surprisingly, many different shapes of protein molecule are found. The proteins that play a structural role, such as keratin, silk fibroin and collagen, have relatively simple fibrous structures. They are bulk materials whose job is to support and strengthen, and the long chains are insoluble in water. We will look at them first because they reveal some of the fundamentals about protein structure.

More complex are the soluble proteins, a class which includes the enzymes, the antibodies and the oxygen-binding proteins such as haemoglobin. These are globular in shape, and unlike the structural proteins, are, of course, soluble in water. Most importantly, however, they have on their surfaces one or more regions that have precise shapes that can interact with other molecules. The result of these interactions will almost certainly be binding of the other molecules, after which some 'event' may happen. In the case of enzymes this 'event' is chemical catalysis of a highly efficient and specific kind. Because enzymes and enzyme catalysis are so important they are given a separate chapter (Chapter 3).

Other types of binding are equally important to life, however. Haemoglobin binds oxygen (O_2) with a high degree of specificity (it does not bind N_2, for example), and binds it tightly or loosely as the circumstances demand so that oxygen is taken up by the blood in the lungs and released in the tissues. When an antibody binds to an invading bacterium in the body, a whole set of further interactions is triggered, resulting ultimately in the destruction of the invader.

The way in which these complex interacting shapes or surfaces in globular soluble protein molecules are generated is now reasonably well understood and will be explained shortly, but we should also not lose sight of that other important class of protein molecules, those that 'sit' in membranes. These have both hydrophobic and hydrophilic surfaces on them and they may also have certain internal shapes and cavities that allow them to act as channels which permit small molecules and ions to pass through the membrane in a controlled way.

First we will look at the structures of the fibrous proteins, which show the basic features of protein structure.

Fibrous proteins

In the early 1950s Pauling and his colleagues, working mainly with molecular models and a set of rules dictated by chemistry, analysed the possible shapes or conformations that polypeptides could adopt. One of the rules was that the peptide bond should be planar and that there could be no rotation of the amide bond (see Figure 2.3), for example. Despite the enormous range of possible protein structures, they found that relatively few conformations satisfied all the rules and were stable. Two major structures they could identify were called the α-helix and the β-pleated sheet (Figure 2.6). In these models the polypeptide chain followed a certain regular course, and non-covalent interactions between atoms either within the same polypeptide chain or between two neighbouring chains stabilized the structures. Keratin, the protein of hair, hoof and feathers, has a great deal of α-helical structure, whereas silk fibroin is almost completely β-pleated sheet in structure with no α-helix at all. The sequence of amino acids in the polypeptide chain is referred to as the primary structure, and so this next level of structure was referred to as the *secondary structure*.

Collagen is different from both keratin and silk fibroin. It has a helix type of structure, but with a different pitch from that of the α-helix. In collagen, three very long polypeptide molecules (e.g. each about 1000 amino acid residues long) are coiled together to form a triple helix which has great mechanical strength. This triple helix is unique to collagen. It will only form if every third residue is a glycine, and it is found that collagen polypeptides have a repeating structure:

-[Gly - Xaa - Yaa-]$_n$

where Xaa and Yaa are other amino acid residues, although these are often formed from the amino acid, proline (Figure 2.7).

Interestingly glycine, the smallest amino acid, is the only one that can 'fit' into the triple helix. When a mutation occurs, such that any other amino acid replaces glycine, the result is disastrous. Collagen triple helices do not form properly and are unstable; they are broken down in the connective tissue leading to weakness. Individuals with such mutations may suffer from a variety of problems from hyperflexible joints ('india rubber man'), to loose, weak skin, or brittle bones. Frequently the disability is so severe that the individual is confined to a wheelchair or bed. An example of such a genetic disease is *osteogenesis imperfecta* where there are weak and brittle bones.

Soluble proteins

Soluble proteins, and indeed membrane proteins too, have further levels of structure. In addition to regions of the protein molecule having, for

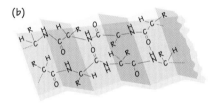

Figure 2.6 α-Helix and β-sheet are the two most common forms of protein secondary structure. **(a)** A right-handed α-helix; the pitch of the helix between two successive turns is 5.4 Å (0.54 nm) and there are 3.6 amino acid residues per turn. The cavity down the centre of the helix is too narrow to allow the penetration by solvent molecules, and this is probably responsible for the stability of α-helices. **(b)** Sketch to show how β-pleated sheets are formed when two polypeptides lie side by side. The polypeptides may run in the same direction ('parallel') or opposite directions ('anti-parallel'). In both α-helices and β-sheets, the forces that hold the structure together are hydrogen bonds. Although these are individually weak, there are many of them.

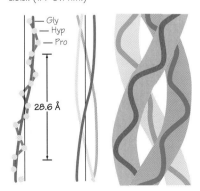

Figure 2.7 Triple-helical structure of collagen. Three polypeptides lie side by side and the pitch of the helix is much greater than in the α-helix. This structure may only be formed if every third residue in each polypeptide is glycine, the smallest amino acid. (1Å=0.1 nm.)

Figure 2.8 Denaturation opens out the tightly folded native three-dimensional configuration of a protein. If the polypeptide has disulphide cross-links, these may be broken with a reducing agent. The process of reduction and denaturation, (**a**) → (**b**) → (**c**), is often reversible. In the reduced, denatured state, biological activity is lost.

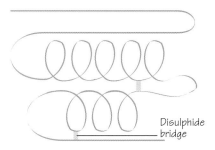

(**a**) Native molecule.

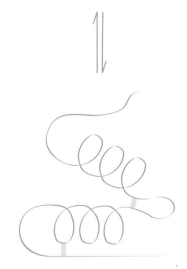

(**b**) Inactive molecule with -S-S bridges still intact.

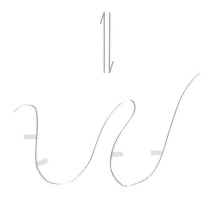

(**c**) Fully denatured, reduced inactive molecule.

example, β-pleated sheet and α-helical structure, the whole long polypeptide is folded to produce a compact, more or less globular, structure. This next level of structure is called *tertiary structure*. It may be stabilized by non-covalent interactions between atoms and groups in the protein, and in some cases by covalent links forming between cysteine thiol groups that become adjacent to one another in the folding process (Figure 2.8). These links may be between regions that are far apart in the amino acid sequence.

The twists and turns of the convoluted polypeptide may appear 'irregular' (at least compared with the regularity of α-helix or β-pleated sheet), but nevertheless they are highly specific. The protein always folds itself in this way and when this structure is disrupted, for example by heating, then the molecule loses its biological activity and is said to be *denatured*.

Haemoglobin is a molecule that has quite a lot (about 70%) of α-helical structure, and very little if any β-pleated sheet. In contrast, the immunoglobulin antibody molecule has a great deal of β-sheet and very little α-helix. Presumably these 'designs' of tertiary structure have evolved to enable the proteins to carry out their biological functions.

Both haemoglobin and the antibody molecule show a further level of structure, *quaternary structure*, which describes the fact that several polypeptides come together to form the biologically active molecule (Spread 2.1). In these cases there are four polypeptides, identical in pairs, although in all other respects the two proteins are as different from one another as could be imagined (Figure 2.9).

Surveying the many known protein structures reveals that they typically show mixtures of α-helix and β-sheet region linked by short sequences of less defined structure including loops and turns (Figure 2.10). More significantly, as we accumulate more and more information on protein tertiary structure, we start to recognize certain patterns or *motifs* from which the soluble protein is built up. Often the same motif will be found in two proteins with apparently different biological functions.

Proteins rarely 'go it alone'

Despite their apparent extensive versatility in being involved in a great number of biological processes, proteins themselves rarely work on their own. The majority of them have help in carrying out their functions, and the 'help' usually comes in the form of a small organic molecule, an inorganic metal ion, or sometimes both. The helper molecule or ion is referred to as a *prosthetic group* (a wooden leg is a prosthesis in medical terminology), but such groups are also called *cofactors*, or in the case of enzymes, *coenzymes*.

These small molecules or ions should be looked upon as means by which the versatility of proteins is increased. They provide sites of action or reservoirs for intermediate products, while the surrounding protein environment provides specificity and other things. To take a very simple example, a number of proteins called *cytochromes* are found in mitochondria and chloroplasts. Obviously the term 'chrome' has the implication that these are coloured and indeed they are a pinkish colour. Cytochromes are proteins that can carry electrons (e^-). Now it is quite difficult for a chunk of protein to carry electrons, but the cytochromes have at their centres an iron atom in a haem group which can exist as Fe^{2+} or Fe^{3+}, the difference between these being one electron. So here the prosthetic group is an electron carrier in the cytochrome (Figure 2.11, page 46).

There are a number of different electron-carrying cytochromes (although all have an Fe). The difference between them is that specific members of the cytochrome family will only accept electrons from certain donors and will only pass them on to certain acceptors. The importance of this will be seen in the section on electron transport in mitochondria and chloroplasts (see section 4.3, page 110). The iron supplies the electron-carrying function; the protein provides the biological specificity.

Haemoglobin is another protein that contains iron. As with the cytochromes, the iron is the centre of a complex organic molecule called a haem group which is linked to the protein. But in this case the surrounding protein and haem group provide the environment so that the iron can reversibly carry oxygen (rather than electrons). A loose complex is formed between the iron and the O_2 which is vital for haemoglobin to be able to pick up O_2 in the lungs and deposit it deep in the tissues. In fact, one molecule of haemoglobin is made up of four subunit polypeptide chains, each of which contains a haem group. Interestingly, the four subunits are non-covalently bound together (although they bind to each other very tightly), and the haem group is non-covalently linked to the protein (Figure 2.9).

We will now consider how this tight but non-covalent binding of one molecule to another can be achieved, and why it is so important in biological systems.

Proteins form surfaces with precise shapes

The special feature of the action of proteins in living systems arises from their unique ability to interact non-covalently with other molecules with very high specificity. For enzymes this means that molecules that are very similar to one another can be clearly distinguished by the enzyme catalyst and only the correct one will be selected (see Chapter 3). In the case of antibodies, 'invading' molecules can be recognized and dealt with whereas closely similar ones belonging to the body itself are ignored. What then are the principles behind this process of biological recognition: how do protein molecules recognize and bind tightly to certain molecules whilst ignoring others? In order to answer this question we have first to remind ourselves that proteins are macromolecules. This makes a great deal of difference when two molecules react. When two small molecules react together we can envisage the process as being one of collision — rather like two billiard balls striking each other — followed, under favourable circumstances, by reaction taking place.

When a small molecule collides with a protein molecule, the process resembles much more a baseball being caught in the glove of a catcher

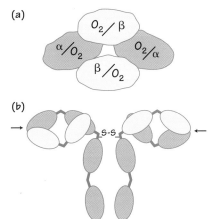

Figure 2.9 Highly simplified diagram to compare haemoglobin with the antibody molecule. Soluble proteins such as these have a more or less globular shape, and each has four subunits. (a) Haemoglobin has two α- and two β-subunits, each about the same size and each capable of binding one molecule of O_2 on its haem group. The four chains fit together very tightly but the binding between them is non-covalent. (b) The antibody molecule also has four polypeptides, but two are large and two small, and they are covalently linked together by disulphide bridges. In the antibody molecule, the polypeptides form a series of more or less globular domains linked together by more open stretches of peptide, two such portions in the so-called light chain and four in the heavy chain. The sites at which antigen binds are arrowed. The relative M_rs are: haemoglobin, 68,000; antibody, 150,000. Haemoglobin is mostly made up of α-helical sections with no β-sheet, but the antibody molecule is practically all β-sheet structure with no α-helix.

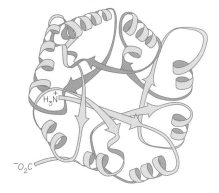

Figure 2.10 'Ribbon diagram' showing the backbone of the enzyme triose-phosphate isomerase. Note that this is highly simplified. It shows that this enzyme has α-helix and β-structure in roughly equal amounts. Folding produces an almost spherical globular protein.

Proteins: solubility, structure, shapes

objectives
- To describe the four levels of structure to be found in proteins (primary, secondary, tertiary, quaternary)
- To recognize that the specific coiling of the polypeptide produces the three-dimensional shape of the protein which gives it its biological function

Solubility: Globular proteins are generally SOLUBLE IN WATER or dilute salt solution because the polar R groups on the surface of the molecule are hydrophilic. Enzymes are globular proteins which function in aqueous solution.

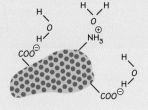

INSOLUBILITY IN WATER, so important to the function of many fibrous proteins, is caused by the many non-polar hydrophobic -R groups on their exterior surfaces.

The solubility of globular proteins is influenced by pH; the pH at which a protein is least soluble is called its ISOELECTRIC POINT. At this pH the protein has NO NET ELECTRIC CHARGE, so that there is no electrostatic repulsion between adjacent molecules and they coalesce and precipitate.

Negatively charged molecules repel one another: PROTEIN IS SOLUBLE ← high pH — PROTEIN IS INSOLUBLE — low pH → Positively charged molecules repel one another: PROTEIN IS SOLUBLE

PROTEINS are made by linking together hundreds or even thousands of amino acid residues in a linear chain. This polypeptide chain folds up in a characteristic way to give a protein with a characteristic shape and solubility properties.

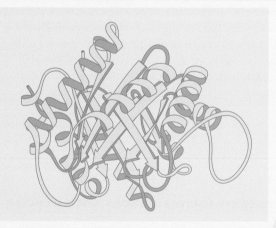

This rather complicated enzyme protein has regions of both α-helix (coils) and β-pleated sheet (broad arrows) linked by loops in the polypeptide. This arrangement produces the three-dimensional shape of the protein including the precise shape of the active site.

Quaternary

Several polypeptide chains may be fitted together to produce the quaternary structure of the protein. The stability of the quaternary structure is maintained by weak interactions between -R groups of adjacent polypeptide chains and by VAN DER WAALS FORCES between subunits. (Not all proteins have quaternary structure.)

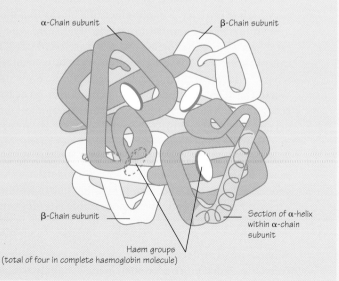

α-Chain subunit, β-Chain subunit, β-Chain subunit, Section of α-helix within α-chain subunit, Haem groups (total of four in complete haemoglobin molecule)

Life chemistry and molecular biology

2.1 Proteins: solubility, structure, shapes

Primary

Individual amino acids are joined together via peptide bonds by condensation reactions catalysed by enzymes.

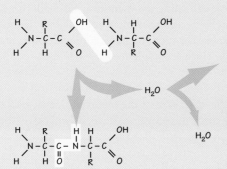

Successive condensations produce a linear chain of amino acids: this sequence of amino acid residues represents the primary structure of the protein. This primary structure is maintained by strong covalent bonds between adjacent amino acids.

Secondary

The polypeptide chains may take on regular arrangements, called the SECONDARY STRUCTURE of the protein. e.g. the α-HELIX. This secondary structure is maintained by hydrogen bonds between the C=O and N-H groups of every fourth peptide link.

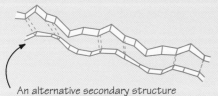

An alternative secondary structure — the β-PLEATED SHEET — has hydrogen bonds between peptide links of adjacent polypeptide chains.

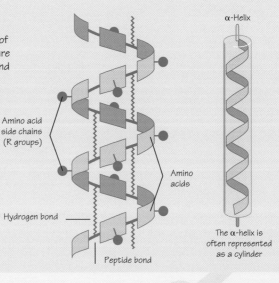

Tertiary

Sections of α-helix may be folded on themselves; this supercoiling of the α-helix represents the TERTIARY STRUCTURE of the protein. This three-dimensional shape or CONFORMATION of the protein is maintained by a series of interactions between -R groups on the polypeptide chain and is vital for the protein's function.

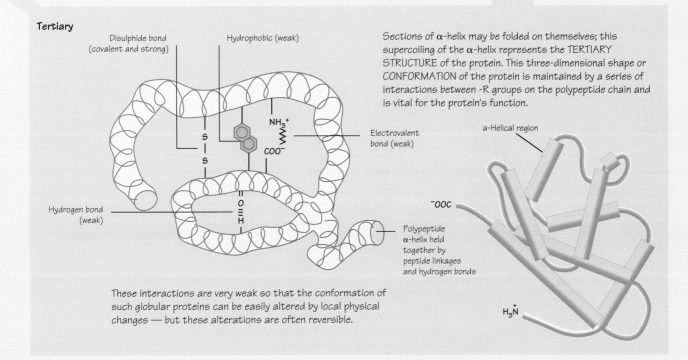

These interactions are very weak so that the conformation of such globular proteins can be easily altered by local physical changes — but these alterations are often reversible.

Chapter 2 • **Biological Molecules**

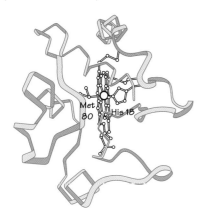

Figure 2.11 Diagram of the protein cytochrome c from tuna. The active part of the molecule is the haem group in the centre whose iron atom can carry an electron [Fe(II) $\rightleftharpoons$ Fe(III)]. In cytochromes the haem group is covalently linked to the polypeptide; in haemoglobin it is non-covalently bound and fits into a hydrophobic pocket in the folded protein.

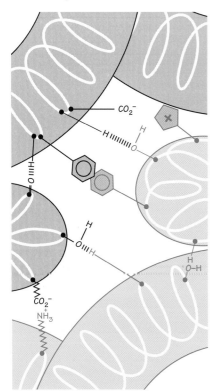

Figure 2.12 Sketch to show how two proteins (or subunits in the same protein) can interact to give highly specific but non-covalent binding. The shapes of the two proteins are complementary, and various weak bonds form (e.g. hydrogen bonds, salt links, van der Waals forces and hydrophobic interactions).

with the glove enfolding the incoming baseball. (One could imagine a spherical ball being accepted but a cubical 'ball' being discarded.) This happens because the protein macromolecule presents a surface to the incoming ball which is complementary in shape to the ball.

When two protein molecules collide the situation is even more complex because of the large size of the macromolecules. Large areas of each protein may come into contact. If — and only if — these areas 'match' very closely in shape will they stick together; in other words, binding of one to the other will take place. Herein lies the secret of biological specificity. Even a small difference in the complementary shape of the interacting areas will cause the binding to fail. This is why a rabbit antibody against horse serum albumin reacts much less strongly with human serum albumin, for example. This is how an enzyme can recognize one substrate but ignore a closely similar compound. Macromolecules make this recognition process possible and provide the basis for biological specificity.

Let us look at the interaction process a little more closely: from what we know of the chemical structures of the amino acid R side chains, we ought to be able to understand specific binding. As we have seen, a polypeptide folds up in a specific way to generate a more or less spherical, solubles, compacts, protein molecule. A close look at the surface of such a molecule will reveal that this surface is irregular and is studded with different types of amino acid side chains. Locally there will be bulges and clefts formed by the way in which the polypeptide has folded up.

We should now recall that the amino acid side chain R groups can have many different characters. A typical area of protein surface that interacts with a small molecule may consist of at least 10 amino acid side chains and these may be hydrophilic or hydrophobic in character, bulky or small, acidic or basic, and so on. When proteins themselves interact, the area of contact may be much larger, and the number of amino acid side chains that can interact with each other is enormous.

There are many ways for side-chain groups to interact. For example, hydrophilic groups will be attracted to hydrophilic groups of a complementary shape but will not interact with hydrophobic groups. The same applies to hydrophobic groups. Positively charged residues ($-NH_3^+$) will be attracted to negatively charged residues ($-CO_2^-$) quite strongly — provided of course that the shape allows this. Within hydrophilic regions hydrogen bonds may form between $-OH$, $-NH_2$ and other groups. Although hydrogen bonds are rather weak individually, many of them may form collectively producing a strong binding effect. If the surfaces can get really close together, then van der Waals forces may come into play. Again, these are individually very weak, but if the juxtaposition of two surfaces allows thousands of them to form, then a strong binding will occur. A very simplified, two-dimensional view of how this happens is shown in Figure 2.12.

Now we can begin to understand how it is that the subunits of, say, haemoglobin stick together so tightly, although non-covalently. The opposing surfaces 'match' each other very precisely. Under normal biological conditions (lowish temperature, pH near neutral) they are very difficult to prise apart. In contrast, heating denatures the three-dimensional shape of the protein and the subunit polypeptides easily separate. We will see that the relative ease of separation is also important in enzyme reactions when substrate is turned into product (see Chapter 3).

Other macromolecules also form three-dimensional shapes and can interact with one another, and with protein molecules, in similar ways to those described above; the principles are the same. It has to be said, though, that proteins are the example *par excellence* of this specific biological recognition by the interaction of surfaces. Probably for this reason proteins fulfil so many of the important functions of life.

We will go on to look at some of the other molecules of life. We have to study their chemistry to some extent, but the approach taken will be to look at biological function and to ask whether it can be understood or explained in terms of chemistry.

2.4 The carbohydrates — storage and structure

'Carbohydrate' is more often than not used almost as a term of abuse in everyday life. Too much carbohydrate is probably a bad thing from the point of view of a balanced diet, but it should be remembered that for the majority of the world's population carbohydrate, as starch, is the major bulk food; it is also the cheapest in the form of grain, roots and tubers. We should also remember that the chief carbohydrate of the body, glucose, forms the basis of cell energy metabolism, is readily used by all cells and is distributed around the body by the bloodstream.

Aside from their role as energy-giving foodstuffs, carbohydrates or sugars have a number of other functions in living organisms (Spread 2.2). They are stored in polymerized forms as starch or glycogen, and carbohydrate polymers form structural materials such as plant and fungal cell walls and the 'ground substance' that makes up a good deal of connective tissue. On the surface of cells, carbohydrates, usually attached to proteins and lipids, allow cells to be recognized as 'friend or foe'. The blood-group substances are examples here, and the mention of 'recognition' brings to mind the interaction of matching or complementary shapes that was discussed above for 'protein shapes'.

Carbohydrates tend to be hydrophilic and water-soluble unless polymerized. The basic monomer unit is the monosaccharide, but this term covers a very wide range of similar compounds.

Carbohydrate chemistry

Sugars, or carbohydrates, are compounds containing an aldehyde or a ketone group together with a number of hydroxy groups (hence their hydrophilic nature). The simplest carbohydrates are glyceraldehyde and dihydroxyacetone (Figure 2.13) which can be seen to be derived from

Figure 2.13 Structure of carbohydrates. The simplest carbohydrates are glyceraldehyde (a triose with an aldehyde group, i.e. an aldotriose) and dihydroxyacetone (a triose with a ketone group, i.e. a ketotriose). It is easy to see how these relate to glycerol. By adding –CHOH– units, four-carbon (tetroses), five-carbon (pentoses), and six-carbon (hexoses) sugars may be constructed. All the aldoses have alternative ketose forms (structural isomers). All (except dihydroxyacetone) have at least one asymmetrical carbon atom, giving the possibility of optical isomerism. Because much of Biology depends on the recognition of three-dimensional structures, the shapes of these molecules are important. (Note that the arrows in this diagram are intended to relate the structures; they do **not** indicate chemical transformations.)

Aldose sugars

$$\begin{array}{c} CH_2OH \\ | \\ CHOH \\ | \\ CH_2OH \end{array} \Rrightarrow \begin{array}{c} CH_2OH \\ | \\ C=O \\ | \\ CH_2OH \end{array} \Rrightarrow \begin{array}{c} CHO \\ | \\ HCOH \\ | \\ CH_2OH \end{array}$$

Glycerol — Dihydroxy-acetone — Glyceraldehyde (triose)

$$\begin{array}{c} CHO \\ | \\ HCOH \\ | \\ HCOH \\ | \\ CH_2OH \end{array}$$

Erythrose (tetrose)

$$\begin{array}{c} CHO \\ | \\ HCOH \\ | \\ HCOH \\ | \\ HCOH \\ | \\ CH_2OH \end{array}$$

Ribose (pentose)

$$\begin{array}{c} CHO \\ | \\ HCOH \\ | \\ HOCH \\ | \\ HCOH \\ | \\ HCOH \\ | \\ CH_2OH \end{array}$$

Glucose (hexose)

Functions of carbohydrates

objectives
- To be able to recognize carbohydrate structures and formulae.
- To list the functions of carbohydrates in micro-organisms, plants and animals
- To recognize the distinction between monosaccharides, disaccharides and polysaccharides

Sugars, or carbohydrates, are one of the major groups of biomolecules and they carry out a wide range of biological functions from the simple storage of energy to participation in the complex process of cell–cell recognition. As their name implies, they consist of carbon, hydrogen and oxygen, but often nitrogen is also present (e.g. in amino sugars such as glucosamine). Their chief characteristic is the possession of a number of hydroxy (-OH) groups (see Spread 2.3), but they usually also have aldehyde or ketone groups.

Carbohydrates can be looked on as relatively 'inexpensive' materials. Plants produce them in abundance by photosynthesis from CO_2 and H_2O (in contrast, molecules containing nitrogen are more 'expensive' because of the plants' difficulty in getting hold of fixed nitrogen, which is why we use fertilizers). For the majority of the people on Earth, carbohydrates such as starch are 'inexpensive', energy-supplying foods; e.g. rice, potatoes, grain, pasta, etc. Only people in the 'rich' nations can afford to eat large amounts of expensive meat protein.

SUGAR DERIVATIVES include: SUGAR ALCOHOLS, e.g. glycerol; SUGAR ACIDS, e.g. ascorbic acid (vitamin C); and MUCOPOLYSACCHARIDES (now called GLYCOSAMINOGLYCANS), which are important components of connective tissues, synovial fluid, cartilage and bone. Heparin (anticoagulant in blood) is a glycosaminoglycan and has a protective function (Figure 2.16).

GLUCOSE is the most common substrate for respiration (energy release) and is used by all cells. FRUCTOSE is a constituent of nectar and sweetens fruits to attract animals and aid seed dispersal. Fructose is about four times sweeter than glucose and about twice as sweet as sucrose. Glucose is often enzymically converted into fructose to get more sweetness per gram of carbohydrate in foods.

SUCROSE (GLUCOSE–FRUCTOSE) is the main transport compound in plants, and is commonly extracted from sugar cane and sugar beet and used as a sweetener. LACTOSE (GLUCOSE–GALACTOSE) is the carbohydrate source for suckling mammals — milk is about 5% lactose. Some adults lack the enzyme needed to split lactose into its component sugars prior to absorption from the intestine (Figure 3.1). They suffer from serious indigestion and diarrhoea because of the action of gut bacteria. MALTOSE (GLUCOSE–GLUCOSE) is a respiratory substrate in germinating seeds.

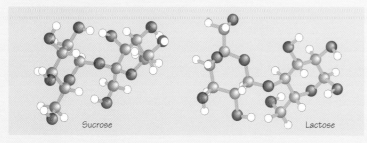
Sucrose — Lactose

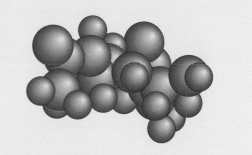
SUGAR DERIVATIVES, e.g. glycerol, ascorbic acid, heparin. Shown here is a model of N-acetylglucosamine.

Structure of glucose, a monosaccharide (aldehyde group at top, primary alcohol at bottom)

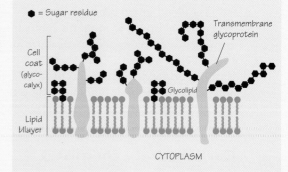
OLIGOSACCHARIDES are short (often 6–12 monosaccharide units) condensation products which combine with protein (GLYCOPROTEIN) or lipid (GLYCOLIPID) and form the outer coat (GLYCOCALYX) of animal cells. They are important in cell–cell recognition and the immune response.

Life chemistry and molecular biology

2.2 Functions of carbohydrates

STARCH (20% AMYLOSE/80% AMYLOPECTIN) is the major storage carbohydrate in plants. CELLULOSE is a major structural component in plants. This book is mostly made of cellulose. CHITIN is a constituent of arthropod exoskeletons and the cell walls in fungi.

 GLYCOGEN is the major storage carbohydrate in mammals and fungi, as well as in oysters and mussels.

DISACCHARIDES include SUCROSE (in sugar), LACTOSE (in milk) and MALTOSE (in germinating seeds).

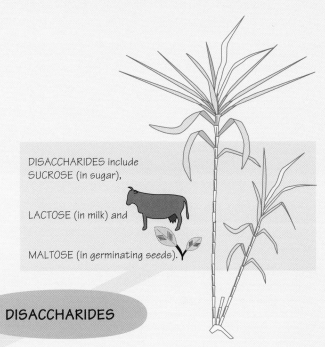

POLYSACCHARIDES

DISACCHARIDES

MONOSACCHARIDES

HEXOSES (C_6)

PENTOSES (C_5)

GLUCOSE and FRUCTOSE.

RIBULOSE 1,5-BISPHOSPHATE (RuBP) is the ACCEPTOR of CO_2 in the Calvin cycle of photosynthesis (see Spread 5.2).

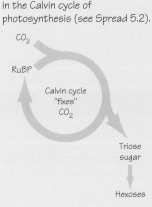

RIBOSE and DEOXYRIBOSE are constituents of NUCLEOTIDES

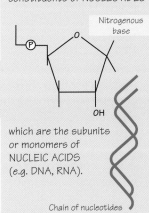

which are the subunits or monomers of NUCLEIC ACIDS (e.g. DNA, RNA).

Chain of nucleotides

Other important roles of pentoses are as part of the molecule in the ELECTRON CARRIERS NAD^+, FAD, and $NADP^+$ and in the 'energy currency' molecule ATP (adenosine triphosphate).

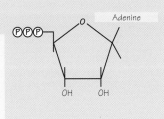

Chapter 2 • **Biological Molecules**

Figure 2.14 (a) Asymmetry generates optical activity when an atom (here E) has four different groups attached to it. This is represented more simply as a perspective formula **(b)** or as a projection formula **(c)**. In the case of the projection formula, this is now a two-dimensional representation of a three-dimensional object. It must not be taken out of the plane of the paper. **(d)** The two optical isomers of glyceraldehyde; **(e)** shows them as mirror images, representing the central carbon as a regular tetrahedron.

(a)
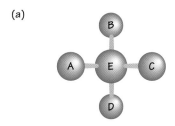

(b)
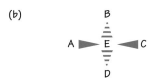

(c)
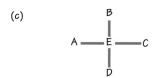

(d)

```
      CHO                CHO
   H—C—OH            HO—C—H
      CH₂OH              CH₂OH
```

(e)
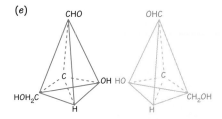

the trihydric alcohol, glycerol. By adding CHOH units more complex sugars may be built up, and indeed the most common sugars in cells are the pentoses (five carbon atoms) and the hexoses (six carbon atoms, e.g. glucose).

Carbohydrate monomers (*monosaccharides*) all have hydrophilic properties because the many –OH groups can form hydrogen bonds with surrounding water molecules, and they have characteristic shapes by which they can be recognized. Many, but not all, monosaccharide sugars also show reducing properties because of the aldehyde or ketone group. Common tests to detect these sugars such as Fehling's or Benedict's test depend upon their reducing properties.

An important feature of carbohydrates is their three-dimensional structure and isomerism (Figure 2.14). Most of the carbon atoms in a carbohydrate are tetrahedral and have four different groupings attached to them. You can easily show by building models that there are always at least two ways of doing this (Spread 2.3) although the isomers that are produced are very similar in chemical and physical properties. Chemically they are almost impossible to distinguish from one another. Physically, solutions of the compounds rotate the plane of polarized light in one direction or the other, that is to say the compounds are optically active. When a carbohydrate has several carbons that can show this asymmetry, the number of possibilities multiplies. In a hexose, for example:

$$CH_2OH(CHOH)_4CHO$$

the four (CHOH) carbons are each capable of having two conformations, giving $2^4=16$ possible compounds, only one of which is the common sugar, glucose.

This is important because living systems deal with shapes. When an enzyme 'recognizes' a carbohydrate molecule, the shape of the carbohydrate molecule fits very precisely into a complementary shape or cleft on the protein. Another carbohydrate that has one of its carbon atoms in the reverse or opposite configuration just will not fit. Although the chemical and physical properties of the two carbohydrates are very similar to each another, the *biological* properties are not, because proteins recognize them as having different shapes.

Incidentally, carbohydrates are not the only biological molecules to show this phenomenon. Amino acids too can exist in two different conformations; only one of these possible conformations is used in building proteins. It is easy to see why this has to be the case. When a complex polymer with three-dimensional shape is being built, inserting one amino acid of the 'wrong' conformation will send the polypeptide chain growing off in precisely the 'wrong' direction!

Another property of the carbohydrates is that those molecules with five or more carbon atoms in them tend to coil up to form rings, and these tend to be the usual form in which the molecules are found in cells. How this happens is shown in Spread 2.3. The link always involves the aldehyde (or ketone) group of the carbohydrate molecule. Glucose tends to form a six-membered ring, while fructose may form five- or six-membered rings. Ribose, a five-carbon carbohydrate which is found in ribonucleic acid (see below), forms a five-membered ring. Although these rings are often drawn as hexagons or pentagons for convenience, it should be remembered that they are not flat or planar rings because each of the carbons maintains its tetrahedral arrangement of bonds, so that the rings are always puckered (Figure 2.15).

The other important property of the carbohydrate molecules is that they can condense together to form multiple units and polymers, and this takes us into an area where sugars fulfil many of their biological roles.

Multiple monosaccharide units

Monosaccharide monomer units can join to give dimers or they can join together in large numbers to produce large polymers, examples of which are starch, glycogen and cellulose (see Spread 2.3).

Several of the dimers, or disaccharides, are important in everyday life. Lactose, which is a disaccharide formed by combination of galactose and glucose, is the sugar found in milk, and is the way in which carbohydrate is fed to infants. Inside the young animal's intestine the disaccharide is hydrolysed back to the two monosaccharide units ready for use by cells to provide energy.

Sucrose is a disaccharide formed by combination of glucose and fructose. It is the sugar we have in our tea, but its biological purpose seems to be for the transport round a plant of sugar formed by photosynthesis (Chapter 5).

The main storage material of many plants is the polysaccharide starch, which is formed by linking together thousands of glucose molecules. A similar polymer found in animal cells is glycogen.

It should be remembered that when two sugars are linked together, there are a large number of possible ways of making the link because of the presence of being several –OH groups on each monosaccharide molecule. In practice, only a few of these possibilities are realized and, in fact, the type of links that are formed have a very profound influence on the properties of the resulting polymer. A striking example of this is the comparison between starch and cellulose, both of which are 'polyglucose' molecules. In starch (and in glycogen) the majority of the links are described as α1,4, that is, they are between carbon 1 of one monosaccharide and carbon 4 of the next, and the hydroxyl group on carbon 1 has the 'α' configuration. In cellulose, the link is β1,4, the only difference being that the hydroxyl group on carbon 1 is in the β configuration. (See Spread 2.3.)

Starch and cellulose are quite different from one another. Starch will dissolve in hot water but cellulose will not, and only starch gives a blue colour with iodine. Because of the way in which long cellulose molecules can pack together, excluding water molecules, the material is extremely stable and resistant, making it an excellent material for forming cell walls

Figure 2.15 Pentose and hexose sugars tend to cyclize to form rings (see Spread 2.3). When the ring is formed from the straight-chain form there are two possibilities (and a new centre of asymmetry is generated). The projections shown in **(a)**, **(b)** and **(c)** are three ways of representing these rings for D-glucose. The puckered rings shown in **(c)** are the closest to the true three-dimensional shapes of these molecules, although they are the most difficult to draw.

(a) Fischer projection formulae.

(b) Haworth projection formulae.

(c) Representation as three-dimensional structures.

α-D-Glucopyranose   β-D-Glucopyranose

Carbohydrates: monosaccharides, disaccharides and polysaccharides

objectives

- To identify the structures of monosaccharides such as glucose and fructose (both hexoses) and ribose and deoxyribose (pentoses)

- To explain how the 'straight-chain' forms of hexoses and pentoses tend to form cyclical structures

- To describe how monosaccharide units may be joined together by condensation reactions to form disaccharides (e.g. lactose, sucrose) and polysaccharides (cellulose and starch)

GLUCOSE and FRUCTOSE are both MONOSACCHARIDES (single sugar units) with the typical formula $C_nH_{2n}O_n$. They each have six carbon atoms and are thus called HEXOSES (PENTOSES have five carbon atoms and TRIOSES have three). Glucose and fructose are isomers of $C_6H_{12}O_6$. The ALDEHYDE and KETO groups of glucose and fructose, respectively, can chemically reduce other compounds. Thus, both these molecules are REDUCING SUGARS.

In naturally occurring di-, tri- and poly-saccharides, monosaccharide rings are joined together by condensation reactions to form GLYCOSIDIC BONDS:

This most usually occurs between the aldehyde or keto group (i.e. the reducing group) of one monosaccharide and a hydroxy group of another monosaccharide (as in LACTOSE), or, more rarely, between reducing groups of adjacent monosaccharides (as in sucrose). Maltose is a reducing disaccharide formed from two molecules of α-glucose.

GLUCOSE occurs in two isomeric forms: α-glucose and β-glucose. Glucose monomers can join together in substantial numbers to produce large polymers, examples of which are starch, glycogen and cellulose. Because there are several -OH groups on each monosaccharide molecule, there are a large number of possible ways of making a link. Although, in practice, only a few of these possibilities are realized, the type of links that are formed have a profound influence on the properties of the resulting polymer. For example, STARCH and CELLULOSE are both 'polyglucose' molecules, but are quite different from one another. In starch, the majority of the links are α1,4 (i.e. between carbon 1 of one α-glucose monosaccharide and carbon 4 of the next), while in cellulose the link is β1,4.

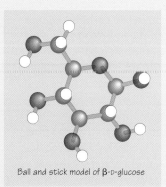

Ball and stick model of β-D-glucose

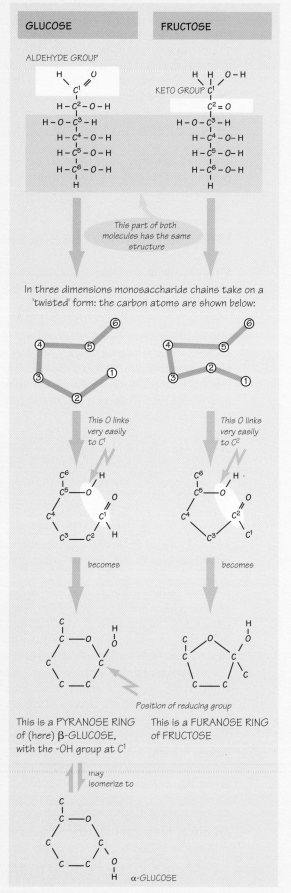

2.3 Carbohydrates: monosaccharides, disaccharides and polysaccharides

GLYCOSIDIC BONDS normally occur between the reducing group of one monosaccharide with the hydroxy group of another.

For example,
LACTOSE is a reducing disaccharide:

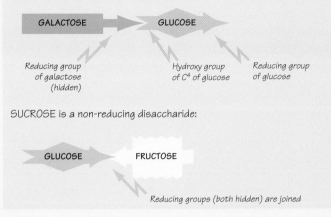

SUCROSE is a non-reducing disaccharide:

CYCLIC GLUCOSE OCCURS IN TWO ISOMERIC FORMS

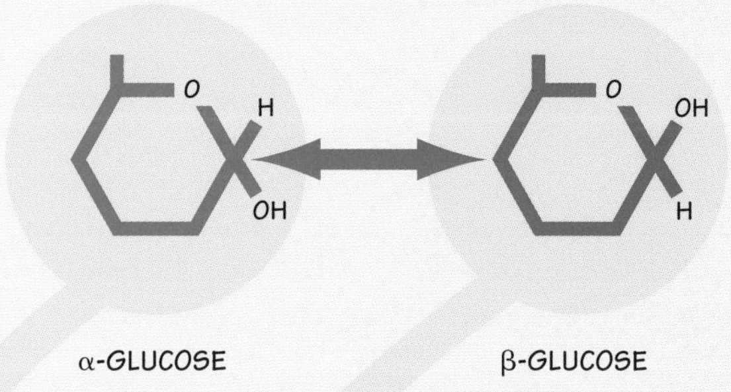

α-GLUCOSE ↔ β-GLUCOSE

STARCH is a mixture of two polymers of α-glucose. AMYLOSE typically contains about 300 glucose units joined by α1,4 glycosidic bonds:

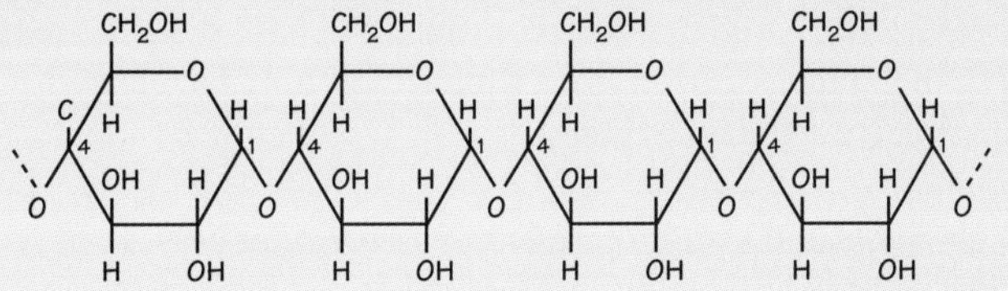

The bulky $-CH_2OH$ side chains cause the molecule to adopt a helical shape (excellent for packing many subunits into a limited space):

Amylose helix
(six glucose units in each turn)

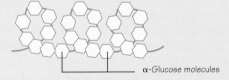

α-Glucose molecules

AMYLOPECTIN is a branched-chain polymer, containing up to 1500 glucose subunits, in which α1,4 chains have branches at α1,6 glycosidic bonds:

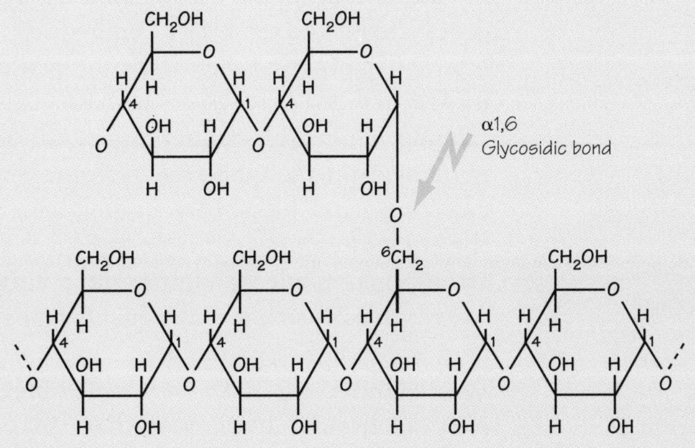

α1,6 Glycosidic bond

CELLULOSE is a polymer of glucose linked by β1,4 glycosidic bonds. The β-conformation inverts successive monosaccharide units so that a straight-chain polymer is formed:

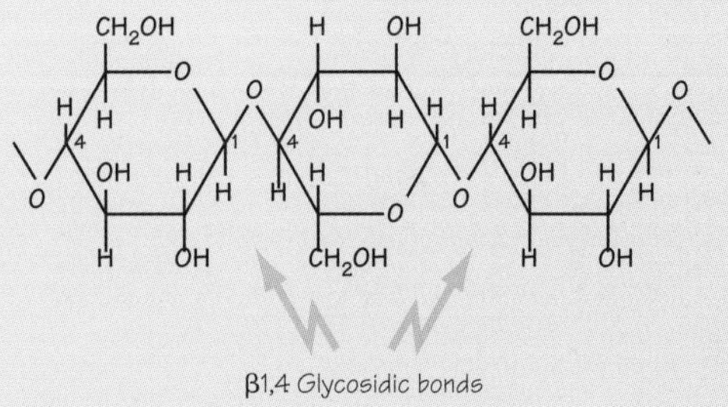

β1,4 Glycosidic bonds

The parallel polysaccharide chains are then cross-linked by hydrogen bonds:

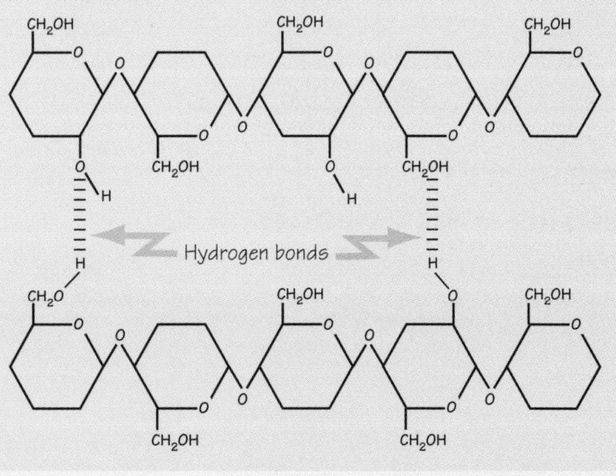

Hydrogen bonds

GLYCOGEN is an α-glucose polymer, very similar to amylopectin, but with very many more α1,6 cross-links and shorter cross-links and

Chapter 2 • **Biological Molecules** 53

in plants. Animals do not have digestive enzymes for breaking down cellulose, whereas they can easily hydrolyse starch in the intestine. Ruminant animals, which consume a lot of cellulose-containing plant material, rely on micro-organisms in the gut to break down the cellulose.

Some types of starch have branch points (see Spread 2.3), making the molecule closely similar to that of glycogen. The tree-like structure formed produces a more or less spherical molecule. In the case of glycogen in animal cells, this macromolecule can grow or shrink by the addition or removal of glucose units, depending whether an animal is well fed (and is storing energy) or is starving (and requires an internal source of energy).

Storing glucose inside cells, as a food storage material, is very convenient. When thousands of glucose molecules are linked into one molecule, the solute pressure (osmotic pressure) exerted is very much reduced. If the cell kept a high concentration of glucose itself in its cytosol, then this would exert a high solute pressure, tending to draw water into the cell which would have to be pumped out — with the expenditure of energy. Solute pressure depends on the colligative properties of matter, that is to say, on the number of particles present (i.e. in this respect one glycogen molecule is equivalent to one glucose molecule). Therefore, having 1000 glucose molecules joined together in one glycogen molecule reduces the osmotic pressure 1000-fold. It is also convenient to have the glucose store tucked away in a single location from which it can only be released at the appropriate time by cellular enzymes.

Cellulose, glycogen and starch represent three forms of glucose polymer that have different biological functions: structural in the first case and foodstore in the second and third. Many other polysaccharides with important biological functions are known. Chitin, for example, is a polysaccharide formed by linking together a modified glucose molecule called N-acetylglucosamine in β1,4 links (Figure 2.16). Chitin forms the exoskeleton of arthropods and crustaceans and in the latter group it may become 'calcified' with crystals of calcium carbonate and calcium phosphate to produce the material we are familiar with in crab shells.

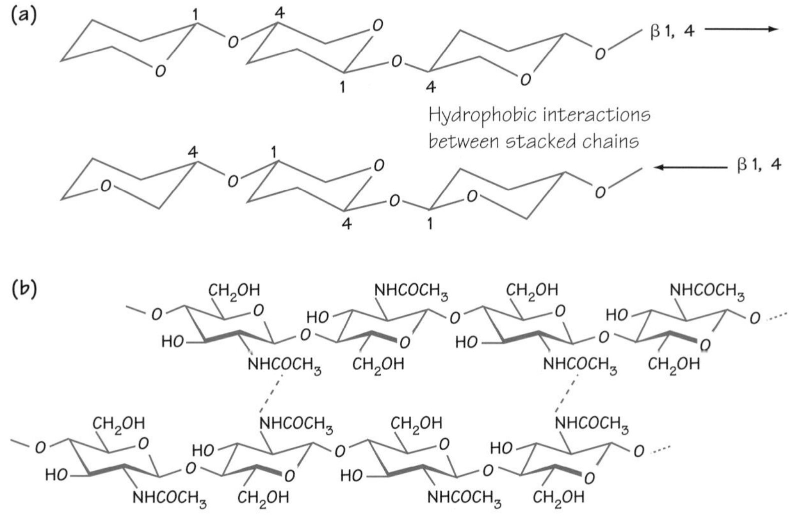

Figure 2.16 Structures of **(a)** cellulose and **(b)** chitin from insect cuticle. The outline structure of cellulose shows how the polysaccharide chains, running in opposite directions, can pack together to give a strong and water-insoluble structure. Cellulose is linear poly(glucose), whereas chitin is poly(N-acetylglucosamine) see Spread 2.2). In both cellulose and chitin the sugar residues are linked together β1,4.

Chitin is very similar to cellulose in its molecular structure. It plays a major role in insects, where the compound eyes have lenses of chitin. In the insect exoskeleton it may be cross-linked and tends to be embedded in a matrix of a protein called sclerotonin which undergoes an oxidative process called tanning. This results from the generation of quinones from the amino acid tyrosine which can form cross-links. Insect wing, which is the envy of engineers for its lightness and strength, is made of almost pure chitin, whereas black beetle cuticle is highly tanned.

In vertebrates there is a very wide range of polysaccharide materials that play a structural or mechanical role. These materials are typically made of modified monosaccharide units, and the most usual structure is a disaccharide repeating unit, $-(A-B)_n-$, which is also found in heparin (Figure 2.17). These materials are now called glycosaminoglycans, and such units composed of about 160 sugar residues are often joined to a central protein core in a test-tube-brush sort of fashion. The resulting huge molecules called proteoglycans are very important in forming the ground substance in bone and cartilage, and also the slimy material in the joints that has a lubricating function.

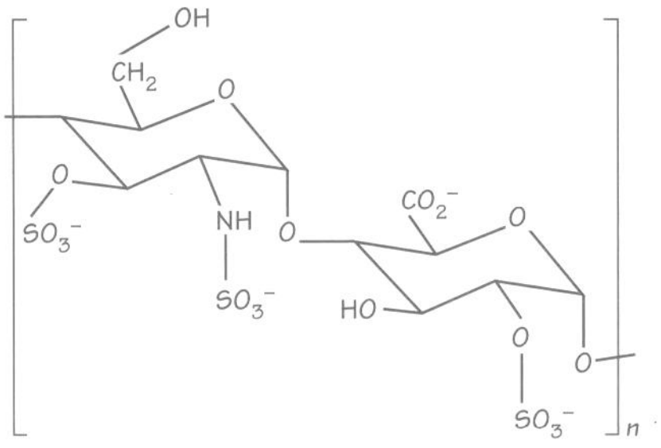

The sulphated sugar residues of the glycosaminoglycan are both highly charged (see Figure 2.17) and hydrophilic, and so they attract water and oppositely charged ions. The sugar chains of the proteoglycans tend to be stiff and stand out from the central protein core. One molecule of proteoglycan can occupy a great deal of space in an aqueous solution because of all the water molecules associated with it. When mechanical pressure is applied to such a (usually rather slimy) solution, the water molecules can be squeezed out. When the pressure is released, water molecules rush back in, and so a solution of proteoglycans can act as a sort of shock-absorber.

In contrast, when calcium salts are deposited in this matrix of proteoglycans and there are collagen fibres present, then the flexible cartilage becomes more and more rigid, to form bone.

To summarize, carbohydrates are a major source of energy-providing food (the metabolic steps will be dealt with in Chapter 4), but they may also be stored as polymers in cells. Polymeric carbohydrates also form an excellent basis for the construction of mechanical elements in an organism, from plant cell walls to insect exoskeleton to bone, not to mention the lubricating and slimy materials that are so important to the life of organisms such as slugs and snails or in the articulating surfaces between vertebrate bones.

Figure 2.17 Heparin is a complex, sulphated polysaccharide which prevents blood from clotting. It is made up of a repeating disaccharide unit, $-(A-B)-_n$, where A is doubly sulphated glucosamine and B is sulphated glucuronic acid. It is produced in granules in the so-called mast cells that line arterial walls, and when released, causes a large increase in antithrombin which binds to thrombin and prevents clotting. Heparin is thought to be released during injury to prevent 'runaway' clot formation and is used clinically for clot prevention.

2.5 The nucleic acids

The third group of macromolecules in living organisms is the nucleic acids, DNA and RNA. DNA stands for deoxyribonucleic acid and RNA stands for ribonucleic acid. They were discovered over 100 years ago and were found to be present in practically every cell. They are acids, because, for example, they can form sodium salts, but their structure and real importance in cells was not realized until the 1950s. In 1953 Watson and Crick showed that a double helix of DNA (see Spread 2.4) was capable of doing two vital things needed by every organism. One was to carry information, the so-called genetic information, that makes a cat a cat and a cabbage a cabbage. The other was that it could duplicate the information accurately, given the appropriate enzymes, so that when a cell divides, each of the daughter cells contains the same set of instructions as the parent cell.

The realization that nucleic acids are *informational* macromolecules was an important step forward in Biology. Previously nucleic acids, from what was known of their structure, were believed to be too simple to carry all the information for making a cell, and it was generally believed

The nucleic acids: deoxyribonucleic acid (DNA)

objectives

- To describe the molecule of DNA, a polydeoxyribonucleotide
- To be able to recognize the nitrogenous bases that occur in DNA which form a linear sequence or code that carries the genetic message
- To explain how the bases pair (A:T, C:G) between two complementary strands (molecules) of DNA to form the double helix

DNA (and RNA; see Spread 2.5) are polynucleotides made by joining together many nucleotide units. A nucleotide contains a nitrogenous base, a sugar (deoxyribose in the case of DNA) and a phosphate. By making a chain of repeating sugar–phosphates, a linear sequence is created of the bases attached to the deoxyribose. This sequence of bases carries the genetic message in a four-letter alphabet. The sequence in the DNA of a human cell is millions of bases long. A gene is part of such a sequence that codes for (carries the instructions for making) one polypeptide chain.

There are FOUR different nucleotides in a DNA molecule: they differ only in the organic base present. There are two different PYRIMIDINE (SINGLE RING) BASES, called CYTOSINE (C) and THYMINE (T), and two different PURINE (DOUBLE RING) BASES, called ADENINE (A) and GUANINE (G). The different dimensions of the purine and pyrimidine bases are extremely important in the formation of the double-stranded DNA molecule.

Nucleotides are linked to form a POLYNUCLEOTIDE by the formation of 3′5′-PHOSPHODIESTER LINKS in which a phosphate group forms a bridge between the C-3 of one sugar molecule and the C-5 of the next sugar molecule.

BASE PAIRING IN DNA was proposed to explain how two polynucleotide chains could be held together by hydrogen bonds. To accommodate the measured dimensions of the molecule, each base pair comprises ONE PURINE–ONE PYRIMIDINE. The double helix is most stable, that is the greatest number of hydrogen bonds is formed, when the base pairs A:T (two hydrogen bonds) and G:C (three hydrogen bonds) are formed. These are COMPLEMENTARY BASE PAIRS.

Because of base pairing, the base sequence on one of the chains in the double helix automatically dictates the base sequence of the other; that is, the chains are COMPLEMENTARY. Another property of the double helix is that the chains are ANTI-PARALLEL; that is, one chain runs from 5′ to 3′ while the other runs from 3′ to 5′.

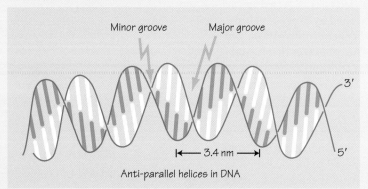

Anti-parallel helices in DNA

NUCLEOTIDES are the subunits of nucleic acids, including DNA. Each of these subunits is made up of:

AN ORGANIC BASE
+
A PENTOSE SUGAR
+
A PHOSPHATE GROUP

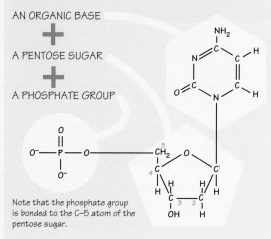

Note that the phosphate group is bonded to the C-5 atom of the pentose sugar.

THE FOUR BASES OF DNA

PYRIMIDINE (single ring) bases: Cytosine, Thymine

PURINE (double ring) bases: Adenine, Guanine

BASE PAIRING IN DNA
Each base pair consists of ONE PURINE–ONE PYRIMIDINE. The COMPLEMENTARY BASE PAIRS are
A = T (two hydrogen bonds) G ≡ C (three hydrogen bonds)

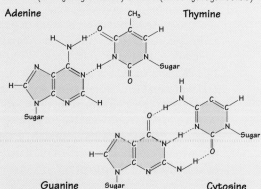

Note that in order to form and maintain this number of hydrogen bonds the nucleotides are inverted with respect to one another so that the phosphate groups face in opposite directions.

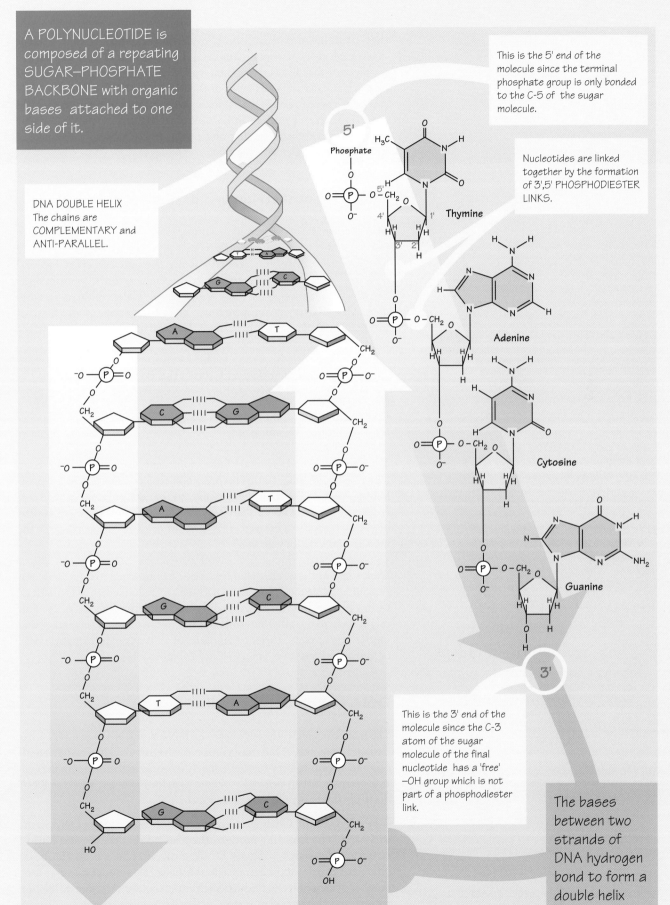

that proteins did this job. It was the realization that a large macromolecule can contain hundreds of thousands of instructions as a simple code that allowed this great leap forward in our understanding of how organisms and cells work at the molecular level. Since the 1980s ways have been found of manipulating this genetic information, and this has led to a great proliferation of technical advances including genetic engineering and the possibility of gene therapy (Chapter 7). This part of Biochemistry is referred to as *molecular biology*.

Fundamental structure of nucleic acids

Information is usually stored in a linear fashion. Perhaps the easiest to understand is the magnetic tape in an audiocassette where regions of magnetism of different intensity along the tape can store musical information, but the same is true of gramophone records, compact disks and diskettes, although the track is now spiral. The medium (tape, vinyl disk, CD plastic) is unimportant: it is the sequence of 'bits' of information that carries the message. This will help us to understand the fundamental structure of nucleic acids.

Chemical analysis of nucleic acids shows that they always contain :

(a) phosphate
(b) a sugar, either ribose or deoxyribose, and
(c) four different organic bases.

The phosphates and sugars are linked together to form a very long chain, (sugar–phosphate)$_n$. In this respect, nucleic acids can be looked on as being rather like polysaccharides except that the polymerizing unit is (sugar–phosphate) rather than (sugar). This backbone is the equivalent of the inert tape in a cassette. It is the four organic bases that carry the message: how do they do this?

Each sugar of the nucleic acid backbone has attached to it one base out of four possible ones (Spread 2.4). These can be looked on as the 'letters' of an alphabet, but this alphabet only has four letters instead of 26. This might seem to be rather limiting, but it is not: we should remember that modern computers store vast amounts of all types of information with an alphabet of just two 'letters', 0 and 1 (Figure 2.18).

DNA — the double helix

The main chemical difference between DNA and RNA is that DNA has deoxyribose and RNA has ribose as the sugar. This rather small difference leads to very fundamental differences in their roles in the cell. Importantly, DNA normally occurs as a double helix, whereas RNA normally does not: why is this so crucial?

When Watson and Crick were carrying out their studies on DNA in the early 1950s by building models, they found that they could build a structure consisting of two DNA molecules in which the bases interacted in pairs at the axis of a double helix. This model accounts for a number of experimental findings such as the molecule's X-ray diffraction pattern and also the relationship between the numbers of bases. In any

Figure 2.18 In computers (and other storage media such as digital audiotape) information is stored as a string of ones and zeros, as in **(a)**. In practice these represent a pulse or no pulse, or magnetism or no magnetism. In DNA **(b)** the genetic information is stored as a four-letter code in a sequence of bases (A, C, G, T). In each diploid human cell the code is about 10^9 bases long.

(a)
0110100100110110101 0010
101010010110101110 10101
10101010010101111 110100
0101011000111101 0100101
101011010101011 11101000
1101001011010001 1110100

(b)
GAGTACGCATCGCGTACGCTAGC
TGCGCGTATGAGATCGAACTAGT
TCGAGCTAGTCGAGCTGATCGAT
TGAGCATGACTAGAAAGCTAGCT
TGATGTACATGTAAGTAGTCGCG
TAAAGGATCGAAGCGCTCGATCG

given sample of DNA, there was always the same amount of one of the bases, adenine, or A, as there was of thymine, T, and similarly there were always the same amounts of the other two bases, cytosine, C, and guanine, G (see Spread 2.4). Watson and Crick argued that if A always matched T, and C always matched G, in the double helix, then a structure could be built in which the two strands were *complementary* (Figure 2.19). This immediately suggested a way in which the DNA double helix could be duplicated (usually 'replicated') at cell division. Simply separate the two molecules (or strands) of the double helix and build up matching or complementary strands on each of them and you've got two identical DNA double helices where you originally had one. (This important process of *replication* is dealt with in Chapter 6.)

In eukaryotic cells the majority of the DNA is in the nucleus in the chromosomes (but there is also a little in mitochondria and chloroplasts). It is packaged in a way that we only partly understand at present, by various complicated coilings (see Spread 6.2, page 168), but this packaging is clearly vital, as every nucleated human cell in the body contains over one metre of DNA double helix. The packaging somehow allows various parts of the DNA to be active at given times. These parts of the DNA are called *genes,* and as a result of Watson and Crick's work in 1953, Biochemistry and Genetics started to move on a common track. The basis of this was that the chemistry of the genes could be understood. This topic is dealt with in Chapter 6.

RNA — the mediator of translation

The information in the DNA of cells (or the genes) is the set of instructions on how to build proteins. To put this another way, the information that says how a sequence of several hundred amino acids is to be strung together to make a particular polypeptide lies in the sequence of bases along a molecule of DNA. RNA provides the mechanism for translating the 'language' of DNA into the 'language' of protein biosynthesis (see Chapter 6).

Chemical analysis of RNA shows that it always contains :

(a) phosphate
(b) a sugar, ribose, and
(c) four different bases.

Three of these bases are the same as those in DNA — adenine, cytosine and guanine (A, C, G) — but the fourth base is uracil (U) rather than thymine (T) (see Spread 2.5). RNA can also join in a double helix with DNA to form what is called a 'DNA–RNA hybrid'. This will only happen if the base sequences are complementary, and in this case the U in RNA matches with A in DNA.

Three types of RNA are found in all cells and they all participate in the process of protein synthesis. A very important experimental finding was that, although the genetic information is stored in the nucleus of eukaryotic cells, the process of making protein goes on in the cytoplasm on the ribosomes. Analysis of ribosomes showed that they contain about 50% RNA (called ribosomal RNA, or rRNA) and 50% protein, and that about 80% of the RNA in cells is rRNA. However, the ribosomes do not contain any specific information for making proteins.

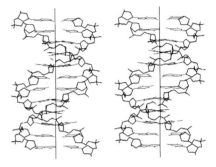

Figure 2.19 Computer-produced stereo image of the DNA double helix showing how the nucleotide bases interact across the axis. The image may be viewed in three dimensions by holding a piece of card between the two images in line with the nose. Many people can view such stereo images without separating them. If you want to try this proceed as follows. Hold the picture 20 cm or so from your eyes in a good, even light (if spectacles are worn, remove them). Try focusing at a distance, 'see' three images, and concentrate on the middle one: the pictures will merge to give you a three-dimensional image. (Courtesy of J.C. Stockert, Universidad Autónoma de Madrid.)

The nucleic acids: ribonucleic acid (RNA)

objectives

- To describe the structure of RNA as a polyribonucleotide based on ribose with four nitrogenous bases: adenine, guanine, cytosine and uracil
- To explain that RNA generally exists as a single-stranded molecule
- To recognize that there are at least three types of RNA in the cell: ribosomal RNA, transfer RNA and messenger RNA, each having its own specific biological function

Although both DNA and RNA are polynucleotides, RNA differs from DNA in three ways:

(a) the sugar–phosphate backbone contains RIBOSE rather than deoxyribose

(b) RNA contains the pyrimidine base URACIL (U) rather than thymine (T); uracil can pair with adenine, but

(c) RNA exists as a single strand rather than as a double strand.

The sole function of DNA is to carry the genetic message. RNA has several functions in the cell as a messenger (mRNA) and as a carrier (tRNA) bringing the amino acids to the site of protein synthesis (rRNA). There will be hundreds or thousands of types of mRNA corresponding to different polypeptides, and at least 20 types of tRNA corresponding to the 20 amino acids.

MESSENGER RNA is single-stranded RNA formed by TRANSCRIPTION of a sense strand of a double-stranded DNA molecule. (The sense strand is the one from which mRNA is made.) Transcription occurs in the nucleus, and permits the flow of genetic information (in the form of triplets of nucleotide bases) from the nucleus to the sites of protein synthesis at the ribosomes in the cytosol.

TRANSFER RNA 'reads' the genetic code in that it is able to pick up specific amino acids and deliver them to the ribosomes. Here, the amino acids may be assembled into proteins according to the instructions carried on messenger RNA. This assembly process is called TRANSLATION.

The RIBOSOMAL RNA molecules are synthesized in the nucleolus, and the many ribosomal RNA molecules are transported via nuclear pores to be assembled in the cytosol. The ribosomes are the sites of the translation stage of protein synthesis and contain both rRNA and protein.

When a length of mRNA is being translated to make protein molecules, there are often a number of ribosomes working their way along one mRNA molecule. Such a structure, shown in the electron micrograph below, is called a POLYRIBOSOME or POLYSOME.

During translation, POLYSOMES are formed when a number of ribosomes are working their way along one mRNA molecule at the same time. Magnification × 260,000.

RNA is composed of nucleotides, but differs from DNA in that ...

The pyrimidine base, URACIL, replaces the base thymine of DNA. When intramolecular (RNA–RNA) or intermolecular (RNA–RNA or DNA–RNA) base-pairing occurs, uracil hydrogen-bonds to adenine.

The pentose RIBOSE, which has an –OH group at the 2' carbon atom, replaces 2'–deoxyribose.

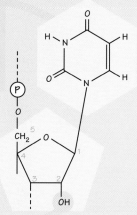

......and RNA is typically single-stranded, although it may assume complex structures depending on its function.

MESSENGER RNA is a single-stranded RNA formed by TRANSCRIPTION of a sense strand of a double-stranded DNA molecule.

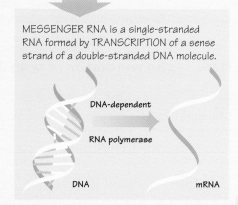

TRANSFER RNA is single-stranded but has a complex three-dimensional shape formed by intramolecular hydrogen bonding.

Simple 'clover-leaf' representation of tRNA:

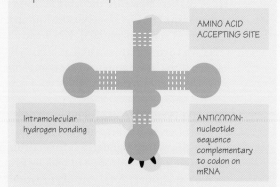

Note that in order to form and maintain this number of hydrogen bonds the nucleotides are inverted with respect to one another so that the phosphate groups (here shown as) face in opposite directions.

2.5 The nucleic acids: ribonucleic acid (RNA)

THE FOUR BASES OF RNA

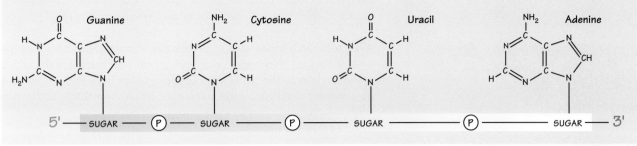

RNA is single-stranded, but contains local regions of complementary base-pairing that can form by a random matching process.

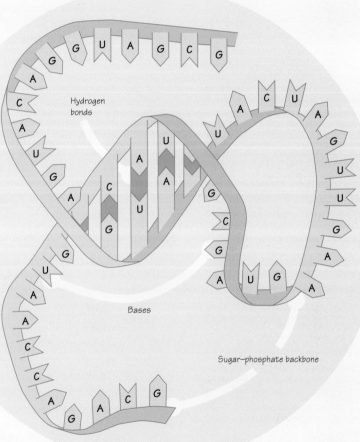

Hydrogen bonds

Bases

Sugar–phosphate backbone

RIBOSOMAL RNA is single-stranded but is folded into a complex series of shapes which form aggregates with ribonuclear proteins in the RIBOSOMES.

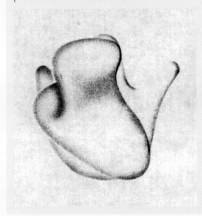

Ribosomes are made up of rRNA and protein, and prokaryotic and eukaryotic ribosomes differ in composition.

PROKARYOTIC, e.g. **Escherichia coli ribosomes (sedimentation coefficient 70 S)**

    Small subunit: one rRNA molecule plus 21 different polypeptides.
    Large subunit: two different rRNA molecules plus 31 different polypeptides.

EUKARYOTIC, e.g. **rat liver ribosomes (sedimentation coefficient 80 S)**

    Small subunit: one rRNA molecule plus 33 different polypeptides.
    Large subunit: three different rRNA molecules plus 49 different polypeptides.

Figure 2.20 Transfer RNA which is about 80 bases long. A is the site where an amino acid may be linked and C is the part that recognizes a sequence in mRNA.

Replication means producing a replica of the DNA double helix. Where there was originally one double-stranded molecule there are now two identical ones. Although the double helix is duplicated, the process is semi-conservative (see Chapter 6) so that the two new double helices each contain one old strand and one new one.

Transcription means 'writing' the information (the sequence of bases) in the DNA into the sequence of mRNA. Only one strand of the DNA is used as the template.

Translation means translating the 'message' (the sequence of bases) in mRNA into the sequence of amino acids in a polypeptide.

Figure 2.21 Some RNA molecules can act catalytically, that is, they are enzymes or 'ribozymes'. In the example shown here, self-splicing takes place in which a stretch of RNA ('intron sequence') catalyses its own removal from a ribosomal RNA precursor molecule.

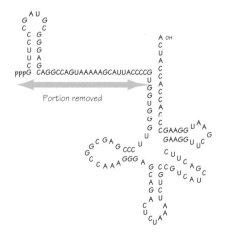

They act as a sort of 'workbench' where proteins are made, but the information to direct building of the sequence of a protein comes from elsewhere. We still do not really know how ribosomes function.

With some difficulty it was eventually shown that there was an intermediary, or 'messenger', that brought the information in the DNA in the nucleus to the ribosomes in the cytosol. This messenger was another sort of RNA, called messenger RNA, or mRNA. Different mRNAs contained the information for making different proteins and so we find a range of sizes.

Messenger RNA makes up only about 1% of the total RNA in a cell, and it is rather short-lived, which made it difficult to identify. This short-lived nature is not really surprising, however, because cells want to make certain proteins at certain times and then at other times they want to make other proteins, so the message must change.

Messenger RNA is made in the nucleus by a process known as *transcription*. This means literally 'writing' the information in the DNA into the language of mRNA. Naturally, the information is complementary, in other words where there is a C in DNA there will be a G in mRNA, and where there is an A there will be a U. When we are talking about DNA and protein synthesis, two words are important: *transcription* and *translation*.

The third type of RNA that is found in cells is transfer RNA, or tRNA (Figure 2.20). The molecules of tRNA are very small (relatively speaking) and contain only about 80 bases or nucleotides (compared with, say, hundreds or thousands in mRNA, and 10^9 in DNA). It is found that there is at least one tRNA corresponding to each amino acid, and this gives us the clue to what tRNA does. Molecules of tRNA act as 'adaptors'. They can recognize, on the one hand, specific amino acids, and on the other hand specific sequences of bases in mRNA. This enables polypeptides to be made by linking together sequences of amino acids based on the sequence of bases in mRNA. How this all happens is explained in Chapter 6.

Transfer RNA molecules are slightly unusual in that a number of non-standard or modified bases are present, such as pseudouridine. It is not really known why this should be. Transfer RNA molecules also tend to form loops within molecules, between internal complementary regions, giving the tRNA molecules characteristic shapes (Figure 2.20). This no doubt has a role to play in their adaptor function.

Some molecules of RNA have recently been found to have catalytic or enzyme activity (Figure 2.21). It has been suggested that in the first primitive life forms that appeared on earth, RNA may have been the key molecule that contained both information and catalytic power.

2.6 The lipids: fats and oils

So far we have looked at three classes of molecules — the carbohydrates, the proteins and the nucleic acids — that make up the bulk of living matter. A characteristic feature of these molecules is their large size; macromolecules play very important roles in life. The fourth class of molecules, the lipids, are equally important but do not form macromolecules. However, their molecules do tend to stay together to form droplets and bilayers and it is this tendency that enables them to carry out many of their biological functions.

We are familiar with the fats (lard, butter, dripping) and the oils (olive oil, sunflower oil) used in the kitchen. These are all lipids, but this term covers other substances as well, such as the phospholipids and glycolipids, that play a vital role in forming biological membranes.

Fatty acids and triacylglycerols

A good place to start thinking about lipids is the fatty acids, although, as we shall see, lipids are a very heterogeneous collection of compounds. A fatty acid has the chemical formula $CH_3(CH_2)_nCO_2^-$, where n can be anything from one to 18 or 20 or more. In the latter case we speak about 'long-chain fatty acids'. Long-chain fatty acids do rather curious things when placed in water, and an understanding of this behaviour enables us to understand biological membranes.

In water, long-chain fatty acids tend to arrange themselves so that their hydrophobic long chains associate together, while their hydrophilic carboxylic acid groups face outwards interacting with the water. In this way a spherical globule called a *micelle* forms (Figure 2.22). Such structures can act as soaps or detergents. Oily materials ('dirt') will find their way into the hydrophobic interior of the micelle because of the hydrophilic carboxy groups pointing outwards. Thus, the whole thing becomes water soluble and may be washed away with water.

Soaps are obtained by boiling animal fats with sodium hydroxide, and a soap is actually the sodium salt of a long-chain fatty acid, $CH_3(CH_2)_nCO_2^- Na^+$. The process of boiling with NaOH is called *saponification* (literally: soap-making), and the animal fat that is saponified is a *triacylglycerol*, or triglyceride, which is an ester formed from the trihydric alcohol, glycerol, and three molecules of long-chain fatty acid. Ester formation is a condensation reaction (with the removal of three molecules of water), and its reverse, saponification, is therefore a hydrolysis (see Figure 2.1 and Spread 2.6).

Triacylglycerols are referred to as 'neutral fats' because, as can be seen from the formula in Spread 2.6, they carry no charge and are very hydrophobic. Molecules of triacylglycerol tend to cluster together to keep away from water and form droplets. Triacylglycerols are also highly reduced compounds [their empirical formula is almost $(CH_2)_n$] and so they have a great potential to be oxidized with the release of energy. In cells they represent a store of energy to be released when needed by enzyme action. The fact that they tend to isolate themselves as hydrophobic droplets makes storage easy. There is another advantage, however. Cells containing droplets of stored triacylglycerol act as thermal and mechanical insulators and, in aquatic animals, supply buoyancy. They also make some humans look more attractive than might otherwise be the case.

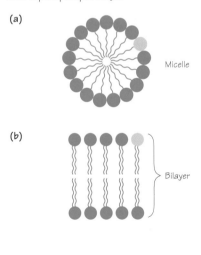

Figure 2.22 (a) Long-chain fatty acids in water tend to form micelles in which the hydrophobic chains are together in the interior of the micelle and the $-CO_2^-$ groups point outwards into the aqueous environment. **(b)** Phospholipids (see Figure 2.23) tend to behave in a similar way (they have two hydrophobic tails instead of one). Imagine a large flattened micelle and you have a phospholipid bilayer.

Structure and function of lipids

objectives
- To state that lipids are a diverse range of molecules characterized mostly by their complete or partial insolubility in water
- To describe the formation of neutral lipid by the formation of a triester of three molecules of long-chain fatty acid with one molecule of glycerol
- To recognize the structures of phospholipids and steroids
- To list the biological functions of lipids

LIPIDS are a very heterogeneous collection of compounds, which include fatty acids and triacylglycerols (triglycerides), fats and oils, phospholipids, sphingolipids and steroids.

Lipid molecules tend to aggregate to form DROPLETS and BILAYERS, and it is this property that enables them to carry out many of their functions, in:

(a) CELL MEMBRANES: phospholipids and glycolipids are found in the plasma membrane and in the membranes around cell organelles.

(b) STORAGE: high energy yield per unit mass and insolubility in water make fats and oils ideal-energy storage compounds, particularly where dispersal or locomotion requires mass to be kept to a minimum, as in some seeds and fruits.

(c) PHYSICAL PROTECTION: the shock-absorbing ability of subcutaneous fat stores and that surround delicate organs such as the kidneys protects them from mechanical damage.

(d) THERMAL INSULATION: fats conduct heat very poorly — subcutaneous fat stores help heat retention in endothermic animals. Incompressible blubber is an important insulator in diving mammals.

(e) ELECTRICAL INSULATION: myelin is secreted by Schwann cells and insulates some neurons in such a way that the conduction of nerve impulses is considerably speeded up.

(f) WATER-REPELLANT PROPERTIES: oily secretions of the sebaceous glands help to waterproof the fur and skin. The preen gland of birds produces a secretion which performs a similar function on the feathers.

Some lipids have vital BIOLOGICAL FUNCTIONS in organisms. For example, an important group of HORMONES, including cortisone, testosterone and oestrogen, are steroids. In NUTRITION, both bile acids (involved in fat digestion) and vitamin D (involved in Ca^{2+} absorption) are also manufactured from steroids. Vitamin A (retinol) and its active derivative, retinoic acid, are long, hydrophobic molecules with a polar group at one end.

Retinol (Vitamin A)

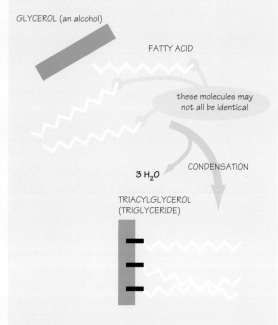

TRUE LIPIDS are esters of fatty acids and alcohols, formed by condensation reactions:

GLYCEROL (an alcohol)

FATTY ACID

these molecules may not all be identical

$3 H_2O$ CONDENSATION

TRIACYLGLYCEROL (TRIGLYCERIDE)

in more detail

GLYCEROL   3 FATTY ACIDS

$3 H_2O$ CONDENSATION

TRIACYLGLYCEROL (TRIGLYCERIDE)

Since the hydrocarbon chains of fatty acids are long (C_{18} in stearic acid; C_{20} in arachidonic acid), most of the weight of the triacylglycerol (triglyceride) is fatty acid.

2.6 Structure and function of lipids

LONG-CHAIN FATTY ACIDS contain a great deal of energy because they represent highly reduced forms of carbon that may be oxidized.

STEARIC ACID is $C_{18}H_{36}O_2$.

Space-filling model of stearic acid

PHOSPHOLIPIDS (PHOSPHATIDES) have a polar 'phosphate-base' group substituted for one of the fatty acids in a triacylglycerol (triglyceride).

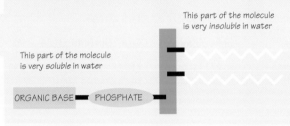

This part of the molecule is very soluble in water

This part of the molecule is very insoluble in water

ORGANIC BASE — PHOSPHATE

STEROIDS are not fatty acids or esters but have similar solubility properties. They have a common basic structure called the **STEROID NUCLEUS** and differ in function as a result of minor changes in the side chains of this nucleus.

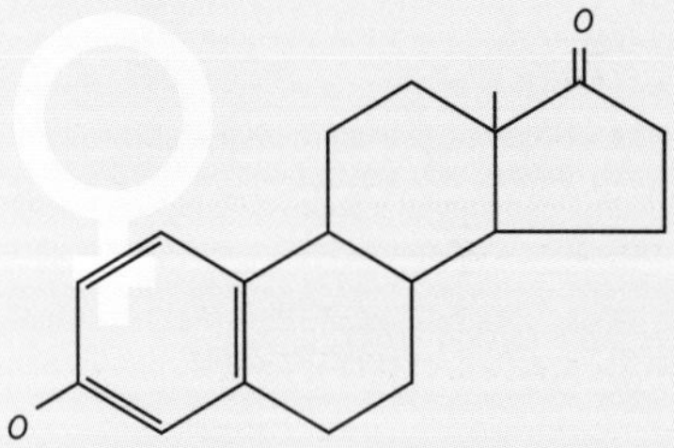

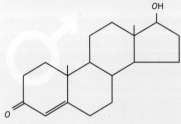

OESTROGEN (female hormone)

TESTOSTERONE (male hormone)

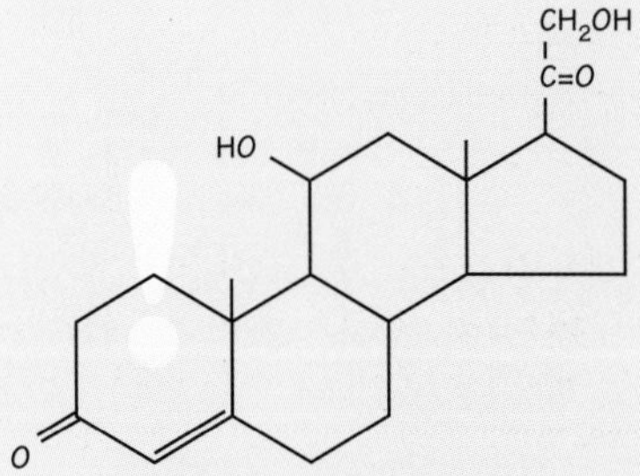

CORTISONE (stress hormone)

SPHINGOLIPIDS are complex lipids composed of one long fatty acid chain linked through a complex polar 'head' to sphingosine which itself has a long-chain structure, e.g. sphingomyelin.

FATS and **OILS** are typical triacylglycerols (triglycerides) which differ chemically in the nature of their hydrocarbon chains — these chains may be saturated ($[-CH_2-CH_2-]_n$) or partially unsaturated (contain some $-C=C-$ bonds).

FATS have a high proportion of **SATURATED HYDROCARBON CHAINS**, and are **SOLID** at room temperature.

OILS have a high proportion of **UNSATURATED HYDROCARBON CHAINS**, and are **LIQUID** at room temperature.

WAXES are esters of higher fatty acids with long-chain fatty alcohols (i.e. not glycerol). They are soft when warm but hard when cold.

Bees use wax (HONEYCOMB) in constructing their larval chambers.

Chapter 2 • **Biological Molecules** 65

The physical properties of long-chain fatty acids and triacylglycerols depend mainly on two factors: (a) the chain length and (b) the degree of unsaturation of the fatty acids. The longer the chain length, the lower the melting point. *Unsaturation* means that some of the carbons in the fatty acid chain are linked by double bonds (–CH = CH–), and the more of these there are, the lower the melting point. Table 2.3 gives some examples.

Table 2.3
Melting points of some fatty acids

Common name	Systematic name	Structure	Melting point (°C)
Lauric	n-Dodecanoic	$CH_3(CH_2)_{10}COOH$	44
Stearic	n-Octadecanoic	$CH_3(CH_2)_{16}COOH$	70
Oleic	cis-9-Octadecenoic	$CH_3(CH_2)_7CH=CH(CH_2)_7COOH$	16
Linolenic	cis-9,12,15-Octadecenetrioic	$CH_3CH_2CH=CHCH_2CH=CH-CH_2CH=CH(CH_2)_7COOH$	–11

Phospholipids

The phospholipids are a group of compounds that are somewhat similar to the triacylglycerols except that they contain a phosphate group instead of one of the long-chain fatty acids. Several other different small molecules may be linked on to the phosphate giving a group of rather complicated but related structures (see Figure 2.23). Phospholipids are vital components of cells because it is these molecules that form the basis for constructing the different types of membranes in cells (plasma membrane, inner and outer mitochondrial membranes, etc.).

The basic structure of two long-chain fatty acids linked via glycerol to a phosphate and then to another molecule is shown in Figure 2.23. This molecule, like long-chain fatty acid molecules, is hydrophobic at one end and hydrophilic at the other. In fact, the hydrophilic end may be negatively charged or neutral (but still hydrophilic), and this enables different types of membrane to be formed, with different properties.

Like long-chain fatty acids, phospholipids will form micelles if dispersed in water. However, if one imagines a very large micelle, the whole structure will become flattened, with the hydrophobic groups in contact with one another, effectively forming a *bilayer*, indeed a phospholipid bilayer. This forms the basis of the phospholipid bilayer of all biological membranes. Different phospholipids will give different properties. Cholesterol molecules may be inserted to increase the fluidity of the membrane (see Figure 1.13, page 19), and proteins are added that give the bilayer the special characteristics of membranes (see Spread 1.4, page 22). It should be remembered that it is important that the membrane bilayer remains fluid in the living cell, and the properties of chain length and degree of unsaturation are also exploited to achieve this for cells that live in a variety of habitats at different temperatures.

Figure 2.23 Phospholipids are made from two molecules of long-chain fatty acid, a molecule of glycerol (or an other similar compound), and a phosphate, to which may be joined a variety of other small molecules. These may be neutral or positively charged. Since the phosphate group carries a negative charge, this hydrophilic end of the molecule may be negatively charged or neutral (net). In this representation the long-chain fatty acids are shown diagrammatically as a zig-zag which represents a sequence of –(CH$_2$)–. However, some of the long-chain fatty acids may have a double bond producing a kink in the chain.

2.7 Other biomolecules

A multitude of other small molecules occur in cells and organisms, but they are difficult to classify. Most are molecules on their way to being built up into something else, or being broken down to provide energy; in other words they are *metabolites,* sometimes called *metabolic intermediates* or just *intermediates.* Some, however, have other biological functions such as acting as hormones and messengers, and yet others have vital roles as vitamins and coenzymes. We shall come across many of these as we proceed. For the time being we have covered the molecules that account for three out of four Ms: *macromolecules, membranes* and *memory.* The next chapter is on enzymes, because the fourth M, *metabolism,* requires the power of enzymic catalysis.

2.8 Further reading

ap Rees, T. (1994) Sucrose and starch. Biological Sciences Review **7(1)**, 13–17

Brown, B. and Gull, D. (1994) Haemoglobin. Biological Sciences Review **6(3)**, 14–16

Brown, T. (1991) DNA unravelled. Biological Sciences Review **4(2)**, 25–26

Bruckdorfer, R. (1993) Cholesterol. Biological Sciences Review **5(5)**, 20–23

Campbell, P.N. (1990) Variations on a protein theme. Biological Sciences Review **2(3)**, 15–19

Cech, T.R. (1986) RNA as an enzyme. Scientific American **255 (Nov)**, 76–84

Doolittle, R.F. (1985) Proteins. Scientific American **253 (Oct)**, 74–83

Felsenfeld, G. (1985) DNA. Scientific American **253 (Oct)**, 44–53

Joiner, A. and Holt, C. (1992) Ice cream and fish antifreeze proteins. Biological Sciences Review **4(5)**, 5–8

Kadler, K. (1994) Collagen in health and disease. Biological Sciences Review **6(4)**, 35–38

Mayes, P. (1991) Lipoproteins. Biological Sciences Review **4(2)**, 37–40

Richards, F.M. (1991) The Protein Folding Problem. Scientific American **264 (Jan)**, 34–41

Sharon, N. and Lis, H. (1993) Carbohydrates in cell recognition. Scientific American **268 (Jan)**, 74–81

Wells, S. and Lowrie, P. (1993) Glucose polymers. Biological Sciences Review **5(4)**, 16–20

2.9 Examination questions

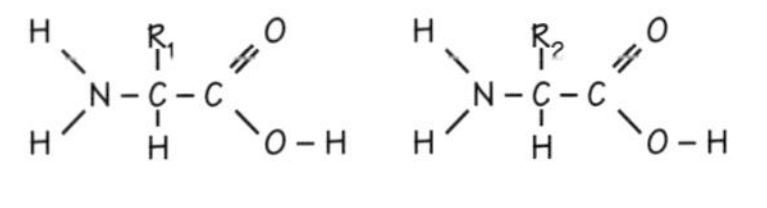

1. The diagrams oppposite show the structural formulae of two amino acids, P and Q.
   (a) Name two elements, other than carbon, hydrogen and oxygen, which may be present in groups R_1 and R_2. (2 marks)
   (b) P and Q may be linked during protein synthesis. In this reaction certain atoms from P and Q combine to form new molecules.
      (i) On the diagram, draw a circle around the atoms that are removed when P and Q are linked together. (1 mark)
      (ii) Draw a line connecting the atoms in P and Q that are bonded together. (1 mark)
      (iii) Name the bond formed by this reaction. (1 mark)
   (c) Explain how the different properties of groups such as R_1 and R_2 are important in the structure and functioning of proteins. (3 marks)

   [ULEAC (AS Biology) June 1991, Paper 1, No. 6]

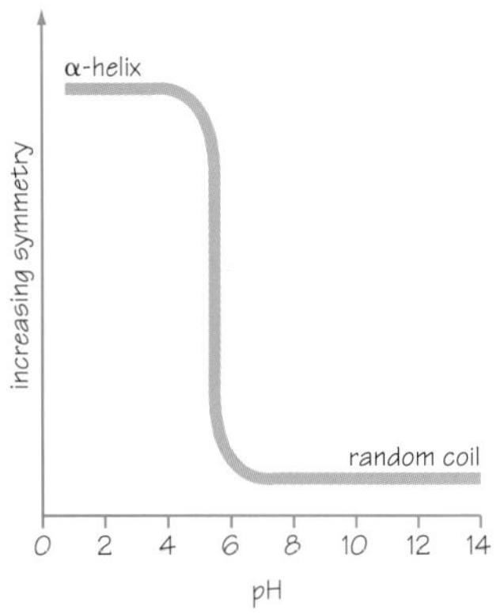

2. The graph shows the effect of pH on the structure of a protein which consists entirely of repeating residues of one amino acid.

   Which of the following statements is true?
   A At high acidity the protein loses its primary structure.
   B At high acidity the protein loses its secondary structure.
   C At high acidity the protein loses its tertiary structure.
   D At low acidity the protein loses its primary structure.
   E At low acidity the protein loses its secondary structure.

3. (a) Name the principal storage carbohydrate of mammals. (2 marks)
   (b) Name two important structural carbohydrates found in plants. (2 marks)
   (c) Name the two carbohydrates found in nucleic acids. (2 marks)
   (d) Name a non-reducing sugar. (1 mark)
   (e) Reducing sugars may be detected by their reaction in the Benedict's (or Fehling's) test. Describe in detail the test you would use to detect a non-reducing sugar. (4 marks)
   (f) Given that the structure of glucose can be expressed as shown in the diagram, draw the structure of a maltose molecule. (2 marks) 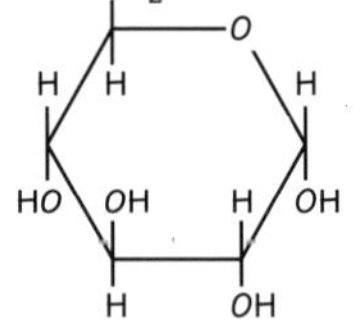
   (g) Of what type of reaction is the formation of maltose from glucose an example? (1 mark)
   (h) Give one advantage that carbohydrate has over fat as a storage material. (1 mark)
   (i) What is a major advantage of fat over carbohydrate as a storage material, particularly in animals? (1 mark)

   [O & C (Biology) June 1991, Paper 1, No. 1]

4. (a) Complete the table below giving the general formula, one named example and one function in living organisms of each of the carbohydrates listed. (6 marks)

Carbohydrate	General formula	Example	Function in living organisms
Pentose			
Disaccharide			

(b) Suggest why the major storage compounds of animals are usually lipids, whereas those of plants are usually carbohydrates. (3 marks)

[ULEAC (Biology) January 1992, Paper 1, No. 8]

5. Describe how you would distinguish each of the following in the laboratory: distilled water, a solution containing glucose and a solution containing sucrose. (5 marks)

6. The diagram opposite represents the molecular structure of part of a DNA molecule.

    (a) Name A–E. (5 marks)

    (b) State three differences between DNA and mRNA. (3 marks)

    [O & C (Biology) June 1994, Paper 1 (adapted) No. 1]

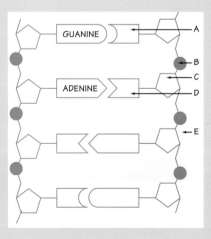

7. Read through the following account of nucleotide structure and then write on the dotted lines the most appropriate word or words to complete the account.

Nucleotides are organic compounds containing the elements carbon, hydrogen, oxygen, ... and .... A molecule of the mononucleotide ATP includes the organic base .... and a sugar called .... which has .... carbon

Chapter 2 • **Biological Molecules** 69

atoms. The organic base in ATP is a double-ringed molecule of the kind called a .... RNA and DNA both contain a nucleotide which includes the organic base guanine, but the nucleotide in RNA contains one more atom of .... In both RNA and DNA the nucleotide containing guanine pairs with the nucleotide containing ...., by means of .... bonding.

(9 marks)

[ULEAC (Biology) January 1989, Paper 1, No. 2]

8. Complete the table with a tick (✔) if the statement is true or a cross (X) if it is not true. (4 marks).

Statement	Substance				
	DNA	Triglyceride	Amino acid	Cellulose	Protein
May contain the element nitrogen					
May contain the element sulphur					
May contain hydrogen bonds					
May act as a substrate for hydrolase enzymes					

[AEB (Biology) Nov 1991, Paper 1, No. 14]

9. (a) Name the three chemical elements found in a fat. (1 mark)

(b) Name the two compounds formed when fats are hydrolysed during digestion in a mammal. (2 marks)

(c) (i) Name one group of lipids other than fats. (1 mark)
(ii) State two functions of lipids in living organisms. (2 marks)

(d) Name two elements which occur in proteins but not in fats or carbohydrates (2 marks)

[ULEAC (Biology) June 1989, Paper 1, No. 2]

3.1 Introduction

The vast majority of the thousands of chemical reactions that go on in living organisms are catalysed by enzymes. Most of these reactions would not proceed at an appreciable rate in the absence of a catalyst, which speeds up the reaction by a factor of a thousand or more. In fact organisms may not want all of their metabolic reactions to continue at a rapid rate all of the time, and by 'switching' enzyme activity on and off they can control their reactions. As we will see later in this chapter, enzyme activity may be controlled in very precise ways. An easy example to understand concerns the cellular enzymes that break down and manufacture glucose. When we are starving we want the cellular enzymes to be breaking down glycogen to release glucose. In contrast, just after a meal we want to be turning glucose into glycogen stores. This control is achieved by controlling the various enzymes that catalyse the metabolic reactions.

Enzymic catalysis has been known about for a long time

Although enzymes in yeast have been used for thousands of years in making bread, beer and wine, the processes by which starch and sugars are broken down to make ethanol were a mystery. The process was called *fermentation*, and in the mid-nineteenth century Pasteur showed that heat-killed yeast cells could not carry out this process. Thus it was believed that their fermentation was a process that could only be carried out by living cells. Some years before this, in 1822, a Canadian doctor, William Beaumont, had treated a fur trapper who had been shot in the stomach. The shot man's stomach lining was exposed through the abdomen and Beaumont carried out a series of experiments in which he lowered pieces of food on a thread into the stomach. He showed that food was digested by the gastric juice and that no mechanical grinding action was necessary. This was another example of enzyme action.

In 1835 Berzelius had found that potatoes contained something that broke down starch, and in 1878 the name *enzyme*, meaning 'in yeast', was used by Kuhne in relation to the process of fermentation. It gradually was realized that biological reactions were catalysed, and 'enzyme' became the accepted name for biological catalysts, although their chemical nature was unclear (Figure 3.1).

A very important step forward came in 1897 when Büchner showed that *living* yeast cells were not essential for fermentation. He ground yeast cells with sand to disrupt their cell walls and then filtered the material to obtain a clear liquid. This liquid was able to ferment glucose to produce alcohol. Later it was shown that the chemicals involved were very large molecules and that their activity was destroyed by heating. Gradually it was realized that these enzymes were proteins, and in 1926 Sumner purified the enzyme *urease* from jackbeans to such a degree that it would crystallize. The crystals appeared to be entirely composed of protein and the idea became accepted that all enzymes are proteins (Figure 3.2).

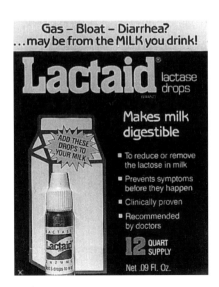

Figure 3.1 Enzymes find many uses. Humans have traditionally used enzymes in yeast to make bread, wine and beer. Present-day usage includes things as diverse as enzyme washing powders and meat tenderizers. The picture shows a sample of lactase purchased in a drug store in the U.S.A. Lactase digests milk sugar, lactose, to give galactose plus glucose. Many American blacks lack this intestinal enzyme. Drinking milk gives them great problems because the lactose is not absorbed and the bacteria in the gut ferment it with great vigour. Adding this enzyme, purified from a micro-organism, to milk can help to overcome this problem.

Figure 3.2 Many purified enzymes will crystallize under the right conditions. The picture shows crystals of trypsin.

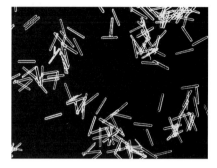

This is generally speaking true, but two things have changed this slightly. One is that in 1975 it was discovered that urease is not quite 100% protein. The enzyme contains less than 0.1% nickel, but this metal is essential for its activity. This was not all that surprising because it had been known for a long time that many enzymes have small molecules or metal ions attached to them that are required for activity. These are called *coenzymes* (see later).

The other thing was that in 1982 Cech found that some RNA molecules have catalytic activity. Although several of these RNA enzymes or *ribozymes* are now known to exist, by far the majority of enzymes are proteins. A typical bacterial cell might contain about 3000 different types of protein enzyme and a human cell contains perhaps 50,000. Nevertheless, some people believe that before proteins evolved there was an 'RNA World' where the majority of the biological activities now carried out by proteins were carried out by RNA molecules. RNA molecules can, of course, carry information too.

3.2 Enzyme-catalysed reactions

A catalyst is a substance that speeds up a chemical reaction whilst itself remaining unchanged at the end of the reaction. Many chemical catalysts are known, including finely divided platinum ('platinum black') which can cause hydrogen and oxygen to combine to form water, a reaction which will only normally occur at a high temperature. Similarly, finely divided nickel is used in the hydrogenation of liquid oils to make the semi-solid margarine (Figure 3.3). The noticeable thing about these catalysts is that the finely divided metal has a large surface area on which the chemical reaction takes place. Furthermore, they tend to be non-specific: finely divided platinum in various forms (platinum black, platinized asbestos, etc.) catalyses a wide range of chemical reactions.

Figure 3.3 Catalysis by metals. Margarine is a solid fat made by the catalytic hydrogenation of seed oils (which are liquid at room temperature). A number of double bonds in the fatty acids are reduced using a finely divided nickel catalyst (0.1%), with hydrogen at high temperature and pressure. The fewer the double bonds, the higher the melting point.

Protein catalysts, in contrast, are generally highly specific for the reactions they catalyse. However, the actual reaction probably takes place on the surface of the protein, or in a sort of cleft, called the *active site*. Usually there is only one of these per protein molecule.

The substance that undergoes catalysis is called the *substrate*. Obviously if an enzyme catalyses the combination of two substances there are two substrates. The compounds resulting from the catalytic action are called the *product(s)* of the reaction. In the case of the enzyme urease the reaction is:

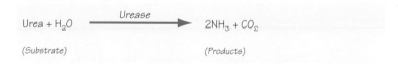

$$\text{Urea} + H_2O \xrightarrow{\text{Urease}} 2NH_3 + CO_2$$

(Substrate)   (Products)

It is normal to use an arrow to indicate the reaction taking place. However, it needs to be remembered that catalysts (including enzymes) simply catalyse *the attainment of equilibrium*. The reaction can go either way depending upon the relative concentrations of substrate(s) and product(s). In this sense an enzyme-catalysed reaction may go 'backwards' and then the product(s) become the substrate(s). An example of this is the enzyme in metabolism that converts pyruvic acid (pyruvate) into lactic acid (lactate) (Figure 3.4). This happens in vigorously contracting muscle, and it is the eventual accumulation of lactic acid that gives us cramp. The reaction turning pyruvate into lactate is a reduction (addition of 2H) and a cofactor is required to 'carry' these hydrogens (see below). After vigorous exercise the accumulated lactate is taken from the muscles by the blood and transported to the liver where it is converted back into pyruvate (an oxidation). Thus in the body, the enzyme responsible — lactate dehydrogenase — works in both directions depending on the body's requirements.

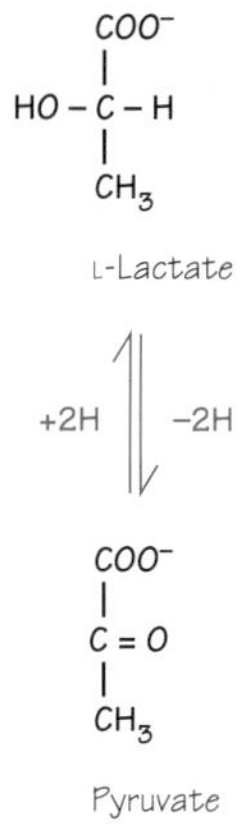

This idea of *equilibrium* is important. Reactions tend to go towards equilibrium; to do other than this would require the input of energy. Neither inorganic catalysts nor enzymes can supply energy, because they are unchanged at the end of a reaction.

Naming and classifying enzymes

A few digestive enzymes have 'old-fashioned' names that end in '-in'. These include pepsin, trypsin, chymotrypsin, rennin and ptyalin, and these names are still used. Sumner used the name 'urease', and this way of naming enzymes by adding '-ase' is the one commonly used today. However, urease, although it tells you that the enzyme works on urea, does not tell you what type of chemical reaction is catalysed (actually it is a hydrolysis). The more recent trend is to use the '-ase' attached to the name of the reaction catalysed. Thus a dehydrogenase catalyses the removal of hydrogens and an oxidase catalyses an oxidation. Obviously lactate dehydrogenase catalyses the removal of hydrogens from lactate. However, when it is working 'backwards' (producing lactate from pyruvate), which is an oxidation, we still call it lactate dehydrogenase.

As with most things in the modern world, enzymes also have numbers (you probably have an UCAS number or a student number, and if you have a bank or savings account, your bank probably thinks of you as '90873754' rather than John Smith). Some years ago the International Union of Biochemistry set up an 'Enzyme Commission' to give unambiguous names and numbers to all enzymes discovered. How this is done is set out in Spread 3.1. The need for numbers is made clear by the fact that tens of thousands of different enzymes are now known. However, the average biochemist working in the research laboratory will tell you that he or she is working on trypsin, not 'EC 3.4.4.4'. Furthermore, because some of the names, like lactate dehydrogenase, are a bit of a mouthful, they tend to get abbreviated, e.g. to 'LDH', in daily parlance. [In case you wanted to know, urease is correctly called urea amidohydrolase, and LDH is (S)-lactate : NAD$^+$ oxidoreductase.]

There is something surprising about the list of enzyme reactions given in Spread 3.1 and this is that there are only six categories of enzyme-catalysed reaction. Thus although tens of thousands of enzymes are known, the number of types of reaction is remarkably small. This illustrates a fundamental point. Although immensely complicated sequences of reactions occur in metabolism in living cells, e.g. turning

Figure 3.4 Structures of L-lactate and pyruvate. These are interconverted by the enzyme lactate dehydrogenase. Note that lactate can exist as two optical isomers. The enzyme recognizes the L-isomer but not the D-isomer. From the optically inactive pyruvate, it only generates L-lactate.

$$\begin{array}{c} COO^- \\ | \\ HO-C-H \\ | \\ CH_3 \end{array}$$

L-Lactate

+2H ↕ −2H

$$\begin{array}{c} COO^- \\ | \\ C=O \\ | \\ CH_3 \end{array}$$

Pyruvate

Naming and classifying enzymes

objectives
- To be able to list the six general types of enzyme-catalysed reaction
- To be able to recognize examples of the types of reaction catalysed by enzymes

OXIDOREDUCTASES catalyse redox reactions involving the transfer of H and O atoms and/or electrons from one substance to another.

e.g. LACTATE DEHYDROGENASE (EC 1.1.1.27)

$$\text{H-O-}\underset{\underset{CH_3}{|}}{\overset{\overset{COO^-}{|}}{C}}\text{-H} + NAD^+ \longrightarrow \underset{\underset{CH_3}{|}}{\overset{\overset{COO^-}{|}}{C}}{=}O + NADH + H^+$$

L-Lactate → Pyruvate

Originally the digestive enzymes of the stomach and the small intestine were given names ending in -*in*: thus peps*in* and chymotryps*in*. Now a more widely used way of naming enzymes has been to add the suffix -*ase* to the reaction catalysed. Thus:

- a hydrolase catalyses a hydrolysis,
- a dehydrogenase catalyses the removal of hydrogen.

A complete 'trivial' (everyday) name for an enzyme would also include the substrate for the enzyme. Thus:

$$\underset{\text{Substrate} \quad \text{Type of reaction}}{\text{RIBULOSE BISPHOSPHATE CARBOXYLASE}}$$

A rigid numerical classification for all enzymes has now been devised and accepted internationally. It recognizes that there are only six basic reaction types catalysed by enzymes and gives an Enzyme Commission number to every enzyme.

Within each group further subdivisions are made so that each enzyme is unambiguously identified by a set of four numbers. For example:

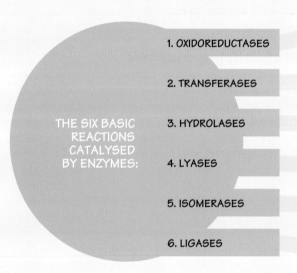

THE SIX BASIC REACTIONS CATALYSED BY ENZYMES:
1. OXIDOREDUCTASES
2. TRANSFERASES
3. HYDROLASES
4. LYASES
5. ISOMERASES
6. LIGASES

```
                EC  1.  1.  1.  27
Enzyme                              L-Lactate
Commission number                   as substrate

   Oxidoreductase              Using NAD+
                               as acceptor
       Acting on a donor -CHOH group
```

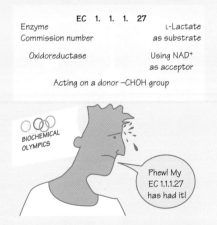

Phew! My EC 1.1.1.27 has had it!

LIGASES catalyse SYNTHETIC reactions, i.e. the joining together of two molecules by the formation of new C–O, C–C, C–N or C–S bonds at the same time as ATP is broken down.

e.g. an AMINOACYL-tRNA SYNTHETASE, such as L-alanine:tRNA ligase* (EC 6.1.1.7)

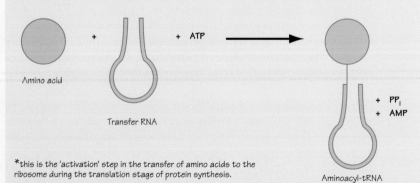

*this is the 'activation' step in the transfer of amino acids to the ribosome during the translation stage of protein synthesis.

74 Life chemistry and molecular biology

3.1 Naming and classifying enzymes

TRANSFERASES catalyse the transfer of groups from one molecule to another.

e.g. HEXOKINASE (E.C. 2.7.1.1)
(systematic name ATP: D-hexose 6-phosphotransferase)

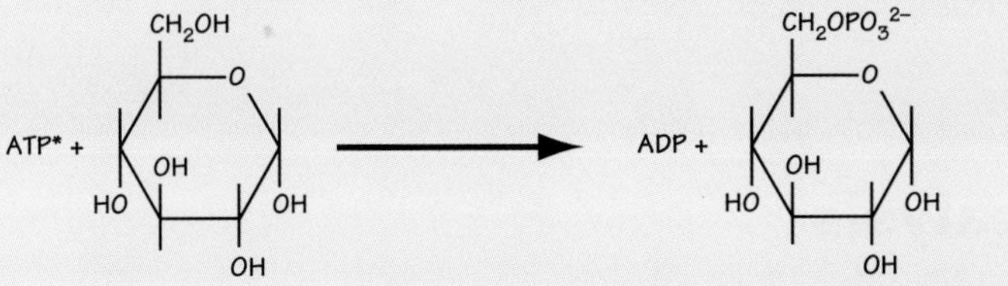

ATP* + D-Glucose → ADP + D-Glucose 6-phosphate

*ATP is an important molecule that can donate a phosphate group; see Chapter 4.

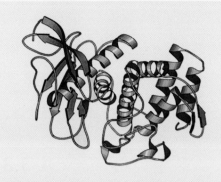

Thermolysin is a protease (note that it still has an old-fashioned name) from a bacterium, *Bacillus thermoproteolyticus*. It has a similar specificity to chymotrypsin, is thermostable and contains zinc. As can be seen it contains both α-helices and β-sheets.

HYDROLASES catalyse hydrolytic reactions involving the breakdown of one molecule into two products by the addition of the elements of water.

e.g. LIPASE (EC 3.1.1.3)

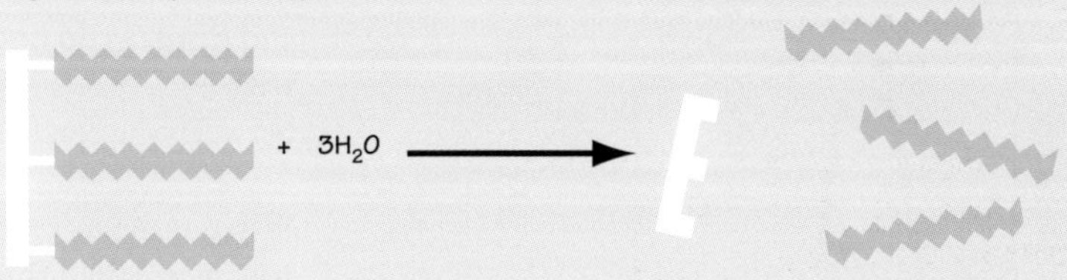

Fat (triacylglycerol) + $3H_2O$ → Glycerol + 3 Fatty acids

LYASES catalyse removal of a group from or addition of a group to a double bond.

e.g. PYRUVATE DECARBOXYLASE (EC 4.1.1.1)

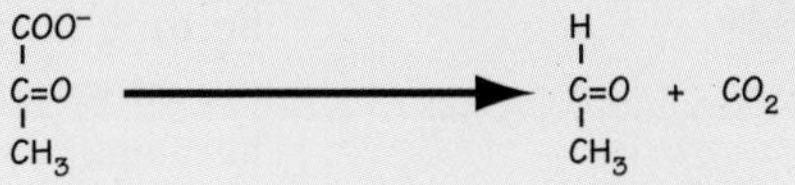

Pyruvate → Ethanal + CO_2

ISOMERASES catalyse isomerizations of many different types.

e.g. GLUCOSE-6-PHOSPHATE ISOMERASE (EC 5.3.1.9)

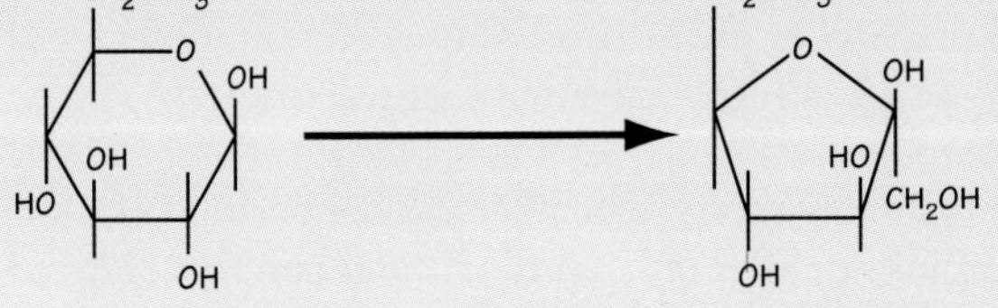

β-D-Glucose 6-phosphate → β-D-Fructose 6-phosphate

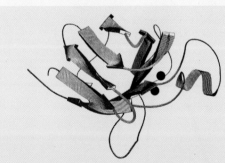

Superoxide dismutase is an enzyme responsible for getting rid of dangerous superoxide radicals that can damage cells. It is found in practically all cells. Shown here is one subunit of a copper- and zinc-containing superoxide dismutase from red cells. Normally the protein exists as a dimer of two identical subunits. As can be seen it is mostly β-sheet structure with almost no α-helix.

Chapter 3 • Enzymes

acetate (ethanoate; C_2) into cholesterol (C_{27}), these overall reactions all take place in very simple steps, each catalysed by one of six general types of enzyme. (You may know that although computers can do apparently very complicated things, at the heart they can only add together one plus one.) Metabolism takes place via a number of very simple chemical transformations or 'steps'.

3.3 Catalysis

Catalysts are substances that increase the rate of a chemical reaction. Reactions proceed because the products have less chemical energy than the substrates, and as we said above, reactions tend to proceed until equilibrium is reached. However, it is a common observation that many substances although capable of reacting do not actually do so. An example is petrol, which contains large quantities of carbon and hydrogen which may be oxidized by oxygen in the atmosphere; and yet petrol can exist at normal temperatures in air, and does not spontaneously burn or explode. The reason for this is that the reactants (petrol and oxygen) do not have enough energy at normal temperatures to react with each other. We say there is an energy barrier, the *activation energy*, between the reactants and the products of the reaction.

When molecules are going to react they must come together to form a complex called the *transition state* which has a higher energy than the mean energy either of the reactants or of the products. Thus the activation energy is the difference in energy between the energy of the reactants and that of the transition state (Figure 3.5).

In the case of the example of petrol, if a spark or a naked flame is applied the energy of a few petrol molecules is increased. The covalent bonds holding the molecules together become less stable, or more energetic, the activation energy is exceeded, and burning starts. This releases some heat energy which is transferred to other petrol molecules whose energy then exceeds the activation energy, and so on, so that the reaction becomes self-sustaining and an explosion is likely.

Now living organisms do not experience changes of temperature like this, and 'burning' would be distinctly hazardous! Nevertheless, practically all of the reactions that go on in cells and organisms have an activation energy. This can be circumvented by enzyme catalysts. In the chemistry laboratory we raise the temperature so that some of the reactant molecules can get over the barrier but, enzymes, like all catalysts, provide an alternative pathway for the reaction. In contrast to the 'normal' (uncatalysed) reaction, this alternative pathway has a different transition state with a lower activation energy (see Figure 3.6). Therefore there is more chance of a higher proportion of the reactant molecules having enough energy to get over the activation energy barrier.

Even in this situation we must remember that the position of equilibrium for the reaction is not affected; equilibrium is simply achieved much more rapidly. The difference in energy between that of the reactants and that of the products is the same whichever route (catalysed or uncatalysed) the reaction takes.

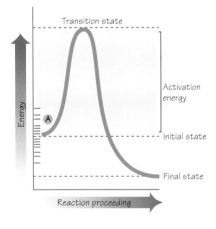

Figure 3.5 In a reaction there is a transition state between the initial and final states which has a higher energy than either of these. Molecules have a range of energies (A), with most of them having approximately the average energy and with very few having enough energy to overcome the activation energy barrier. Heating the reaction gives more molecules enough energy to get over the barrier.

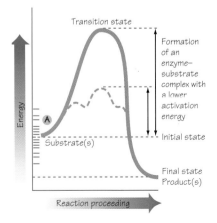

Figure 3.6 In an enzyme-catalysed reaction the formation of an enzyme–substrate complex of a lower activation energy means that many more molecules have sufficient energy to react (compare with Figure 3.5).

3.4 Enzymes and rate enhancement

The idea that enzymes can increase the rate of a reaction implies that we have some way of measuring rate. This is simply the rate at which substrate is converted into product, and so it will be in, say, moles converted per minute, under standard conditions. (It is necessary to define the conditions because the rate, as we shall see, varies with enzyme concentration, substrate concentration, temperature and pH.) It is equally possible, instead of measuring the rate of transformation of the substrate, to measure the rate of appearance of the product; it usually depends on which is the more convenient and accurate.

As an example, we might take the case of the enzyme amylase from saliva, which digests (i.e. catalyses the hydrolysis of) starch into smaller carbohydrate units such as maltose. Now starch gives a blue colour with iodine whereas maltose does not. So we could take samples out of the starch-plus-amylase mixture from time to time and test them with iodine. As time goes on we would find less and less blue colour as the reaction proceeds.

Because enzyme reactions to measure rate are usually carried out in small volumes at low concentration, it is usual to define the unit in terms of 10^{-6} moles (or µmoles) converted per minute. The *international unit* (SI unit) of enzyme activity is defined as the amount of enzyme that will convert 1 µmol of substrate into product in 1 minute under defined conditions of pH and temperature. The equivalent SI unit of enzyme activity, the *katal*, is defined as the amount of enzyme that catalyses the conversion of 1 mole of substrate into product in 1 second. However, this unit has never really caught on (although the conversion from µmol min^{-1} is not all that difficult!). Most people use the 'old' SI unit and have not come to terms with using katals (or really 10^{-9} katals or nanokatals, which is what would be used in typical experiments).

If a fixed amount of enzyme is added to a fixed amount of substrate (usually in great excess), and the amount of product formed determined at suitable time intervals, we get a *progress curve* (see Spread 3.2). This is a curve rather than a straight line because as the enzyme works the substrate concentration falls, eventually lowering the rate of reaction. As equilibrium is approached the rate slows down, eventually to reach zero.

The rate of reaction is greatest at the outset. This is called the *initial rate* and this is the value that should be used in calculations. If the initial rate of reaction is measured at a series of substrate concentrations (with a fixed amount of enzyme), we get a graph like that in Spread 3.2. It shows that at high substrate concentrations the rate approaches a maximum value. This is called V_{max} (for maximum velocity), and at this substrate concentration each enzyme molecule is working as fast as it can under the prevailing conditions. This graph is an important one as will be seen from Spread 3.2.

The V_{max} of the reaction may be used to calculate the *turnover number* of the enzyme. This is obtained by expressing the V_{max} in terms of moles (or µmoles) of protein rather than per milligram. It gives a measure of the efficiency of the enzyme. It represents the number of

Measuring enzyme activity

objectives
- To define the *unit* of enzyme activity
- To describe how enzyme activity may be measured
- To explain how the rate versus substrate concentration graph may be transformed

Enzyme activity is measured as 'units', where a unit is that amount of enzyme required to transform 1 μmole of substrate into product per minute under stated conditions (temperature, pH, etc.). Such units are called 'international units' or simply 'units'. The SI unit is the *katal* (amount catalysing the transformation of 1 mole of substrate to product per second). This unit is rarely used.

For the lactate dehydrogenase-catalysed reaction, enzyme activity may be determined by measuring the amount of lactate disappearing per unit time or the amount of pyruvate appearing per unit time:

Lactate + NAD$^+$ ⟶ Pyruvate + NADH + H$^+$

(It is assumed that the amount of NAD$^+$ present is not limiting.)

In 1913 Leonor Michaelis and Maud L. Menten proposed a general theory of enzyme action:

ENZYME + SUBSTRATE ⇌ ENZYME–SUBSTRATE COMPLEX
ENZYME + PRODUCT

If a fixed amount of enzyme is added to a fixed amount of substrate (usually in great excess), and the amount of product determined at suitable time intervals, we get a PROGRESS CURVE (see opposite). This is a curve rather than a straight line because as the enzyme works the substrate concentration falls, until eventually the rate of reaction is lowered. As equilibrium is approached, the rate slows down to zero.

The rate of the reaction is greatest at the onset and this is what is known as the INITIAL RATE OF REACTION. On the graph opposite, the initial rate of reaction is measured at a series of enzyme concentrations (with a fixed amount of enzyme) and plotted against substrate concentration. At high substrate concentration, it can be seen that the rate approaches a maximum value (V_{max} for MAXIMUM VELOCITY). At this substrate concentration, each enzyme molecule is working as fast as it can under the prevailing conditions.

The substrate concentration necessary to achieve half-maximal velocity, K_m, is known as the MICHAELIS CONSTANT. The V_{max} of a reaction may be used to calculate the turnover number of the enzyme, and the K_m may give an indication of enzyme–substrate affinity.

In 1934 H. Lineweaver and D. Burk devised a convenient way of evaluating K_m and V_{max} (see Lineweaver–Burk plot opposite).

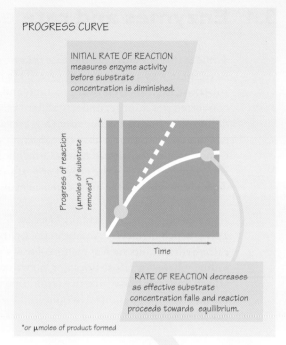

PROGRESS CURVE

INITIAL RATE OF REACTION measures enzyme activity before substrate concentration is diminished.

RATE OF REACTION decreases as effective substrate concentration falls and reaction proceeds towards equilibrium.

*or μmoles of product formed

This indicates that rate of reaction depends on substrate concentration, a relationship expressed in the MICHAELIS–MENTEN EQUATION:

$$v = \frac{V_{max}[S]}{K_m + [S]}$$

which may be plotted as:

The rate of an enzyme reaction depends on the amount of enzyme present (more enzyme, more catalysis).

Life chemistry and molecular biology

3.2 Measuring enzyme activity

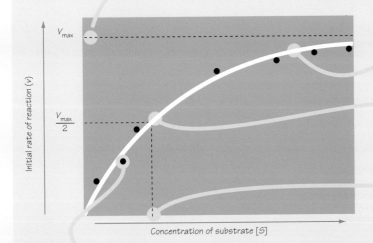

V_{max} is the maximum velocity of the reaction at a given enzyme concentration; it is difficult to measure (theoretically it occurs when $[S] = \infty$).

Some other RATE-LIMITING FACTOR affects the number of E–S complexes formed, and hence the rate of reaction. This factor is commonly the number of available active sites, i.e. the concentration of the enzyme.

Rate of reaction is proportional to the number of substrate molecules available to form enzyme–substrate complexes.

K_m is the substrate concentration necessary to achieve half maximum velocity: it may give an indication of E–S affinity, i.e. low K_m = high affinity.

K_m is called the MICHAELIS CONSTANT.

Each of these plotted points represents an initial rate determined from a progress curve at a given substrate concentration.

V_{max} and K_m are useful to know but difficult to measure. They may be determined from the LINEWEAVER–BURK EQUATION:

$$\frac{1}{v} = \frac{K_m}{V_{max}} \frac{1}{[S]} + \frac{1}{V_{max}}$$

which may be plotted graphically

as $\frac{1}{v}$ versus $\frac{1}{[S]}$

Lineweaver–Burk plot

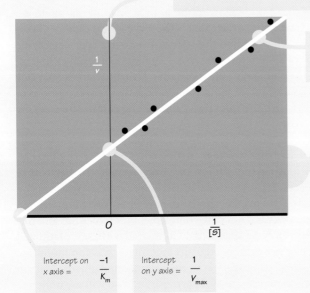

Straight line obtained by taking reciprocals

Slope = $\frac{K_m}{V_{max}}$

Intercept on x axis = $\frac{-1}{K_m}$

Intercept on y axis = $\frac{1}{V_{max}}$

Chapter 3 • **Enzymes**

molecules of substrate 'turned over', i.e. converted into product, by 1 molecule of enzyme in 1 minute when there is excess substrate present. Obviously it is necessary to know the M_r of the enzyme protein in order to calculate the turnover number (Table 3.1).

Table 3.1
Turnover numbers for some enzymes

Enzyme	Turnover number (given here per second)
Acetylcholinesterase	1.4×10^4
Carbonic anhydrase	1.0×10^6
Catalase	4.0×10^7
Chymotrypsin	1.0×10^2
Lactate dehydrogenase	1.0×10^3
Lysozyme	5.0×10^{-1}

3.5 The active site

In most cases, enzymes are very much larger than their substrates. Enzyme proteins typically have M_rs of at least 25,000 and often very much more, while substrates such as glucose etc. have M_rs of less than 200. This statement is not completely true because, for example, a proteolytic enzyme such as trypsin has an M_r of 25,000 but may act on protein substrates much much larger than this. Nevertheless, at any one time trypsin will be acting on one peptide bond alone and so the true size of the 'substrate' may be regarded as two amino acids bonded in a dipeptide (M_r about 200).

So why are enzymes so large in comparison with their substrates? The answer to this is that enzymes are globular proteins and that there is a small area or cleft on their surface that is the place where catalysis takes place (Figure 3.7). This is called the *active site* and in a typical enzyme protein it may be of a size equivalent to 6–12 amino acid residues (out of, say, 200–400 in the whole enzyme molecule). Usually there is only one active site per enzyme protein molecule although there are exceptions to this. We will deal with the question 'what is the rest of the enzyme protein for?' shortly.

The presence of active sites was suggested over 100 years ago by Fischer, who was investigating the ability of yeast extracts to hydrolyse derivatives of sugars. He suggested that enzymes catalysed reactions by binding to substrates in a manner similar to how a key (*substrate*) fits into a lock (*enzyme*). Locks and keys are complementary structures and this would also explain enzyme specificity. Compared with, say, platinized asbestos, most enzymes exhibit exquisite specificity. It is commonplace that an enzyme can distinguish between galactose and glucose (which differ in the position of just one –OH group), and also will use only one

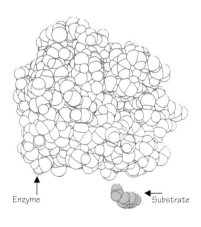

Figure 3.7 Enzymes are generally much larger than their substrates. The active site occupies only a small area or cleft on the surface of the enzyme protein. The picture shows a computer-generated model of trypsin and a small substrate.

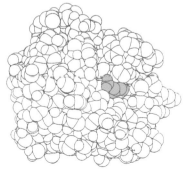

optical isomer of galactose (see Figure 3.4, page 73). Thus only the 'correct' substrate will bind to the enzyme and only then can catalysis take place. Catalysis takes place on surfaces or in clefts where molecules are brought together in an especially favourable orientation for reaction to take place (Figure 3.8). In solution in the absence of a catalyst, in contrast, substrate molecules will collide with one another at random, and only very occasionally will the collisions be correctly orientated such that a successful reaction takes place.

Fischer's *lock and key hypothesis* was a remarkable idea at the time and was accepted largely unchanged for about 70 years. It is the more remarkable since Fischer did not really know that enzymes were very much larger than their substrates, and indeed the hypothesis or theory is still essentially correct. It has only had to be modified slightly in the light of our modern ideas on protein structure.

Fischer's lock and key hypothesis explained many of the actions of enzymes but there were a few problems. Enzymes may be inhibited by a number of compounds (see below) which can get into or on to the active site and block the enzyme's catalytic action. Compounds with rather similar structures to that of the true substrate can act as inhibitors. It was found that some inhibitors were of the same size or were smaller than the true substrates and this was not really surprising. A smaller key can be placed in a lock and jam it. What was surprising was that some inhibitors were larger than the true substrate. Such a false 'key' should not be able to fit into the lock. These observations gradually changed our perception of enzyme and substrate as a lock and key. It became clear that both enzyme and substrate were flexible. When combination takes place there is a mutual fitting together: the substrate induces the enzyme protein molecule to fit around it. This model, which is now generally accepted, is called the *induced fit model* of enzyme and substrate combination (Spread 3.3). This explains why inhibitors larger than the substrate can, indeed, inhibit.

The process of inducing fit is very subtle and may, at least in some cases, explain the process of catalysis. The story is something like this. The active site of the relatively flexible enzyme protein molecule is only an approximate match for the substrate. When enzyme and substrate form a *complex*, structural changes occur which allow the substrate to occupy the active site. This process of 'forcing' the substrate into the active site imposes a strain on the substrate molecule which makes it less stable and more likely to react with nearby molecules. This, together with participation by specific amino acid side chains in the active site, promotes the reaction being catalysed.

This idea also explains, at least to some extent, why protein enzymes have to be large. The rest of the protein molecule not only provides the 'scaffolding' that holds the amino acids in the active site in precise position with respect to each other. Also when combination takes place very subtle changes occur in the whole protein molecule resulting in a perfect match between the enzyme active site and the substrate (Figure 3.9, page 84). The changes in shape undergone by the protein molecule (so called conformational changes) are only small ones. They typically involve fairly major 'swings' by a few amino acid side chains and more subtle shifts in the rest of the protein molecule.

Figure 3.8 Both chymotrypsin and trypsin catalyse the hydrolysis of peptide bonds. However, chymotrypsin (**a**) cleaves bonds after an aromatic (hydrophobic) amino acid residue whereas trypsin (**b**) cleaves after a basic one. The specificity pocket that recognizes these is close to the active (catalytic) site. (**c**) Shows a sketch of the complete active site; the serine, histidine and aspartic acid residues are the chief ones responsible for catalysis.

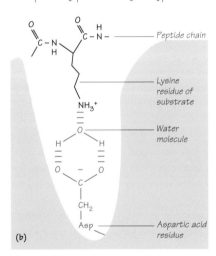

The specificity pocket of chymotrypsin

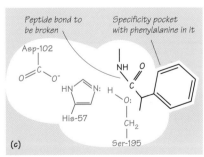

The specificity pocket of trypsin

Complete active site of chymotrypsin

Chapter 3 • **Enzymes** 81

Binding of substrate to enzyme

objectives

- To be able to describe the active site of an enzyme as a region of the surface or a cleft in a polypeptide that has a complementary shape to that of the substrate(s)
- To recognize that, when enzyme and substrate combine, a subtle change of shape takes place leading to a better fit between substrate and enzyme (the 'induced fit' theory).

Enzymes are GLOBULAR PROTEINS, which are normally very large in comparison to their substrate. The ACTIVE SITE of an enzyme is the region that binds to the substrates and induces the 'strain' which leads to the making or breaking of bonds.

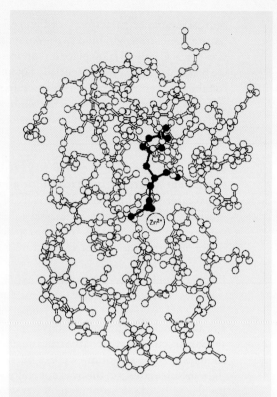

Enzyme–substrate complex of carboxypeptidase with its substrate (black). Note the single zinc atom in the active site. Note also that the active site occupies only a small area of the surface of the protein. The shape of the active site results from precise folding of the polypeptide. (In a space-filling model there would be no spaces in the protein molecule). [Reproduced with permission from Lipscombe, W.N (1971), Proceedings of the Robert Welch Foundation Conferences on Chemical Research, **15**, 140, Houston, Texas.]

SOME FEATURES OF ACTIVE SITES are as follows:

1. Active sites take up a relatively small part of the enzyme's total volume.
2. Active sites are three-dimensional, made up of chemical groups from the side chains of amino acids from different parts of the polypeptide chain.
3. During enzyme–substrate (E–S) complex formation the substrate molecules are bound to the active site by relatively weak forces — typical free energies of interaction are in the order -12 to -50 kJ mol^{-1} (compared with -200 to -500 kJ mol^{-1} for covalent bonds).
4. Active sites are commonly hydrophobic clefts. Water is usually excluded from the cleft unless it is a substrate. There are usually some polar side chains important for binding and catalysis, but binding is considerably enhanced by the non-polar nature of the cleft.

Any factor which alters the conformation (three-dimensional nature, dependent on tertiary and quaternary structure) of the enzyme will alter the shape of the active site, affect the frequency of E–S complex formation and thus influence the rate of the enzyme-catalysed reaction.

COFACTORS are essential for enzyme activity. Some, such as Zn^{2+} or Mg^{2+}, or porphyrin groups such as the HAEM in catalase, may form part of the active site and cannot easily be separated from the enzyme protein: these are commonly called PROSTHETIC GROUPS.

Others, such as NAD$^+$ (nicotinamide–adenine dinucleotide) bind temporarily to the active site and actually take part in the reaction as a substrate. For example:

$$\text{Lactate} + \text{NAD}^+ \underset{}{\overset{\text{Lactate dehydrogenase}}{\rightleftharpoons}} \text{Pyruvate} + \text{NADH} + \text{H}^+$$

Such coenzymes shuttle between one enzyme system and another — most are formed from dietary components called VITAMINS (e.g. NAD$^+$ is formed from niacin, one of the B vitamin complex).

CATALASE is abundant in red cells and catalyses the breakdown of

$$H_2O_2 \longrightarrow H_2O + \tfrac{1}{2}O_2$$

It is one of the most active enzymes known (see Table 3.1, page 80).

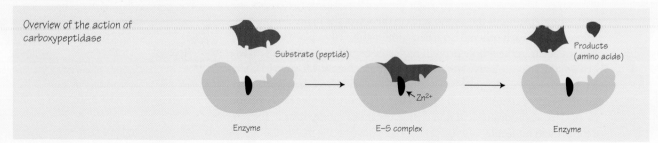

Overview of the action of carboxypeptidase

82 Life chemistry and molecular biology

3.3 Binding of substrate to enzyme

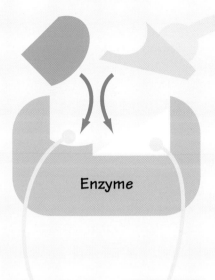

Active site

COFACTORS (e.g. prosthetic groups, coenzymes) are essential for enzyme activity.

STEREOSPECIFICITY: RELATIONSHIP OF SUBSTRATE(S) TO ACTIVE SITE

Emil Fischer's LOCK AND KEY HYPOTHESIS suggested that the active site and the substrate were exactly complementary.

but more recent work allowed Koshland to propose the INDUCED FIT HYPOTHESIS, which suggests that active site and substrate are only fully complementary after the substrate is bound.

This latter process of DYNAMIC RECOGNITION is now the more widely accepted hypothesis.

STEREOSPECIFICITY

Because of the matching of the shape of the substrate with that of the enzyme active site, enzymes normally act on only one isomer of a pair of stereoisomers. The three-point attachment shown here illustrates how this stereospecificity is achieved.

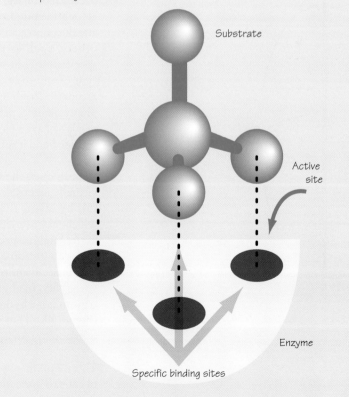

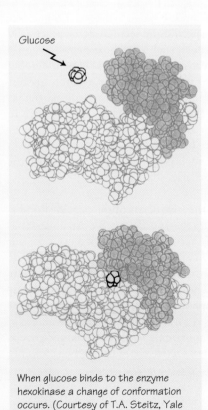

When glucose binds to the enzyme hexokinase a change of conformation occurs. (Courtesy of T.A. Steitz, Yale University.)

Chapter 3 • Enzymes

Figure 3.9 (a) The overall structure of an enzyme protein provides a sort of scaffolding to hold the amino acid residues in the active site (dark blue) in precise positions. (b) When the polypeptide chain is opened out, the active-site amino acid residues may be quite widely separated along this polypeptide.

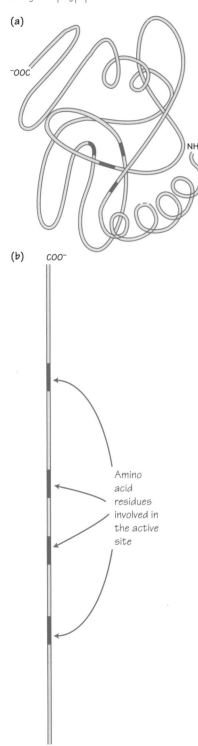

3.6 Specificity of enzyme action

It has been mentioned that, for all the many thousands of types of enzymes that exist, they can all be divided into six categories in respect of the reactions catalysed. This means that there are only a few ways in which enzymes work; so how can we explain their enormous range of specificities?

Table 3.2
Many different kinds of protease enzymes are known

Enzyme	Origin	Amino acid residue(s) involved in active site	Specificity
Chymotrypsin	Intestine	Serine, histidine	Near to aromatic residues
Trypsin	Intestine	Serine, histidine	Near to basic residues
Papain	Papaya latex	Cysteine	Non-specific
Pronase	Bacterial	Cysteine?	Non-specific
Carboxypeptidase	Intestine	Arginine, glutamic acid, tyrosine plus a zinc atom	Carboxy-terminal
Pepsin	Stomach	Aspartic acid	Non-specific

To give but one example, there are many different types of protease (Table 3.2). They come from animals, plants and micro-organisms, and they all catalyse the cleavage of a peptide bond (a hydrolysis). Some, like bacterial pronase, are rather non-specific and seem to attack practically all peptide bonds in proteins. Others, such as carboxypeptidase, only attack amino acid residues at the carboxy-terminal end of peptides. Yet others, such as trypsin and chymotrypsin, have a much greater specificity. They only attack the peptide bond between one particular type of amino acid and another. Thus chymotrypsin needs to have an aromatic (hydrophobic) amino acid residue, such as tyrosine or phenylalanine, next to the peptide bond it cleaves. In contrast, trypsin needs to have a basic amino acid, such as arginine or lysine, next to the peptide bond it splits (see Figure 3.8, page 81).

In both cases the mechanism of the *hydrolysis* of the peptide bond is virtually identical; it is the *specificity* that is different. We can understand this by saying that the active site actually contains two regions, but which may overlap. One region contains the amino acids that participate in the catalytic process (in the case of trypsin and chymotrypsin, these include a histidine and a serine). The other region contains amino acids that make a complementary shape that provides for 'recognition' or specificity. In chymotrypsin this is a hydrophobic pocket, whereas in trypsin it is a hydrophilic pocket containing negatively charged groups (Figure 3.8). Thus we can speak of the active site as having two (possibly overlapping) parts, the *catalytic site* and the *specificity site*.

This is building up a picture of the active site of enzymes and enables us to start to understand how they work. We can envisage a large protein molecule folded into a more or less globular shape. A small region of its surface is the active site which has amino acid side chains arranged in a very precise way (a) to give a mechanism that carries out catalysis, and (b) to provide an accurate shape into which the substrate fits, giving a high specificity. The bulk of the protein is a sort of scaffolding to hold these shapes in place, but, as we have seen, the whole thing is flexible so that *induced fit* is possible. Because of the way in which the polypeptide chain is folded up, amino acid residues from various parts of the polypeptide are brought together in the active site (Figure 3.9).

3.7 Cofactors, coenzymes and prosthetic groups

Some enzymes consist of protein alone, but in fact this category of enzymes is rather small as a fraction of the total. Many, or even most, enzyme molecules have additional small molecules or atoms, that 'help' them to carry out the catalysis. Such groups may be as simple as a metal ion, e.g. zinc in carboxypeptidase (see Spread 3.3), or they may be very complex organic molecules, such as haem groups. There are various names for these additional bits. 'Prosthetic group' is a good name because it has the idea of helping: a *prosthesis* is a wooden leg — it helps you to walk. But other names such as *cofactor* and *coenzyme* are also used.

Some cofactors are covalently attached to the enzyme protein and some are not. Some actually act as substrates (or 'co-substrates') in that they participate in a cycle of events during catalysis. An example of this is the coenzyme called nicotinamide–adenine dinucleotide, or NAD for short. This is an important molecule that participates in many reactions that involve a dehydrogenation (removal of hydrogen), effectively an oxidation, although oxygen itself is not involved. NAD (see Figure 3.10) is closely associated with certain dehydrogenase enzymes but is not covalently bound to the enzyme protein. It can exist in an oxidized (NAD^+) or a reduced (NADH) form, and this is how the hydrogens are carried. NAD^+ is actually a modified form of one of the B group of vitamins called niacin or nicotinamide. An example of a dehydrogenase with NAD^+ as a coenzyme is lactate dehydrogenase already mentioned above:

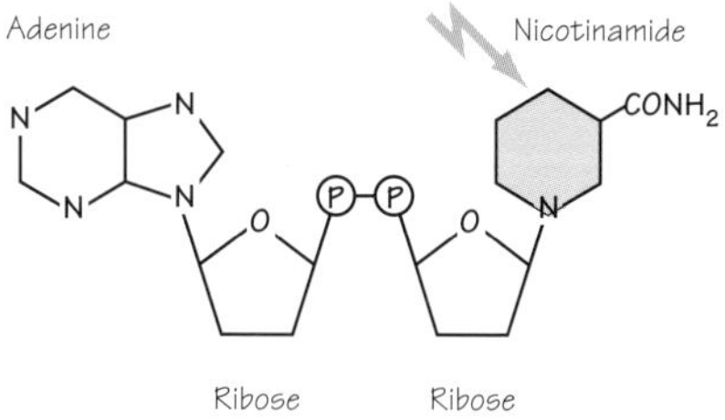

$$\text{Lactate} + NAD^+ \xrightarrow{\text{LDH}} \text{Pyruvate} + NADH + H^+$$

(Oxidized form) (Reduced form)

In this reaction lactate becomes oxidized to pyruvate, but the reaction is easily reversed. In reality lactate dehydrogenase has two substrates, lactate and NAD^+, and there are two adjacent regions in the active site where these are recognized (by their shapes) and bind. The reaction can only be kept going if the NADH product that forms is continually recycled (by another reaction) to regenerate NAD^+.

To summarize, many protein enzymes rely on ions or smaller molecules to help them in the catalytic process (Table 3.3). Many of these small

Figure 3.10 Skeletal structure of the coenzyme nicotinamide–adenine dinucleotide (NAD). This complicated organic molecule can exist in oxidized or reduced forms, that is, it can carry hydrogens. The 'business end' of the molecule (*) is where the hydrogen is added. The rest of the molecule is 'recognized' by the enzyme molecule because of its characteristic shape, but has nothing to do with carrying hydrogens. It is not clear why such a complicated molecule should have evolved for carrying out this function. Even so the lactate dehydrogenase molecule, i.e. the protein, is much much larger than this.

molecules are derived from the B group vitamins, and some are attached to the enzyme covalently and some are not. If they are attached covalently they are often called prosthetic groups, and loosely attached ones like NAD^+ are often called coenzymes. It is difficult to give a definite rule, however, and some of these have been named on a rather arbitrary basis. Above all, these small molecules provide *chemical diversity*, enabling enzymes to do things that otherwise they would not be able to do.

Table 3.3
Some coenzymes are derived from water-soluble (B group) vitamins

Vitamin	Coenzyme derivative	Enzyme (example)
Thiamin (B_1)	Thiamin pyrophosphate (TPP)	Pyruvate dehydrogenase
Riboflavin (B_2)	Flavin–adenine dinucleotide (FAD)	Succinate dehydrogenase
Niacin	Nicotinamide–adenine dinucleotide (NAD^+)	Lactate dehydrogenase
Pyridoxine (B_6)	Pyridoxal phosphate	Transaminase
Folic acid	Tetrahydrofolate (THF)	Glycine synthase
Cobalamin (B_{12})	Cobamides	Methylase

3.8 Effects of temperature and pH

Enzymes are globular proteins and their catalytic power depends upon the active site having a precise arrangement of amino acid residues to give the shape that carries out the catalysis and controls specificity. Therefore factors that change this precise arrangement will affect activity. In general anything that disrupts the complex folded shape of protein molecules will decrease the effectiveness of catalysis (see Figure 3.9).

One such influence is temperature which, by increasing the thermal motions of a polypeptide, eventually causes denaturation (Spread 3.4). The effect is complex, however, because increasing temperature normally makes reactions go faster because molecules have a greater chance of colliding and hence of reacting. What we find is that at low temperatures this is in fact the case. An increase in temperature increases the rate of reaction. At higher temperatures the 'opposing' effect of thermal denaturation starts to become more important and activity is lost. The shape of the curve is more or less the same for most enzymes (Spread 3.4) but the actual temperature where denaturation becomes significant varies enormously. The extreme example is of bacteria that live in hot springs ('extremophiles') whose enzymes may be active at 80°C or more, a temperature which would completely denature practically all mammalian enzymes.

Recently an enzyme from an extreme thermophile *Pyrococcus furiosus* has been purified and its structure determined. This organism lives at 100°C in hot springs. The enzyme protein contains both iron and vanadium, but its thermal stability seems to reside in having a low surface area and a large number of ion pairs and 'buried' atoms relative to other proteins.

It is sometimes said that enzymes have a temperature optimum but this is not really true; it depends upon how the measurement of activity is done. One could measure enzyme activity at a given temperature, or one could expose the enzyme to different temperatures for different time periods but then carry out the measurement of enzyme activity at 25°C. Different results will be obtained in each case, partly because the process of denaturation is wholly or partially reversible. This latter factor also varies a great deal between different enzymes.

Enzyme activity is also dependent upon pH. Extremes of pH (high acidity or high alkalinity) will denature proteins by disrupting the polypeptide chain's precise three-dimensional arrangement. Even near to neutral pH (pH 7) small changes in pH can affect the ionization state of amino acid side chains in the active site. If enzyme activity depends on particular residues being charged, or not charged, then a shift of even one pH unit (remember that this represents a ten-fold change in hydrogen ion concentration) can change the amount of enzyme activity measured very significantly.

Most enzymes are active over a fairly narrow pH range and can be said to have an optimum pH. This may be different for different enzymes, depending on their normal working environment. An extreme example here is, of course, the stomach digestive enzyme pepsin, which normally works at pH 1–2 in the stomach. Enzymes in 'biological' washing powders work at alkaline pH (Figure 3.11).

This dependence on temperature and pH shows us why, when we measure enzyme activity, it is very important to be precise about the conditions under which the measurement was made, including the temperature and the pH.

Figure 3.11 'Biological' washing powders contain proteases that break down protein stains such as food and blood. They are bacterial enzymes that are stable in the alkaline conditions necessary for detergent activity. They retain their activity for an hour or so at temperatures up to 65°C.

3.9 Inhibition of enzyme activity

When we considered the lock and key hypothesis above, we said that enzyme activity could be inhibited, as if the wrong key were jammed in the lock. It is easy to see that anything that alters or gets in the way of the precise three-dimensional shape of the active site is likely to stop catalysis taking place or at least slow it down. Indeed, many substances are known that can inhibit enzymes in a variety of ways and the study of these compounds is very important for a number of reasons. One reason is that if we compare the shapes of the true substrate with those of the inhibitors we can build up a picture of what the shape of the active site is like. Another reason is that many drugs, toxins, pesticides and herbicides act as enzyme inhibitors and hence it is important to know how they work. A valuable insecticide would be a compound that strongly inhibited a given enzyme in insect species but had no effect on humans and domestic animals. Similarly, knowledge of the shapes of active sites and inhibitors enables more effective drugs and pesticides to be 'designed'.

Factors that affect enzyme activity: temperature and pH

objectives
- To explain how temperature affects the rate of enzyme action
- To explain how pH affects the rate of enzyme action
- To recognize that some enzymes may have evolved to have certain pH and temperature profiles

Enzymes usually have a distinct optimum pH, although some enzymes are equally active over a broad range of pH values. However, not all enzymes have the same pH optimum. PEPSIN, which hydrolyses proteins to peptides in the very acidic conditions of the stomach, and TRYPSIN, which hydrolyses peptides to amino acids in the alkaline conditions of the duodenum, are enzymes with pH optima adapted to their environment. Pepsin is denatured as the chyme which contains it is squeezed through to the duodenum.

SUCRASE and ARGINASE are enzymes with pH optima which differ from the pH of their surroundings. This illustrates that local pH may be an important factor in regulation of enzyme activity.

It is possible to determine a temperature optimum for an enzyme, but the value obtained depends very much on how the experiment is carried out, e.g. for how long the enzyme is exposed to the given temperature. A brief exposure to high temperature may do little harm, whilst a long exposure may denature the enzyme.

INCREASING TEMPERATURE increases the rate of enzyme activity since the molecules involved in the reaction [enzyme, substrate(s) and cofactors, where appropriate] have greater kinetic energy as temperature increases. This means that:

(a) individual molecules move more quickly, increasing the frequency of collision;
(b) a greater proportion of molecules possess the minimum energy of activation.

This effect (Q_{10} approaches 2*) continues up to an OPTIMUM TEMPERATURE.

HIGH TEMPERATURES often irreversibly inhibit enzyme activity since the thermal motion of regions of the enzyme causes disruption of the weak interactions responsible for maintenance of enzyme conformation. As the enzyme changes in shape the active site(s) may be altered so that the number of E–S bindings is severely reduced and the enzyme loses its catalytic activity — the enzyme has undergone THERMAL DENATURATION. Even brief exposure to such high temperatures can be very damaging to metabolism — this is one of the consequences of high fever (a body temperature of 44°C, 112°F, generally results in death).

HIGH TEMPERATURE ADAPTATIONS are shown by some thermophiles in hot springs whose enzymes are THERMOSTABLE. Such enzymes may have their conformation stabilized by an increased number of intramolecular covalent bonds (disulphide bridges) or by prosthetic groups. Human enzymes are more thermostable than those of many infective agents — raising of the body temperature during mild fever can reduce infection by selectively denaturing microbial enzymes.

The effect of temperature on enzymes is utilized in FOOD PRESERVATION, which may depend upon FREEZING (low temperatures completely inhibit enzyme-induced deterioration of food) or COOKING (high temperatures denature enzymes in food or in microbes). Many fish contain an enzyme called THIAMINASE which degrades vitamin B_1 — animals on a diet of raw fish risk B_1 deficiency, and humans should typically boil raw fish to denature the enzyme and thus avoid the problem.

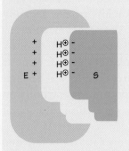

EFFECT OF pH

An enzyme has an OPTIMUM pH which reflects the effects of hydrogen ion concentration on the three-dimensional shape of the enzyme or the charge relationships of the enzyme's active site and the substrate molecule.

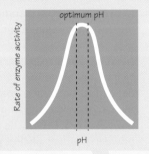

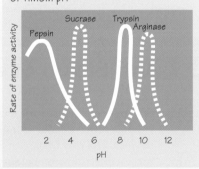

NOT ALL ENZYMES HAVE THE SAME OPTIMUM pH

*The effect of temperature on the rate of a reaction is expressed as Q_{10}. This represents the ratio of the rate of reaction at two temperatures, T°C and (T + 10)°C, in other words the effect on the rate of increasing the temperature by 10°C. For most enzyme reactions the value of Q_{10} is about 2.0.
Examples: hydrolysis of urea catalysed by urease, Q_{10} = 1.7; hydrolysis of casein catalysed by pepsin, Q_{10} = 2.3.

3.4 Factors that affect enzyme activity: temperature and pH

The formation of an ENZYME–SUBSTRATE COMPLEX is essential to the catalytic activity of an enzyme.

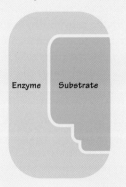

EFFECT OF TEMPERATURE

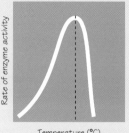

Enzymes have an OPTIMUM TEMPERATURE which represents a compromise between activation due to increased rate of collision between E and S and loss of activity due to denaturation of E molecules.

NOT ALL ENZYMES HAVE THE SAME TEMPERATURE OPTIMUM

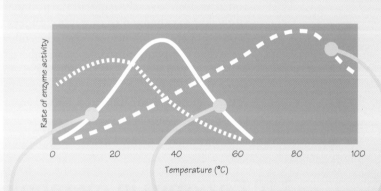

INCREASING TEMPERATURE increases the rate of enzyme activity until the optimum temperature is reached.

HIGH TEMPERATURES can bring about THERMAL DENATURATION of enzymes.

In hot thermal springs, some Cyanobacteria show HIGH TEMPERATURE ADAPTATIONS, e.g. the possession of THERMOSTABLE enzymes.

Minor fluctuations in pH may modify enzyme activity by alteration of the charge relationship between substrate and active site.

e.g. A positively charged substrate would normally bind to negatively charged side chains at an active site . . .

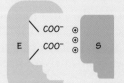

. . . but a fall in pH (increased H⁺ concentration) may 'mask' the active site and prevent E–S complex formation.

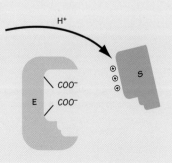

Chapter 3 • Enzymes

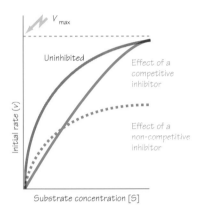

Figure 3.12 Effects of competitive and non-competitive inhibitors on rate of reaction in a plot against substrate concentration. The amount of enzyme is constant. In the presence of a competitive inhibitor the curve approaches V_{max} provided that the substrate concentration is sufficiently high. For non-competitive inhibitors it does not.

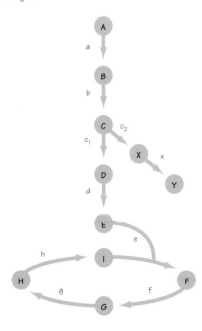

Figure 3.13 Metabolic pathways may be straight, branched or even circular. The compounds or 'intermediates' are shown here as A, B, C, etc. undergoing reactions catalysed by enzymes a, b, c, etc.; the arrows represent enzyme-catalysed transformations. Each of these transformations represents only a small chemical change catalysed by a single enzyme.

Several types of enzyme inhibition are known but we will concentrate on two types here to illustrate this important phenomenon (Spread 3.5).

Inhibitors which resemble the substrate in shape cause inhibition because they compete with the substrate to occupy the active site. When they are present in the active site, of course, no catalysis is taking place. We call this *competitive inhibition* because there is, indeed, a competition between substrate and inhibitor to occupy the active site. The result of this competition depends on the relative concentrations of substrate and inhibitor. If the concentration of the substrate is increased to a sufficiently high value then reaction can still attain the maximal rate (V_{max}). This effect can be evaluated by studying the rate of reaction under different conditions of substrate and inhibitor concentration. The study of the rates of enzyme reactions and the effects of inhibitors is called *enzyme kinetics* (Figure 3.12).

The other type of inhibition is *non-competitive inhibition*. In this case it seems that the inhibitor molecule binds to the enzyme protein at a position other than the active site (but probably near to it). This either prevents the substrate gaining access to the active site or alters the conformation (overall shape) of the polypeptide such that the active site's shape is changed so that it no longer recognizes its substrate. Non-competitive inhibitors bear no resemblance to the substrate and there is no competition; adding excess substrate to the reaction mixture will therefore not overcome the inhibition. Again, the enzyme kinetics show this (Figure 3.12). V_{max} is reduced because a fraction of the enzyme molecules are inhibited. However, any uninhibited molecules present behave normally.

3.10 Enzymes in metabolism

As we have seen, there are relatively few types of enzyme activity although there is a great range of specificity. When we look at what enzymes actually do, we find that they only catalyse small chemical changes — a hydrolysis, the addition of a double bond, an isomerization (see Spread 3.1, page 74). When changes occur in metabolism they are often major changes, such as when glucose is completely oxidized to CO_2 and H_2O. What we find in metabolism, then, is that changes take place in a number of small steps, each catalysed by a specific enzyme (Figure 3.13). The cumulative effect of these small changes produces the major conversion during metabolism. We call a series of small enzyme-catalysed steps *metabolic pathways*, and pathways may be linear, or branching, or even circular (Figure 3.13).

In fact, this is no different from how chemists carry out complicated chemical syntheses, namely in a number of relatively simple steps. The difference is that, in a series of chemical steps, yields for each step are typically a lot less than 100% and it is often necessary to purify the product of one reaction before going on to the next. In cells there is not the facility for 'purifying' at each step and enzymes need to produce high yields.

3.11 Allosteric enzymes

Cells and organisms need different metabolic pathways to operate at different times. Glycogen in the liver needs to be built up in times of plenty and broken down when starving, and the same applies to fat. How are the metabolic pathways controlled? It is really not feasible to make more enzyme protein when it is needed and to break it down when not. Apart from the inefficiency of this type of process, it cannot happen very quickly; it takes time to synthesize protein. What we need is minute-to-minute control so that after a rich meal, for example, the appropriate enzymes come into play to catalyse the metabolic steps that are needed in cells. These sorts of controls will be discussed later, but for the time being we have to ask whether it is possible to turn enzyme activity on and off at a minute's notice.

There are in fact two general ways of doing this. One is to add phosphate groups covalently to the enzyme molecule in a reaction known as *phosphorylation*. Many examples of this are known where the addition of a single phosphate can completely activate, or inactivate, that enzyme molecule. Some of the enzymes involved in glycogen metabolism are modified in this way to switch them on and off.

The other way is to have an enzyme that subtly changes its overall shape under certain conditions so that its activity is switched on or off. Such enzymes are called *allosteric enzymes* ('*allo*' is from a Greek word meaning 'other'). In general, allosteric enzymes have, in addition to the active site, another site or sites where other molecules can bind. It seems that such enzymes have two conformations or slightly different shapes, one active and the other inactive. Some allosteric enzymes are normally active and the binding of a given molecule to the 'extra' site (the allosteric site) switches activity off. Others are inactive until a compound binds to the allosteric site. If proteins were rigid molecules this would obviously not be possible, but we know that they are not rigid. One could imagine that the binding of a compound to the allosteric site can cause a subtle change in the conformation of the protein, and therefore of the active site, such that it becomes active (or inactive). This is illustrated in Figure 3.14. This is another reason why a rigid lock and key model does not explain what we observe.

What are these compounds that turn enzymes on and off? There are a great variety of them but mostly they are compounds that are intermediates in a pathway. One common way of controlling a metabolic pathway is by *negative feedback*. Here the product of a sequence of enzyme-catalysed reactions inhibits the first enzyme of the pathway allosterically. When there is 'enough' final product for a cell's needs, the concentration of this product rises and becomes high enough to bind to the allosteric site of the first enzyme of the pathway and inhibit its action (Figure 3.15). Thus there is no more throughput to the pathway until the concentration of the final product falls. At this point the first enzyme switches on again. Many examples of this type of control are now known, and some enzymes are switched on or off by a wide variety of compounds in this way, providing for very delicate control of metabolism. You will come across examples in future chapters.

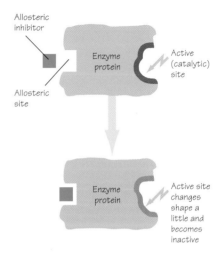

Figure 3.14 When an allosteric inhibitor binds to the allosteric site on the enzyme, the protein molecule undergoes a small conformational change so that the active site ceases to have catalytic activity. The binding of an allosteric inhibitor to its site is non-covalent and is easily reversible if the concentration of inhibitor falls.

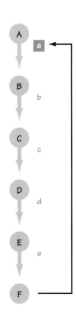

Figure 3.15 Feedback inhibition. Substrates or intermediates in a metabolic pathway are shown as A, B, C, etc. and the enzymes catalysing the steps as a, b, c, etc. Enzyme a is an allosteric enzyme that is inhibited by F. When sufficient F has been formed, accumulation of F stops the initial step happening.

Effect of inhibitors on enzyme activity

objectives
- To explain that enzyme action may be inhibited by a variety of compounds by a variety of mechanisms
- To realize that many drugs and pesticides are, in fact, specific enzyme inhibitors

The study of enzyme inhibitors is a very important part of biochemistry. A study of competitive inhibitors reveals the shape of the active site of an enzyme and gave rise to the INDUCED FIT HYPOTHESIS (see Spread 3.3). Many drugs and pesticides are also enzyme inhibitors.

Much of the chemical therapy used to combat cancer utilizes compounds structurally designed to compete with substrates needed for the enzymatic processes of DNA replication in cancer cells, and, in this manner, cell division can be prevented.

Sulphanilamide is an antibacterial agent because it effectively competes with p-aminobenzoic acid, which is required for synthesis of the essential metabolite folic acid by some pathogens.

Sulphanilamide can inhibit the bacterial enzyme selectively because humans require the B-vitamin folic acid but cannot synthesize the vitamin from p-aminobenzoic acid and, therefore, do not have the comparable enzyme.

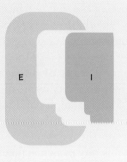

COMPETITIVE INHIBITORS compete for the active site with the normal substrate. These inhibitors therefore must have a similar structure to the natural substrate.

The success of the binding of I to the active site depends on the relative concentrations of I and S, and such inhibition is therefore REVERSIBLE BY AN INCREASE IN SUBSTRATE CONCENTRATION.

ETHANOL CAN PREVENT POISONING BY ANTI-FREEZE (ETHYLENE GLYCOL)

In the body ethylene glycol is converted to the lethal product oxalic acid — the first step is catalysed by alcohol dehydrogenase.

The reaction is inhibited by administration of a large dose of ethanol; ethanol competes for the active site on alcohol dehydrogenase and binds more tightly than ethylene glycol. It is, of course, also a substrate, but it competitively inhibits the binding of ethylene glycol. Eventually the ethylene glycol is lost via the kidneys.

Succinate dehydrogenase shows competitive inhibition by malonate. The reaction catalysed is:

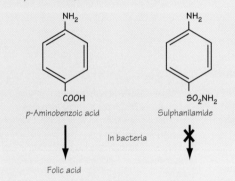

(FAD is a flavin compound that acts as a cofactor here, carrying hydrogens).

This enzyme is competitively inhibited both by oxaloacetate and by malonate:

$$\begin{array}{l} COO^- \\ | \\ CH_2 \\ | \\ C=O \\ | \\ COO^- \end{array} \qquad \begin{array}{l} COO^- \\ | \\ CH_2 \\ | \\ COO^- \end{array}$$

Oxaloacetate (inhibitor)   Malonate (inhibitor)

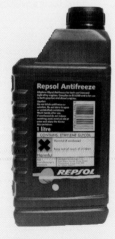

(Reproduced with permission of Repsol Petroleum Limited.)

3.5 Effect of inhibitors on enzyme activity

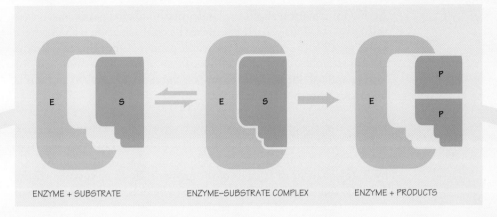

ENZYME + SUBSTRATE ENZYME–SUBSTRATE COMPLEX ENZYME + PRODUCTS

NON-COMPETITIVE INHIBITORS reduce enzyme activity by distortion of enzyme conformation caused by binding to some site other than the active site. If the binding is non-covalent the inhibition may be reversible if the inhibitor concentration is diminished.

If the binding is covalent the distortion of enzyme conformation may be permanent: IRREVERSIBLE INHIBITION.

AN IRREVERSIBLE INHIBITOR becomes covalently linked to the enzyme.

The Times of Tuesday 21 March 1995 reported a horrific attack on the Tokyo underground in which Sarin was used. A dozen people died and hundreds were hospitalized. Sarin is very similar to DFP.

$$H_3C-CH-O-\underset{\underset{O}{\overset{F}{|}}}{P}-CH_3$$
$$CH_3$$

Sarin, a nerve gas

IODOACETAMIDE is an irreversible inhibitor of enzymes with cysteine residues at or close to the active site.

$$\boxed{E}-CH_2SH \;+\; ICH_2-\overset{O}{\underset{}{C}}-NH_2$$

$$\downarrow$$

$$\boxed{E}-CH_2S-CH_2-\overset{O}{\underset{}{C}}-NH_2 \;+\; HI$$

This compound is often used in studies of active site composition and function.

DFP (di-isopropyl phosphofluoridate) is a nerve gas which exerts its effects by binding irreversibly to a serine residue in the active site of acetylcholinesterase.

$$\boxed{E}-CH_2OH \;+\; \boxed{DFP}$$

$$\downarrow$$

$$\boxed{E}-CH_2O-DP \;+\; HF$$

Many very useful insecticides including Malathion and Parathion were developed from highly toxic compounds such as Sarin and DFP. They kill insects by the same mechanism (i.e. acetylcholinesterase inhibition) but insects seem to be more susceptible than humans.

$$H_3C-O \underset{H_3C-O}{\overset{S}{\underset{}{\overset{\|}{P}}}}-S-CH-\overset{O}{\underset{}{C}}-O-C_2H_5$$
$$CH_2-\overset{O}{\underset{}{C}}-O-C_2H_5$$

Malathion

Chapter 3 • Enzymes 93

Actual information about the conformational changes themselves is rather sparse. This is partly because the conformational changes are rather small in the context of the whole protein. One thing that almost always stands out is that allosteric enzymes have 'unusual' or 'non-standard' kinetic behaviour (see Spread 3.2, page 78) and this is frequently how they are identified.

Discovering the precise nature of the conformational changes is a goal and has been achieved for a few allosteric enzymes. However, the usual method for determining the three-dimensional structure of proteins is by X-ray analysis of crystals of the proteins. By their very nature the crystals, and the protein in them, have to be fixed in shape rather than flexible. Consequently X-ray analysis typically only gives information about the shape of one of the forms of the enzyme protein.

3.12 The role of enzymes in metabolism

In the next chapter we will start to understand the myriad of enzyme reactions that control metabolism in cells. The various processes of making new proteins and new cellular materials, obtaining energy in various ways, movement, and all the other life activities, are only possible because of the action of enzymes.

3.13 Further reading

Bickerstaff, G.F. (1987) Enzymes in Industry and Medicine. Edward Arnold (New Studies in Biology)

Buttle, D.J. and Knight, C.G. (1993) Proteins that digest proteins. Biological Sciences Review **5(4)**, 10–13

Cech, T.R. (1986) RNA as an enzyme. Scientific American **255 (Nov)**, 76–84

Cheetham, P. (1993) Enzymes at work — commercial uses of bio-catalysts. Biological Sciences Review **5(5)**, 34–36

Gull, D. and Brown, B. (1993) Enzymes — fast and flexible. Biological Sciences Review **6(2)**, 26–29

Vadgama, P. (1989) Biosensors — artificial sniffer dogs. Biological Sciences Review **1(5)**, 36–39

Wymer, P. (1993) Abzymes and ribozymes — new kinds of enzyme. Biological Sciences Review **5(4)**, 40–42

Wynn, C. (1989) Enzymes — nature's accelerators. Biological Sciences Review **1(5)**, 16–21

3.14 Examination questions

1. (a) Indicate which of the graphs on the right represents the usual relationship between enzyme activity and the following: (i) temperature; (ii) pH; (iii) enzyme concentration; (iv) substrate concentration. (4 marks)

   (b) Explain what is meant by: (i) coenzyme (3 marks); (ii) competitive inhibition (3 marks); (iii) enzyme inactivation at high temperatures. (2 marks)

   [ULEAC (Biology) June 1983, Paper 1, No. 2]

2. Penicillin is hydrolysed and therefore inactivated by a digestive enzyme (penicillinase) resulting from a mutation in some bacteria. The amount of penicillin hydrolysed in 1 minute in 10 cm^3 of solution containing 10^{-3} g of purified enzyme was measured.

   (a) If the molecular weight of the enzyme is 29,600, what is its molar concentration in the incubation mixture? (3 marks)

   (b) When the experiment above was performed, the rate of hydrolysis of penicillin was found to vary with its concentration as follows:

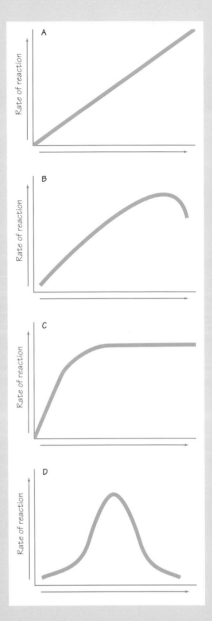

Penicillin added (µM)	Amount hydrolysed (nmol)
1	0.11
3	0.25
5	0.34
10	0.45
30	0.58
50	0.61
90	0.61

   Plot these results on graph paper and label the curve 'curve A'. (4 marks)

   (c) The experiment was repeated in the presence of a constant amount of inhibitor and the following results were obtained.

Penicillin added (µM)	Amount hydrolysed (nmol)
1	0.03
3	0.10
5	0.18
10	0.26
30	0.48
50	0.58
80	0.61

Chapter 3 • Enzymes

Plot these results on the same axes as the previous graph and label the curve 'curve B'. (1 mark)

(d) What is this type of inhibition called? (1 mark)

(e) Why can high substrate concentrations overcome this type of enzyme inhibition? (1 mark)

(f) The enzyme was briefly heated to 80°C and rapidly cooled. The experiment was repeated in the absence of inhibitor and the following results were obtained.

Penicillin added (µM)	Amount hydrolysed (nmol)
1	0.08
3	0.15
5	0.19
10	0.25
30	0.32
50	0.38
90	0.38

Plot these results on the same axes as the previous graphs and label the curve 'curve C'. (1 mark)

(g) Why was the maximum rate of enzyme activity reduced in this last experiment while the gradient of the graph, which is a measure of affinity between enzyme and substrate, was relatively little affected? (2 marks)

(h) What other procedure might produce the same effect on the enzyme as that seen in curve C? (1 mark)

(i) Describe the steps linking the mutation mentioned in (a) to hydrolysis of the penicillin. (2 marks)

(j) State two factors which must be kept constant in this experiment. (2 marks)

(k) Describe the experiment that must be performed to show that penicillin does not hydrolyse spontaneously. (2 marks)

[O & C (Biology) June 1991, Paper 1, No. 7]

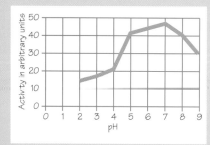

3. (a) What is meant by an enzyme? (3 marks)

(b) The graph opposite shows the effect of pH on the activity of the enzyme catalase in potato.
    (i) Write an equation in words or symbols to show the action of the enzyme catalase. (2 marks)
    (ii) Comment on the effect of pH on the activity of catalase. (3 marks)

(c) Catalase is required for detoxification of a metabolic by-product. There is generally a higher concentration of catalase in animal cells than in plant cells. Suggest a reason for this difference. (2 marks)

[ULEAC (Biology) January 1992, Paper 1, No. 3]

4. (a) The figure illustrates the effect of pH on the activity of an enzyme which catalyses the hydrolysis of proteins.
   (i) In what region of the digestive tract might this enzyme be found? (1 mark)
   (ii) Explain how pH affects the activity of an enzyme. (4 marks)

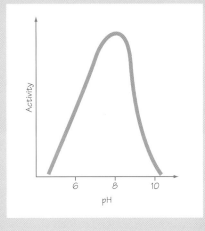

(b) (i) Using the lower axes opposite, sketch the expected effect of temperature on the activity of an enzyme extracted from mammalian tissue. (3 marks)
   (ii) What is meant by the term optimum temperature of an enzyme? (1 mark)
   (iii) The optimum temperature for an enzyme-catalysed reaction measured over a 5-minute period was found to be 45°C. However if measured over a 15-minute period the optimum temperature was 30°C and when measured over a 30-minute period it was 25°C. Comment on these results. (3 marks)
   (iv) Would you expect enzymes extracted from plant tissues to have the same optimum temperature as those extracted from mammal tissues? Give a reason for your answer. (3 marks)

(c) Explain why the presence of an enzyme increases the rate of a reaction. (3 marks)

(d) Cell organelles are often described as multi-enzyme units. For each of the following organelles, state one of their major functions: (i) smooth endoplasmic reticulum; (ii) Golgi apparatus; (iii) lysosomes; (iv) ribosomes. (4 marks)

[O & C (Biology) June 1993, Paper 1, No. 6]

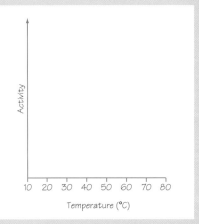

5. A student was given four unlabelled flasks containing solutions and told that one flask contained starch and the other three contained different concentrations of an enzyme which catalyses the hydrolysis of starch.

   (a) (i) Name the enzyme given to the student. (1 mark)
   (ii) What would have been formed when the starch was hydrolysed in this experiment? (1 mark)

   (b) Describe how you would determine which of the flasks contained starch. (2 marks)

   (c) Describe an experiment you could carry out to determine which of the enzyme solutions was the most concentrated. (5 marks)

[ULEAC (AS Biology) June 1991, Paper 1, No. 7]

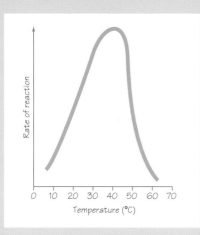

6. (a) State what is meant by each of the following terms in relation to enzymes.
   (i) Active site. (2 marks)
   (ii) Competitive reversible inhibition. (2 marks)

(b) The graph shows the relationship between the rate of an enzyme-controlled reaction and temperature.
   (i) Give a reason for the increase in the rate of reaction between 15°C and 37°C. (1 mark)
   (ii) Give a reason for the decrease in the rate of reaction between 45°C and 60°C. (1 mark)

[ULEAC (Biology) June 1992, Paper 1, No. 5]

7. The turnover number of an enzyme is the number of substrate molecules that one molecule of enzyme can convert to product in one second under optimum conditions.

The table gives the turnover numbers of some different enzymes.

Enzyme	Turnover number as molecules converted per second
Carbonic anhydrase	600,000
Catalase	200,000
Acetylcholinesterase	25,000
Lactate dehydrogenase	1000
Chymotrypsin	100
Lysozyme	1

(a) In measuring the turnover number of a particular enzyme, there must be an excess of substrate present. Why is this so? (1 mark)

(b) (i) What effect would a drop in temperature of 10°C have on the turnover numbers of the enzymes?
   (ii) Explain your answer. (2 marks)

(c) Where precisely in the body of a mammal would you expect the following enzymes to be active:
   (i) acetylcholinesterase?
   (ii) carbonic anhydrase? (2 marks)

(d) What is the action of carbonic anhydrase? (1 mark)

[AEB June 1992, Paper 1, No. 2]

4.1 Introduction

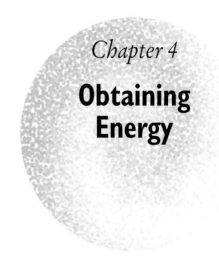

Life represents an ordered state amongst the chaos of the universe, and to maintain this order requires the expenditure of energy. Living cells need energy to make more cellular substance as they grow and divide, and as they move and engage in other life activities. In addition, just to maintain cells in a viable state requires that energy be available for pumping substances across membranes, in and out of cells. One of the first signs that a cell is damaged or dying is that its membrane becomes leaky because the membrane pumps are not working; lack of energy will have just this effect. Cells actually spend quite a lot of their daily energy requirement in this maintenance process.

It may be asked what 'energy' is in biological terms and how living cells get hold of it. Obviously they cannot 'make' energy; they need to obtain it by transformation from some source in the same way that oxidation of petrol converts chemical energy into kinetic energy in a car. Ultimately, for practically all life on Earth, the source of energy is light from the sun, trapped by green plants and photosynthetic bacteria during photosynthesis (Figure 4.1). Everything else is dependent upon this process and, sooner or later, the light energy trapped on earth is released as low-grade energy or heat. This happens when coal formed from living matter millions of years ago is burned, but the process is going on all the time.

Figure 4.1 Conversion of light energy into other forms of energy. The solar-powered calculator traps light energy (photons) on a silicon wafer to produce electrical energy to run the electronic circuits. The green plant traps photons and converts the light energy into chemical energy which is used to drive the chemical synthesis of all the organic molecules of which the plant is made.

The major questions to be asked in this chapter are: what is 'biologically useful energy'? how is it generated from light energy? and how do non-photosynthetic organisms get their energy? It will also be helpful to think how energy is stored (for example, petrol in a fuel tank), for even plants can carry out photosynthesis only when the sun is shining.

ATP is the energy currency of the cell

Money, or currency, is very useful stuff. It does not actually matter what it is made of — iron bars for the Romans, Cowrie shells in the South Pacific, gold for the Spaniards, or pieces of paper at the present time — what matters is that it can be generated in a variety of ways and used in a number of different ways in transactions. We can think of an energy currency in the same way, and, as anyone who has travelled in Europe will know, having a single currency would make life a lot easier than having several as at present.

For the majority of energy transactions in practically all organisms, the universal energy currency is *adenosine triphosphate* or ATP. This is a molecule that may be generated in a number of ways, including by photosynthesis, and used in a variety of ways, such as in muscle contraction. The molecule is sometimes said to be 'energy rich' or to have 'high-energy bonds', but these phrases are really examples of biochemical jargon or shorthand; there is nothing chemically unusual or unorthodox about the ATP molecule. Its formation and hydrolysis obey the laws of chemistry. Its key property lies in the fact that it is easily hydrolysed within cells and readily donates its phosphate group to other molecules. When ATP is hydrolysed to *adenosine diphosphate* (ADP)

Adenosine triphosphate: ATP

objectives
- To describe the structures of ATP and ADP
- To explain why ATP hydrolysis rather than its formation is favoured, releasing energy for metabolic use

BALL AND STICK MODEL OF THE ATP MOLECULE (hydrogen atoms not shown). At the pH values found in the cell the phosphate groups will be almost completely ionized.

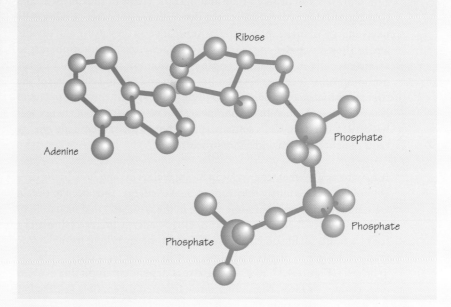

The universal energy currency for the majority of energy transactions in organisms is adenosine triphosphate (ATP). ATP is generated primarily by the energy-yielding reactions of photosynthesis in chloroplasts, and by oxidative metabolism in mitochondria. The hydrolysis of ATP to adenosine diphosphate (ADP) releases the energy that is required by cells to engage in life activities, such as muscle contraction and protein synthesis.

The hydrolysis of ATP (see formulae on opposite page) may be summarized as:

$$ATP^{4-} + H_2O \longrightarrow ADP^{3-} + P_i^{2-} + H^+$$

This hydrolysis occurs readily in aqueous solution, although not necessarily very rapidly in the absence of an enzyme catalyst. To reverse the above reaction will require the bringing together of ADP^{3-} and $phosphate^{2-}$, and since like charges repel one another this will require energy. Consequently, the reverse reaction (ATP hydrolysis) can be said to release energy.

The equilibrium constant for the reaction

$$ATP + H_2O \longrightarrow ADP + P_i$$

at pH 7.4 (the pH inside the cell) and 25 °C is about 1×10^{-7} mol dm^{-3}. This says that at equilibrium there is about 10,000,000 times as much ADP as ATP present in the solution!

PYROPHOSPHATE, two phosphates joined together (usually represented as PP_i), may be hydrolysed as follows:

$$PP_i + H_2O \longrightarrow 2P_i$$

and the equilibrium for this reaction is strongly to the right. Sometimes ATP may be hydrolysed to give AMP and PP_i. Subsequent hydrolysis of the PP_i will 'pull' the overall reaction to the right.

ROLE OF ATP AS AN ENERGY CURRENCY

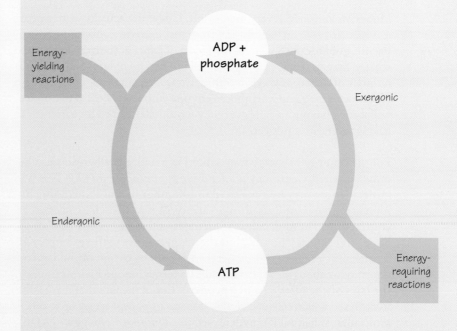

4.1 Adenosine triphosphate: ATP

HYDROLYSIS OF ATP TO ADP

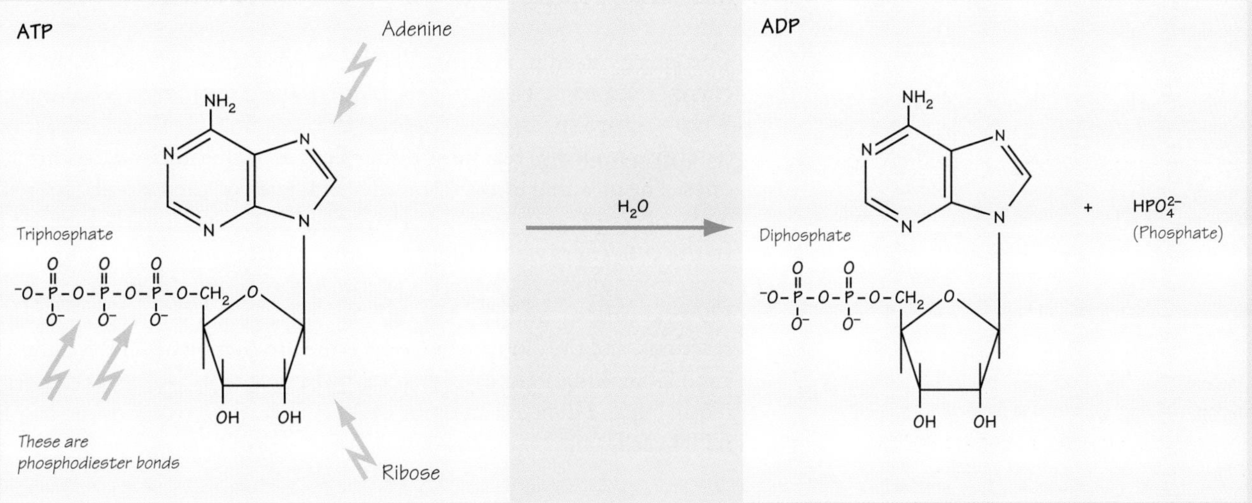

ATP — Adenine, Triphosphate, Ribose
These are phosphodiester bonds

$\xrightarrow{H_2O}$

ADP — Diphosphate + HPO_4^{2-} (Phosphate)

Other nucleosides also form triphosphates with very similar free energies of hydrolysis, e.g.:

Guanosine triphosphate (GTP)
Cytidine triphosphate (CTP)
Uridine triphosphate (UTP)

Adenine–ribose = a nucleoside (adenosine)
Adenine–ribose–phosphate–phosphate–phosphate = adenosine triphosphate
[base–sugar–phosphate = a nucleotide]

ADP can be hydrolysed to produce AMP (adenosine monophosphate) in a similar reaction:
ADP + H_2O $\longrightarrow$ AMP + phosphate

GENERATION OF ATP

CHLOROPLASTS (light energy)

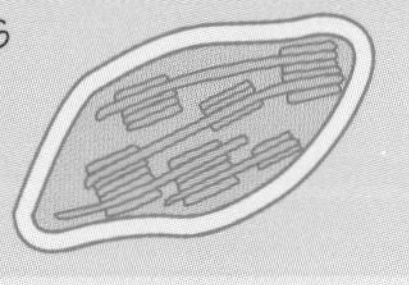

MITOCHONDRIA (oxidation energy)

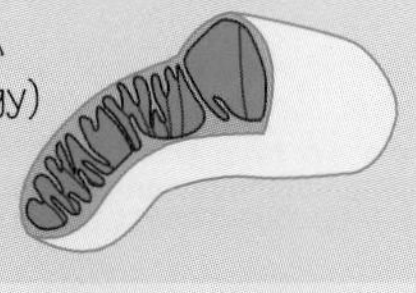

CYTOPLASM (rearranging molecules)

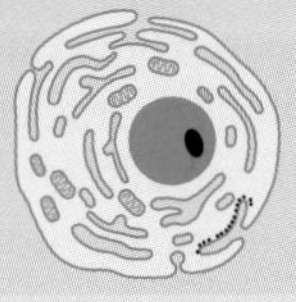

→ **ATP** →

MUSCLE CONTRACTION
ATP hydrolysis changes the position of myosin 'head' relative to actin.

PROTEIN SYNTHESIS
ATP energy is used to 'load' amino acids on to transfer RNA.

'DARK' PHASE OF PHOTOSYNTHESIS
ATP is used to drive the reactions by which carbohydrate is synthesized.

ACTIVE TRANSPORT ACROSS MEMBRANES
is driven by ATP.

BIOLUMINESCENCE
The light production of fireflies and glow worms depends on ATP.

Chapter 4 • Obtaining Energy

Figure 4.2 Coupling reactions together.

The reaction:

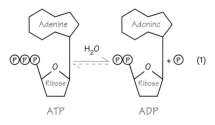

has a strong tendency to proceed from left to right. At equilibrium there is ten million times more ADP than ATP.

The reaction:

$$\text{Glucose} + \text{P} \underset{H_2O}{\rightleftharpoons} \text{Glucose phosphate} \quad (2)$$

has little tendency to proceed from left to right and indeed 'prefers' to go backwards. At equilibrium there is much more glucose than glucose phosphate present.

If one imagines the two reactions going on in the same test tube, the tendency is for reaction (1) to increase the local concentration of phosphate. By the Law of Mass Action this 'pushes' reaction (2) to the right. In other words, glucose becomes phosphorylated at the expense of ATP hydrolysis.

In cells we cannot speak of the two reactions going on in the same test tube; they take place on the active site of an enzyme:

Glucose + ATP → glucose phosphate + ADP

The enzyme catalysing this reaction is hexokinase. Notice that ATP hydrolysis as such, to produce ADP and phosphate, does not occur.

Reactions (1) and (2) are said to be coupled and this is typically how ATP 'hydrolysis' is used to drive reactions that would not normally proceed.

and inorganic phosphate, the phosphate is usually shown as P_i rather than HPO_4^{2-}. Spread 4.1 explains the role of ATP in cells.

The energy used in making ATP from ADP and phosphate, and the energy released in its hydrolysis, is not heat energy but chemical energy. We therefore speak of *exergonic* (energy-yielding) and *endergonic* (energy-requiring) reactions rather than endothermic and exothermic ones. The meaning of the term 'chemical energy' can be understood in terms of coupling two reactions together. This is explained in Figure 4.2. In the cell, energy-yielding reactions (of catabolism) are coupled to energy-requiring reactions (of anabolism) through the agency of ATP. Photosynthesis and the breakdown of sugars are energy-yielding reactions, and the ultimate outcome is the production of ATP. Syntheses, such as making new proteins, carbohydrates or lipids, are energy requiring and demand an input of ATP. Spread 4.1 explains the role of ATP in cells.

Where and when is ATP formed?

ATP is synthesized in cells from ADP and inorganic phosphate (P_i), and by far the largest proportion of the ATP manufacture takes place in the mitochondria and, in green plants, in the chloroplasts. These are both membrane-enclosed organelles with a special structure that is required for ATP synthesis. Prokaryotic organisms make ATP by similar processes, but in this case it is at infoldings of their plasma membranes. A small amount of ATP is produced in the cytosol of cells by reactions that take place in the soluble phase. This ATP production is important when cells, and especially muscle cells, are operating anaerobically, but under normal circumstances it represents only a minor proportion of the total energy needed by eukaryotic cells.

Since practically all life on Earth is dependent on the light energy from the sun, it is the chloroplasts of green plants, and the analogous structures in cyanobacteria and photosynthetic bacteria, that are the key organelles in ATP production. We will deal with chloroplasts first, but, as we will see, the mitochondria are very similar in the way that they make ATP; it is just the source of the energy that is different, namely oxidation of organic compounds rather than absorption of sunlight.

4.2 Photosynthesis

Photosynthesis is the process of trapping the sun's light energy and converting this into chemical energy in the form of ATP and other organic molecules. The first part of the reaction is a photochemical reaction. When light interacts with photographic film in a camera, the light energy is converted into chemical energy which drives a reaction that eventually produces an image. Solar-powered calculators provide an even more relevant example, because here the light energy causes movements of electrons in wafers of silicon and after the calculation is complete some of the energy is used to drive the liquid crystal display with the answer (see Figure 4.1). These are all energy transformations.

Organisms which use light energy to furnish their energy needs are the basic trophic level of nearly all ecosystems (Spread 1.2, page 8) and are called autotrophs. The photosynthetic process requires pigments to absorb the light energy, and the pigment in question is *chlorophyll* in its various forms, along with some accessory pigments (Figure 4.3). The process is basically very similar in green plants, cyanobacteria and photosynthetic bacteria, but the details are different, presumably reflecting the fact that these separate lines evolved differently over millions of years.

The result of light energy absorption is that the plant or cell can combine very simple compounds — CO_2, H_2O, NH_3 — to make all the complicated organic compounds of which plant cells are made. The first phase of the process is the conversion of light energy into chemical bond energy, not only in ATP but also in cofactors that have 'reducing power' (Figure 4.4). After this, a myriad of reactions takes place to carry out the biosynthesis of complex organic molecules, including carbohydrates. The first part is often called the *light reaction,* and the second part the *dark reaction.* This is rather an oversimplification because 'reaction' actually means many reactions. Also the name 'dark reaction' is intended to indicate that, given ATP and 'reducing power', this phase could go on in the dark (Figure 4.4, page 104). Normally, however, both processes go on simultaneously in the plant cell and the activities of many of the enzymes of the dark reaction are controlled by light. In very simple terms, plants use the energy of ATP to drive the synthesis of new chemical bonds and reducing power (NADPH) to reduce carbon dioxide to compounds more reduced than carbon dioxide:

$$CO_2 + 2[H] \longrightarrow CH_2O + \tfrac{1}{2} O_2$$

Figure 4.3 The chlorophylls *a* and *b* are the pigments in chloroplasts responsible for trapping light energy. They are helped by accessory pigments, such as carotenes and phycocyanobilins, which trap light energy at wavelengths where chlorophylls do not absorb light, and pass the energy on to chlorophyll.

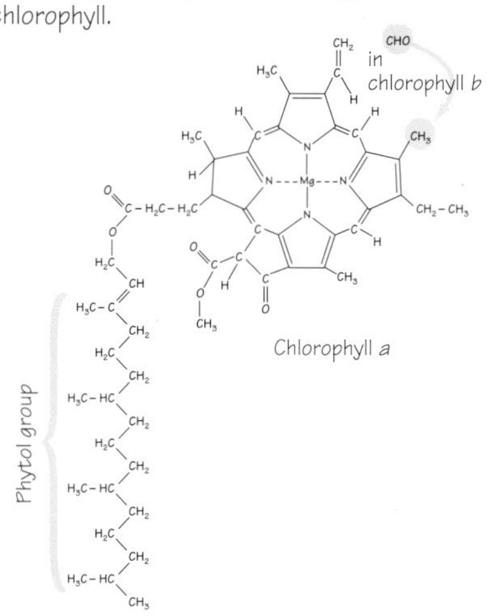

The light reaction of photosynthesis

Chlorophylls are the pigments responsible for capturing light energy and starting the process of converting it into chemical energy. They are helped by other pigments, such as the carotenoids (Figure 4.3). The relationship between which wavelengths of light are effective and the rate at which photosynthesis takes place may be measured. The resulting graph is called an *action spectrum* and is helpful in telling us which wavelength and which pigments are involved in the process (Figure 4.5, page 105). This is how we know that chlorophylls and carotenoids are involved.

Chlorophylls may be extracted from plant tissues using acetone. A freshly prepared solution of chlorophylls fluoresces strongly. The phenomenon of fluorescence means that light is being absorbed but then re-emitted at a longer wavelength. Thus, in this case, the light has not been trapped. In plants, instead of being re-emitted, the absorbed light energy is passed from compound to compound.

During fluorescence, when a molecule absorbs light energy, some of the electrons jump to a higher orbital (and have a higher energy level); later

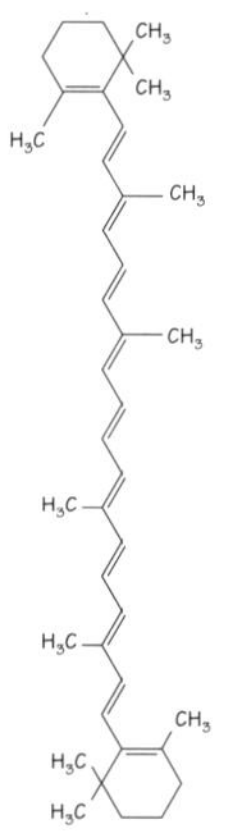

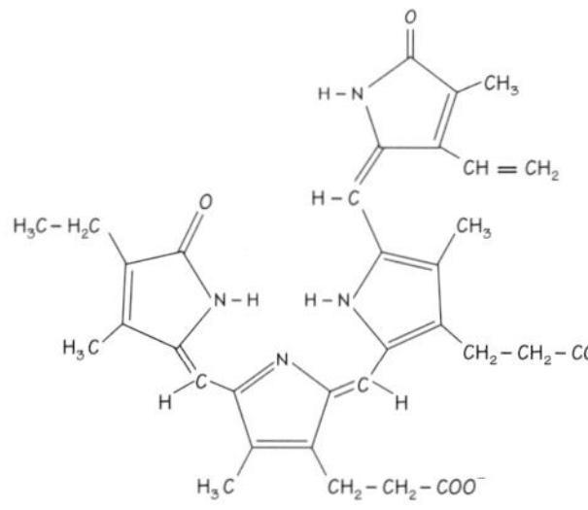

Figure 4.4 (a) In order to make organic compounds, all of which are more reduced (i.e. contain more hydrogen and electrons) than CO_2, 'reducing power' and chemical energy (ATP) are needed. (b) Many different compounds, usually coenzymes, can act as reducing agents in cells, but in photosynthesis the key compound is NADPH, which stands for nicotinamide–adenine dinucleotide phosphate. A nucleotide is base-sugar-phosphate; this dinucleotide is two nucleotides joined tail-to-tail. We can think of the molecule as having a 'recognition end', a complicated organic molecule that can be recognized specifically by enzyme proteins, and a 'business end' that can carry electrons and hydrogens. (c) Because the nicotinamide ring already carries a positive charge, the reversible oxidation reaction should be written:

$NADP^+ + 2H^+ + 2e^- \longrightarrow NADPH + H^+$
Oxidized form   Reduced form

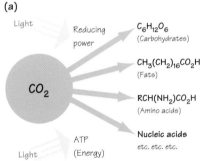

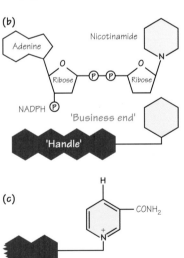

they fall back. In photosynthesis, instead of these electrons dropping back to their original energy level, they are captured by other compounds and are subsequently used to drive the synthesis of ATP and NADPH. The chlorophyll molecules consequently become electron deficient and these electrons need to be replaced. This is done by using electrons produced by the splitting of water (H_2O) molecules:

$$H_2O \longrightarrow 2H^+ + [O] + 2e^-$$

The oxygen produced in this way is released into the atmosphere. Exactly how this all happens is still not fully understood despite an enormous amount of research work. Nevertheless, we have a pretty good idea of how it all happens, and recent work has shown in detail how some of the (rather simpler) bacterial systems work.

Chloroplast structure

Chloroplasts have three sets of membranes. An outer and an inner membrane enclose the whole organelle (Spread 4.2). The interior is a protein-rich gel called the *stroma*, where the dark reaction of photosynthesis takes place in solution (see Chapter 5). Also in the interior there is a third membrane system consisting of stacks of flattened bags forming a network. This is the *thylakoid network* and the thylakoid stacks are called *grana*. These are connected by intergranal thylakoid *frets*. The volume enclosed by the thylakoid membrane is called the *lumen*. The chloroplast is a complex organelle — but then it has a complicated job to do. The membrane systems are vital for the proper functioning of chloroplasts.

The parts of the chloroplast that function in light trapping and ATP production are located in the thylakoid membranes; the actual process of light trapping is carried out by *photosynthetic units*. Each of these contains 200 – 400 chlorophyll molecules plus associated carotenoids, in a very precise arrangement with proteins. One particular chlorophyll molecule acts as the *reaction centre* for the unit, and all the other pigment molecules form a sort of 'antenna' that catches photons of light and passes these on to the reaction centre (Figure 4.6). Cyanobacteria and green plants have two types of photosynthetic unit called *photosystems I* and *II* (often abbreviated to PSI and PSII). These have different reaction centres and absorb light at two different wavelengths: 700 nm for PSI and 680 nm for PSII. The reaction centres are sometimes referred to as P700 and P680 and this refers to the single chlorophyll molecule in the 'centre' of the antenna in each case. Photosynthetic bacteria have only one type of reaction centre and do things somewhat differently although 'overall' the mechanism of light trapping is the same.

When a photon of light strikes any of the pigment molecules of PSI, the energy captured is eventually channelled to the reaction centre P700 which represents the lowest energy 'sink' of all the pigment molecules. The energy of the absorbed light allows P700 to pass an electron having a raised energy level to an acceptor molecule. This therefore becomes *reduced* and a positively charged 'hole' is left in the chlorophyll molecule (for the time being). The reduced acceptor molecule then passes its electron to a small iron-containing protein called *ferredoxin*, which thereby becomes reduced. Eventually ferredoxin passes this electron to NADP$^+$, reducing it to NADPH using a proton (H$^+$) from the medium. The result of this photochemical reaction and associated electron transfers generates reducing power but not ATP. Before we get to this we have the pressing problem of a positively charged 'hole' left in the electron-deficient chlorophyll P700. (Another way of thinking about this is that the chlorophyll molecule has become oxidized.) The hole is filled, or the reduction completed, by using electrons taken from the splitting of water. This is achieved using energy collected in a similar way by PSII, which is channelled to P680, which can then pass on an electron. This electron passes through a series of electron-carrying compounds until it is finally delivered to the electron-deficient P700. A chain of electron carriers is simply a series of compounds that can exist in either the reduced (electron-rich) or oxidized (electron-poor) states and so can pass electrons from one to another. What has happened here is quite a remarkable process. Water, a notoriously poor reducing agent, has been induced to supply electrons. The 'inducement' comes from the trapped light energy.

This process is shown diagrammatically in Figure 4.7 (page 108), which also explains how ATP is made.

ATP production during photosynthesis

Schemes of the type shown in Figure 4.7 are called 'Z schemes' because the diagram showing the flow of electrons was originally drawn like a letter Z. Normally, these days, the diagram is rotated through 90° to give a letter N, the logic being that things at the 'top' have a higher energy level than those at the 'bottom'. The flow of electrons through the various 'carriers' in the scheme is like an electric current and can be harnessed to do work — in this case making ATP. Many different types of compound can act as electron carriers, from organic compounds that can exist in either the oxidized or reduced state, to metalloproteins containing iron or copper atoms, which can switch valencies (Figure 4.8, page 109).

The driving force to produce chemical energy is now electrons flowing from a high potential to a low potential. Remember that the energy was originally light energy. It has to go through another form before it can be transformed into chemical bond energy. The various electron carriers are set in the thylakoid membrane and the protein carriers are arranged in a fixed orientation. As electrons flow through the carriers, the orientation of the carriers ensures that hydrogen ions (protons, H$^+$) are pumped across the membrane, so generating a higher concentration on one side than the other. (The higher concentration is on the inside of the thylakoid units.) What we have, therefore, is a *proton gradient*, and, like water behind a dam, a gradient represents potential energy. In fact it is not only a gradient of hydrogen ion concentration, but also of positive charge, very similar to an electric battery. It has the tendency to dissipate with the release of energy.

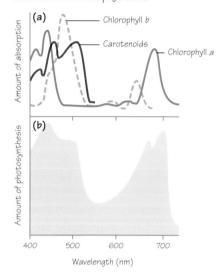

Figure 4.5 (a) Absorption spectra of chlorophylls *a* and *b* and of carotenoids, compared with (b) the so-called action spectrum, which relates amount of photosynthesis to wavelength. Light energy can only be trapped if there are pigments to absorb it; the accessory pigments enable light to be absorbed across a greater range of the solar spectrum than could be achieved with chlorophyll alone.

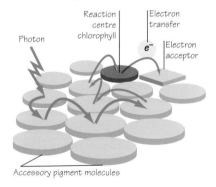

Figure 4.6 The arrangement of pigment molecules in a chloroplast (antenna) allows captured light energy to 'wander' from one molecule to another by an efficient process called resonance transfer. Finally, the energy arrives at the reaction centre chlorophyll molecule. [Redrawn with permission from C.K. Mathews and K.E. van Holde (1990) *Biochemistry*, Benjamin/Cummings Publishing Co., California.]

Chloroplast structure and photosynthesis

objectives
- To describe the structure of higher plant chloroplasts
- To explain how light energy is trapped as usable energy in ATP and reducing power in NADPH
- To summarize how chloroplast structure is related to function

The STRUCTURE of chloroplasts is closely related to their function in photosynthesis and illustrates their possible endosymbiotic origins. The 'double membrane' chloroplast envelope, the 70 S ribosomes and the circular DNA all suggest that chloroplasts may have been derived originally from invasive algae — this is the SYMBIOTIC INVASION HYPOTHESIS.

ELECTRON MICROGRAPH OF A PLANT CELL CHLOROPLAST

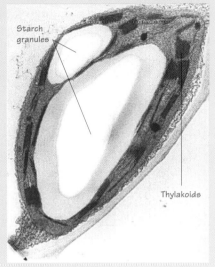

The CHLOROPLAST ENVELOPE comprises two membranes: the inner one is the true 'chloroplast membrane' which is freely permeable to carbon dioxide and oxygen but selectively permeable to sugars (photosynthetic product) and ions such as K^+ and Mg^{2+} (used in chlorophyll synthesis) and iron (used in the electron transfer system).

THYLAKOID MEMBRANES are the site of the LIGHT DEPENDENT REACTIONS of photosynthesis. They have on their surface, or embedded within them, the necessary molecules of chlorophyll and accessory pigments, together with electron transport systems and enzymes. The light-absorbing molecules are arranged in units called PHOTOSYSTEMS.

Green plants and cyanobacteria have two types of photosynthetic unit called PHOTOSYSTEMS I and II. These have different reaction centres and absorb light at different wavelengths: 700 nm and 680 nm respectively (see Figure 4.7).

GRANA are stacks of thylakoid membrane. They are sac-like structures enclosing a thylakoid space which is the hydrogen ion reservoir used in the CHEMIOSMOTIC SYNTHESIS OF ATP.

The STROMA (CHLOROPLAST MATRIX) contains the enzymes which catalyse the LIGHT-INDEPENDENT REACTIONS of photosynthesis. These include ribulose bisphosphate carboxylase, which catalyses the reaction:

$$CO_2 + \text{ribulose bisphosphate (RuBP)} \longrightarrow \text{2-phosphoglyceraldehyde (PGA)}$$

and the starch synthase complex which catalyses the reaction:

$$(\text{Glucose})_n + \text{glucose 1-phosphate} \longrightarrow \text{'starch'}$$

Within the stroma, multi-layered STARCH GRANULES are the insoluble storage carbohydrate product of photosynthesis.

In addition, LIPID GLOBULES are often associated with the breakdown and regeneration of chloroplast membranes.

The CIRCULAR DNA found within the chloroplast is not contained within a nucleus, nor is it organized into a chromosome. This is similar to the arrangement of bacterial DNA.

The RIBOSOMES are also similar to those found in bacteria, being of the same size (70 S). Protein synthesis in the chloroplast can supply some, but not all, of the chloroplast proteins.

GRANA enclose a thylakoid space which is the hydrogen ion reservoir used in CHEMIOSMOTIC SYNTHESIS OF ATP

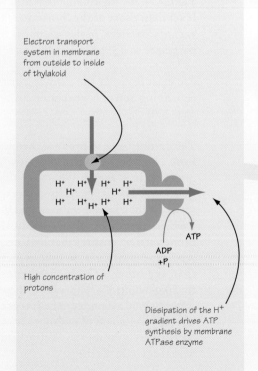

Life chemistry and molecular biology

4.2 Chloroplast structure and photosynthesis

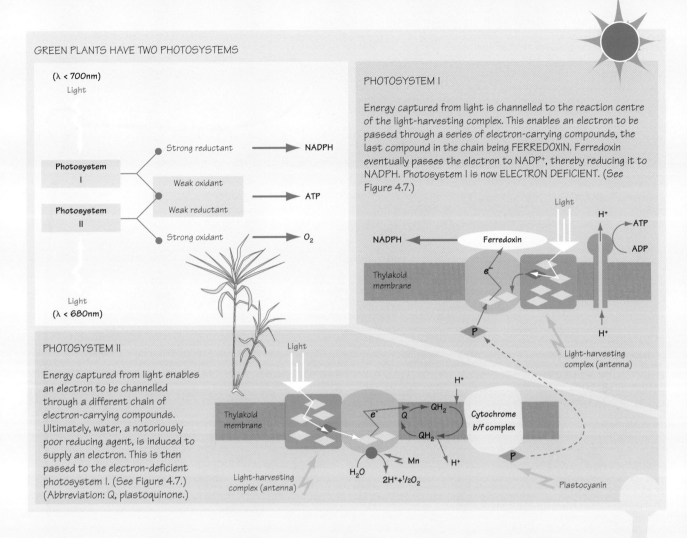

GREEN PLANTS HAVE TWO PHOTOSYSTEMS

($\lambda < 700$nm) Light → Photosystem I → Strong reductant → NADPH; Weak oxidant → ATP

Photosystem II → Weak reductant → ATP; Strong oxidant → O_2

Light ($\lambda < 680$nm)

PHOTOSYSTEM I

Energy captured from light is channelled to the reaction centre of the light-harvesting complex. This enables an electron to be passed through a series of electron-carrying compounds, the last compound in the chain being FERREDOXIN. Ferredoxin eventually passes the electron to $NADP^+$, thereby reducing it to NADPH. Photosystem I is now ELECTRON DEFICIENT. (See Figure 4.7.)

PHOTOSYSTEM II

Energy captured from light enables an electron to be channelled through a different chain of electron-carrying compounds. Ultimately, water, a notoriously poor reducing agent, is induced to supply an electron. This is then passed to the electron-deficient photosystem I. (See Figure 4.7.) (Abbreviation: Q, plastoquinone.)

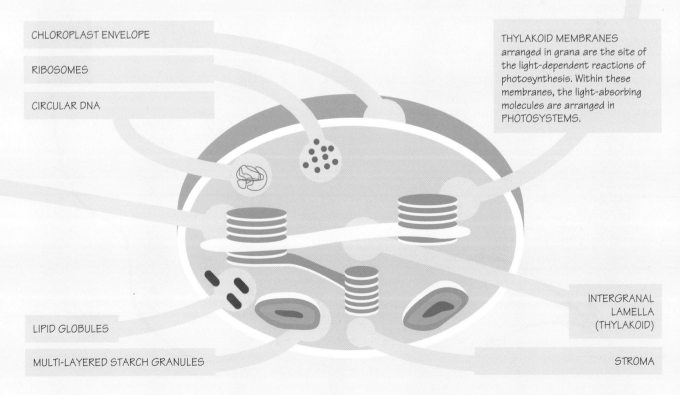

CHLOROPLAST ENVELOPE

RIBOSOMES

CIRCULAR DNA

THYLAKOID MEMBRANES arranged in grana are the site of the light-dependent reactions of photosynthesis. Within these membranes, the light-absorbing molecules are arranged in PHOTOSYSTEMS.

LIPID GLOBULES

MULTI-LAYERED STARCH GRANULES

INTERGRANAL LAMELLA (THYLAKOID)

STROMA

Chapter 4 • **Obtaining Energy** 107

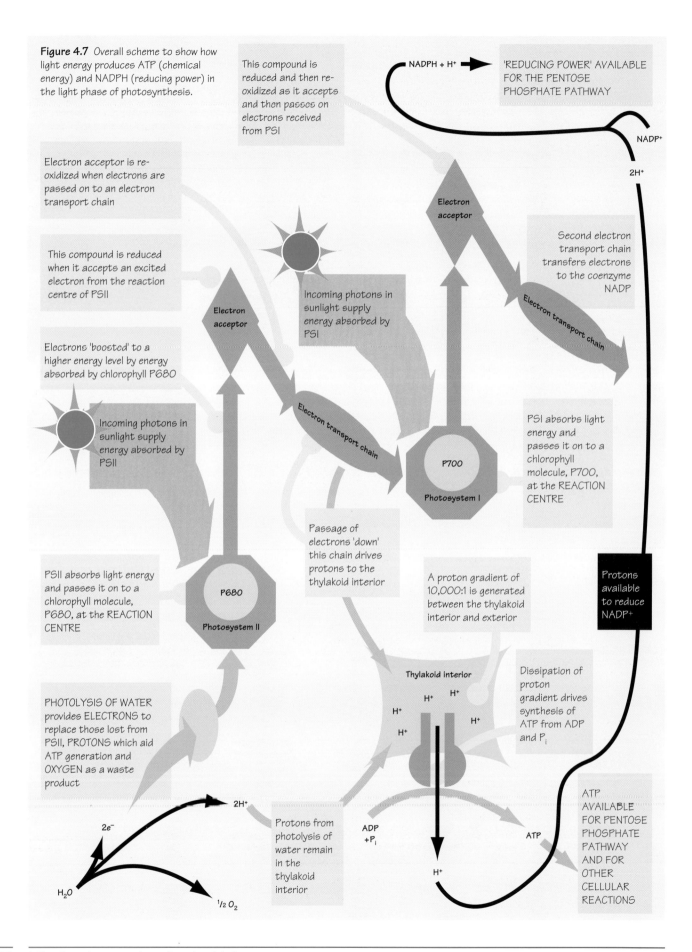

Figure 4.7 Overall scheme to show how light energy produces ATP (chemical energy) and NADPH (reducing power) in the light phase of photosynthesis.

It is the energy in this gradient that is used to generate ATP from ADP and P_i; remarkably, the process by which this is done is very similar to that occurring in mitochondria (see below). The thylakoid membrane is impermeable to protons except at certain places where rather peculiar membrane proteins are inserted into the membrane. These proteins have a 'stalk' in the membrane and a 'head' pointing out into the stroma. The stalk has a proton channel through its centre and the head has so-called ATPase activity. 'ATPase' activity means that this 'head' portion is an enzyme which can catalyse the reaction:

$$ATP + H_2O \longrightarrow ADP + P_i$$

However, enzymes can catalyse both the forward and reverse reactions, and when this enzyme is properly situated in the membrane it can catalyse the synthesis of ATP, but only when protons flow through the proton channel in the stalk. The flow of protons dissipates the proton gradient and the energy thereby released drives the synthesis of ATP (Figure 4.9):

$$ADP + P_i \longrightarrow ATP + H_2O$$

Exactly how this happens is not yet certain, but recent studies on the detailed structure of ATPase give us some clues (Figure 4.9, page 110). Whatever the true mechanism, it is also likely to apply to mitochondrial ATP production as well as that in bacteria with only minor differences.

Cyclic and non-cyclic electron flow

The route taken by electrons in Figure 4.7, as described above, is one-way: electrons originally in water end up in NADPH. However, there is an alternative way of passing electrons, via a cyclic route, the so-called cyclic pathway, which generates only ATP, because the electrons are returned to the chlorophyll molecules. The cyclic and non-cyclic pathways are shown in Figure 4.10 (page 110). It is quite difficult to prove experimentally that cyclic electron flow actually occurs, because all that can be observed is the production of ATP, which also occurs in non-cyclic electron flow, which is going on simultaneously. PSII does not participate in cyclic electron flow, which only occurs when $NADP^+$ is unavailable to accept electrons from reduced ferredoxin.

Bacterial photosynthesis

Bacterial photosynthesis differs from that in green plants and cyanobacteria in that oxygen is not involved. Bacteria do not use water as the reductant or supplier of electrons, but rather they use compounds such as hydrogen sulphide (H_2S) or a variety of organic acids (H_2A) which in general are better reducing agents than water. Bacteria only have one photosystem, which is similar to the PSII of green plants.

Figure 4.8 Electron carriers of chloroplasts. (a) Ferredoxin is a small protein containing clusters of iron atoms co-ordinated with sulphur atoms, held in the protein by linkage to cysteine residues. Iron can carry electrons ($Fe^{3+} + e^- \rightarrow Fe^{2+}$). The complete structure of a ferredoxin as a 'ribbon diagram' is shown with the iron cluster set in the protein. (b) Plastocyanin is a copper protein, and here again copper can carry electrons ($Cu^{2+} + e^- \rightarrow Cu^+$). (c) Plastoquinone is a hydrophobic organic molecule that can exist in either an oxidized (quinone) or a reduced (quinol) state.

Figure 4.9 Simplified diagram to show arrangement of the components of the green plant photosynthesis system in the thylakoids. The two photosystems consist of a number of proteins that sit in the membrane. Mn represents a cluster of manganese atoms that is associated with splitting water. Electrons originally present in water end up in NADPH, and ATP is produced by the ATPase working in reverse.

Bacterial photosynthesis was important in the discovery of the mechanism of photosynthesis. For green plants we can write an overall reaction:

$$CO_2 + 2H_2O \longrightarrow (CH_2O) + O_2 + H_2O$$

and (for example) for the so-called green sulphur bacteria:

$$CO_2 + 2H_2S \longrightarrow (CH_2O) + 2S + H_2O$$

(CH_2O) is intended to represent carbohydrate or cell material more reduced than CO_2. What this means is that the O_2 released from green plant photosynthesis all comes from water, and this was shown many years ago by using isotopes of oxygen.

It is believed that the oxygen in the Earth's atmosphere is there because cyanobacteria and green plants evolved the mechanism for aerobic photosynthesis, releasing oxygen from water. The atmosphere had previously been a reducing environment. This paved the way for the evolution of many types of heterotrophic organisms.

Another peculiarity of bacterial photosynthesis is that bacteria possess bacteriochlorophylls, which are similar to chlorophylls except that their small chemical differences mean that they absorb light in the infrared region rather than the visible. For example, the photosynthetic centre of *Rhodopseudomonas viridis* has a dimer of bacteriochlorophyll *b* molecules that absorbs maximally at 960 nm.

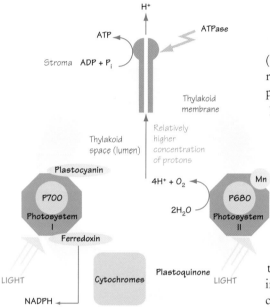

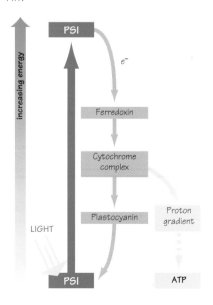

Figure 4.10 Electron flow in cyclic photophosphorylation. Light excites an electron in the pigments of PSI, but instead of passing from ferredoxin to NADP⁺ it returns to the chlorophyll via the cytochrome complex, generating a gradient of protons which leads to the production of ATP.

4.3 ATP production in heterotrophs

In cells other than photosynthetic ones, and indeed in photosynthetic cells in darkness, ATP is synthesized as a result of oxidative processes going on mainly in the mitochondria. Instead of light being the energy supplier to raise electrons to a level such that they can flow through a chain of transfer carriers, the energy is already locked up in organic compounds — the sugars, fats and proteins that are already present in the cell or what we eat. These compounds are more reduced than CO_2 and can therefore supply electrons to oxygen (i.e. reduce it) with the formation of water. This process is analogous to burning petrol and is the reverse of the photosynthetic process:

$$O_2 + (CH_2O) \longrightarrow CO_2 + H_2O$$

The sites of this process are the *mitochondria* (Spread 4.3, page 114), which have a double membrane system and in which the synthesis of ATP from ADP and P_i takes place through an ATPase enzyme in a very similar way to that occurring in chloroplasts.

The electrons required to run this process in mitochondria are derived from the various organic fuel molecules, such as starch and fat, that we regard as foods or foodstores. They are broken down in a series of reactions, but at various stages reactions take place by which reduced cofactors are generated (Figure 4.11). Some of the breakdown takes place outside the mitochondria and some inside. Ultimately, electrons from these carriers pass down an electron-transfer chain, falling in energy all the time, and generating a gradient of protons across the inner mitochondrial membrane.

The electrons are finally passed on to oxygen to produce water at the very last stage of the process. Thus the cycle on Earth is complete: photosynthesis evolves oxygen; mitochondrial respiration uses it (Figure 4.12). Interestingly, when someone is poisoned with cyanide, the cyanide blocks the last step in the mitochondrial electron-transfer pathway to oxygen. This knowledge is presumably small consolation during the minute or so it takes to die. Nevertheless, it illustrates the importance of the process. It really is the major supplier of energy for the body and, if it is blocked, then death follows rapidly.

The proton gradient generated across the inner mitochondrial membrane by the passage of electrons is used to generate ATP in a very similar way to that in chloroplasts. The membranes, proteins, carriers and enzymes are distinct and different, but at the heart they work in almost precisely the same way. The process occurring in mitochondria is called *oxidative phosphorylation*: reduced cofactors get oxidized and ADP gets phosphorylated to produce ATP. The analogous process occurring in chloroplasts is called *photophosphorylation*.

Mitochondrial oxidative phosphorylation

Mitochondria take on many different forms depending upon the cells they are in, from spherical to elongated or even branched structures (Spread 4.3). The outer membrane is smooth, but the inner membrane, where oxidative phosphorylation takes place and across which the protein gradient is generated, is infolded, increasing its surface area. The infoldings are called *cristae*. In general, the more active a cell is metabolically, the more mitochondria it has and/or the more cristae there are. The inner membrane is relatively impermeable, especially to protons. The enclosed central space is called the *matrix* and is a rich soup of enzyme proteins. This is where fatty acids are degraded and where the Krebs cycle takes place (see p.113).

The electron carriers of the inner mitochondrial membrane are different from those in the chloroplast thylakoid membrane, but the general pattern will be recognizable. Thus there are organic compounds, such as quinones, that can exist in either oxidized or reduced states, and iron and copper proteins (Table 4.1 and Figure 4.13, page 112). It is almost inconceivable that mitochondria and chloroplasts did not evolve from a common origin, albeit millions of years ago.

Whereas in chloroplasts the ATPase enzyme heads point into the thylakoid lumen, in mitochondria they point into the matrix. Although

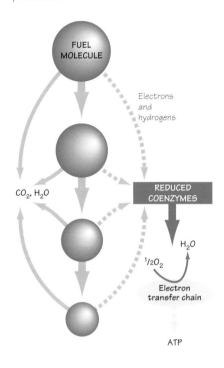

Figure 4.11 Fuel molecules are more reduced (i.e. contain more hydrogen and electrons) than CO_2. Their stepwise oxidation generates reduced coenzymes and electrons from these pass through the electron-transfer chain leading to ATP production.

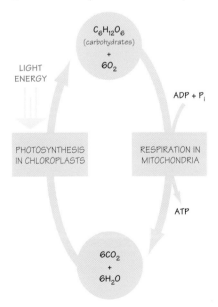

Figure 4.12 Light energy from the sun is trapped in photosynthesis and is used to synthesize all the organic materials of plants. When organic materials are broken down by heterotrophs (respiration in mitochondria), some of the energy is conserved in ATP (some is lost as heat).

Figure 4.13 Some of the electron-carrying coenzymes of mitochondria. **(a)** Nicotinamide–adenine dinucleotide (NAD^+); the only difference between this and $NADP^+$ (see Figure 4.4b) is the absence of a phosphate group on one of the riboses (arrowed). Enzymes recognize this apparently small difference and are generally specific for either NAD^+ or $NADP^+$. **(b)** The outline structure of flavin–adenine dinucleotide (FAD). Here the flavin unit is the 'business end' of the molecule and can carry two electrons and two hydrogens ($FADH_2$). **(c)** Coenzyme Q, also called ubiquinone, CoQ or just 'Q'. Notice that it is very similar to plastoquinone (Figure 4.8c). In addition, other electron carriers are involved, including iron–sulphur proteins, cytochromes (iron-porphyrins) and copper atoms (see Table 4.1).

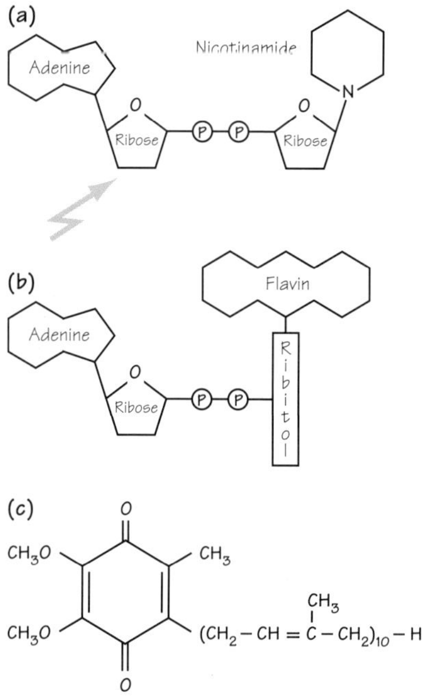

at first sight this might seem 'backwards way round', a simple diagram showing the origin of the thylakoids shows that it is not (Figure 4.14).

In the course of electron transfer, the chain of carriers in the inner mitochondrial membrane has to deal with input of electrons at different energy levels. It does this by using NADH and $FADH_2$ as intermediate carriers. NAD^+ accepts electrons from some compounds and is, relatively, a better reducing agent than FAD, which accepts electrons from other compounds. Electrons from NADH produce more ATPs than do electrons from $FADH_2$. As electrons pass from the reduced coenzymes to oxygen, so protons are pumped out. These protons pass back through the membrane ATPase units and ATP is generated.

Table 4.1
Coenzymes involved in carrying electrons in photosynthesis and oxidative phosphorylation

	Photosynthesis	Oxidative phosphorylation
Electron and hydrogen carriers	$NADP^+$, plastoquinone	NAD^+, FAD, ubiquinones (Coenzyme Q)
Electron carriers		
Iron–sulphur proteins	Ferredoxin	Iron–sulphur proteins in mitochondrial membrane
Cytochromes (iron–haem)	Cytochromes *b, f*	Cytochromes *a, b, c*
Copper proteins	Plastocyanin	Cytochrome *c* oxidase

The chemiosmotic theory

The idea that ATP is generated from ADP and P_i using the energy stored in a gradient of protons across a membrane (in both chloroplasts and mitochondria) originated from a British biochemist, Peter Mitchell, in the early 1960s. In those early days his ideas were not well received and proved controversial. This so-called *chemiosmotic theory* was slow to gain acceptance and, indeed, it was difficult to find experimental evidence to support it. One problem with both mitochondria and chloroplasts is that to work properly they have to have an intact membrane system. This is fairly obvious if the mechanism depends upon having a gradient of protons *across* an impermeable membrane, but it means that if you try to take the organelles apart to find out how they work, they stop working! Valuable evidence was eventually obtained by using inhibitors and uncoupling agents (Table 4.2).

In the end the chemiosmotic theory became very widely accepted and Mitchell was awarded a Nobel Prize in 1978 in recognition of his contribution to science. Some of the ideas connected with mitochondrial electron transfer and the chemiosmotic theory are explained in Spread 4.4 (page 118), but it is important to remember that photosynthetic phosphorylation works in a very similar way.

Mitochondrial oxidative phosphorylation will only work if it is supplied with electrons carried by NADH and FADH$_2$ produced by the catabolism of food materials in the cell. The so-called Krebs or tricarboxylic acid cycle acts as the final common pathway in the oxidation of the majority of compounds that are metabolized and we shall begin by looking at how this supplies NADH and FADH$_2$.

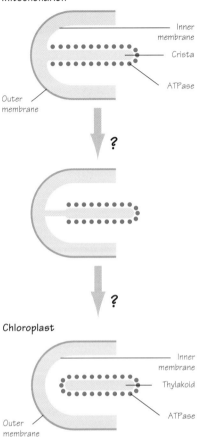

Figure 4.14 Possible relationships between mitochondria and chloroplasts. Cristae are infoldings of the inner mitochondrial membrane. If these were to 'break away' as shown, a structure similar to a thylakoid would be produced.

Table 4.2
Inhibitors and uncouplers of oxidative phosphorylation

Inhibitors of electron transport block the passage of electrons along the electron transfer chain and thus oxidative phosphorylation cannot take place

Examples:

ROTENONE blocks the transfer of electrons from NADH into the electron-transfer chain.

ANTIMYCIN A blocks transfer of electrons between cytochromes.

CYANIDE blocks cytochrome oxidase so that electrons cannot be transferred to oxygen.

Uncoupling agents (or 'uncouplers') prevent the formation of a proton gradient across the inner mitochondrial membrane, hence ATP cannot be formed. The energy that would have been used to form ATP from ADP and P$_i$ is released as heat.

Examples:

2,4-DINITROPHENOL (DNP)

TRIFLUOROCARBONYLCYANIDE PHENYLHYDRAZONE (FCCP)

Both can dissolve in membrane lipids and ferry H$^+$ across the inner mitochondrial membrane, allowing the proton gradient to dissipate.

4.4 The Krebs tricarboxylic acid cycle

It will be recalled from Chapter 3 that enzymes, although highly specific and very efficient catalysts, can only catalyse simple chemical transformations. If metabolism requires complicated chemical change, then a series of enzymes work, one after the other, in what is called a *metabolic pathway*. Pathways may be branched or unbranched, and some of them are even circular, as is the case with the Krebs cycle.

It may be thought that it is rather inefficient to do things this way, in little steps, but we should remember that all that computers can do is add one plus one (although they can do it very rapidly). Metabolic pathways 'work' because enzymes are so highly specific that each step yields very few by-products. In a typical series of organic chemical reactions, yields are a lot less than 100% and purification procedures are required before continuing, but this is not a problem for enzyme-catalysed sequences. This is perhaps just as well, since living cells do not have the facilities for purification.

Mitochondria and ATP production

objectives
- To be able to describe the structure of mitochondria
- To explain the process of oxidative phosphorylation
- To relate mitochondrial structure to function (biological energy production)

MITOCHONDRIA are the powerhouses (or power-stations) of the cell where oxidation of food materials culminates in the production of ATP from ADP + P_i.

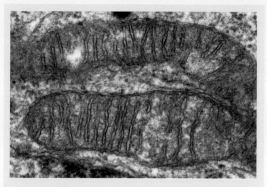

Electron micrograph of mitochondrion. Magnification × 28,000.

The MITOCHONDRIAL MATRIX (M COMPARTMENT) contains the enzymes of the tricarboxylic acid (TCA) cycle and is thus the site of oxidation of pyruvate to carbon dioxide and the reduced coenzymes NADH and $FADH_2$. Since the TCA cycle is the principal metabolic hub of the cell, the mitochondrial matrix is also the site of many syntheses (e.g. steroids) and interconversions (e.g. transaminations). It also contains the mitochondrial DNA, ribosomes, tRNAs and the enzyme systems necessary for the synthesis of some of the mitochondrial proteins.

The OUTER MITOCHONDRIAL MEMBRANE (OMM) is permeable to pyruvate, oxygen, ATP and many other molecules. The membrane also contains proteins which convert lipids into compounds which can cross the inner mitochondrial membrane and be catabolized in the mitochondrial matrix.

In the 'O' COMPARTMENT or INTERMEMBRANE SPACE, a high proton concentration is built up by the action of the three proton pumps.

The INNER MITOCHONDRIAL MEMBRANE (IMM) is impermeable to NADH so that shuttle systems are necessary to transfer 'reducing power' (H^+ and e^-) from glycolysis into the matrix. The membrane is also impermeable to ATP, so that ATP produced by oxidative phosphorylation must be transported to the cytosol by an energy-demanding translocase system. The membrane *is* permeable to pyruvate, the product of glycolysis.

The IMM contains the ELECTRON TRANSPORT SYSTEM (which generates a proton gradient), and the ATPase COMPLEX (which uses the proton gradient to generate ATP from ADP and P_i).

Thus, the IMM is the site of OXIDATIVE PHOSPHORYLATION.

The CRISTAE are infoldings of the IMM, which increase the surface area available for the redox and phosphorylation reactions leading to ATP generation. The number and form of the cristae may vary from tissue to tissue.

The MITOCHONDRIAL DNA carries the genes for the synthesis of many, but not all, mitochondrial proteins. The DNA is loosely coiled and is very prone to mutation by oxidation. This leads to a decrease in mitochondrial efficiency and may be a contributory factor in cell aging.

The INNER MITOCHONDRIAL MEMBRANE contains the ELECTRON TRANSPORT SYSTEM (which generates a proton gradient), and the ATPase COMPLEX (which uses the proton gradient to generate ATP from ADP).

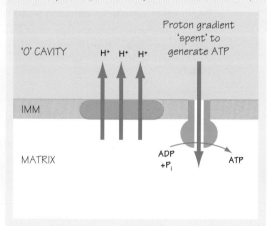

Thus the IMM is the site of OXIDATIVE PHOSPHORYLATION

ELECTRON TRANSPORT SYSTEM

ROUTE FOLLOWED BY ELECTRONS

114 Life chemistry and molecular biology

4.3 Mitochondria and ATP production

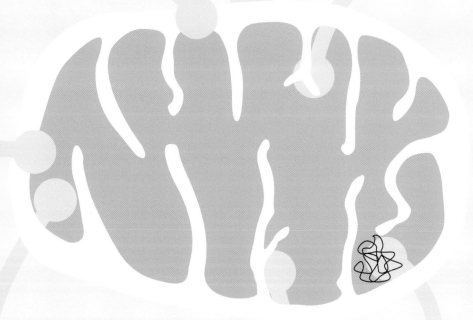

OUTER MITOCHONDRIAL MEMBRANE (OMM)

'O' COMPARTMENT/INTERMEMBRANE SPACE

CRISTAE

MITOCHONDRIAL MATRIX ('M' COMPARTMENT)

MITOCHONDRIAL DNA

OXIDATIVE PHOSPHORYLATION couples electron flow (oxidation) to ATP synthesis (phosphorylation) via a series of proteins in the inner mitochondrial membrane.

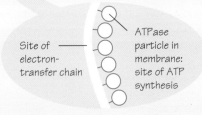

Site of electron-transfer chain

ATPase particle in membrane: site of ATP synthesis

SIMPLIFIED MODEL OF ELECTRON FLOW AND ATP PRODUCTION IN THE INNER MITOCHONDRIAL MEMBRANE
The enzyme–coenzyme complexes are shown as 'chunks' of protein in the membrane. These pump protons out when electrons flow to oxygen. When these protons flow back through the membrane ATPase, ATP is generated (see Spread 4.4).
(Abbreviations: Q, ubiquinone; cyt c, cytochrome c.)

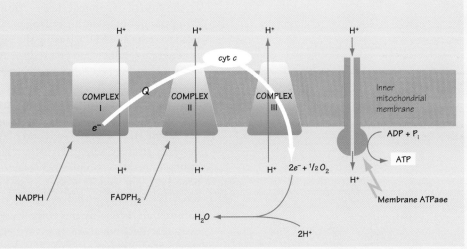

Chapter 4 • **Obtaining Energy**

The majority of the foods that we eat consist of reduced forms of carbon, i.e. carbohydrates, fats and proteins, and the same is true of the nutrients that cells receive. These highly reduced carbon compounds undergo catabolism by a variety of metabolic pathways, ultimately to produce a few simple compounds that can then be completely oxidized to produce CO_2 and water (Figure 4.15). It is these final steps that concern us here because it is in these last metabolic reactions that the bulk of the NADH and $FADH_2$ resulting from oxidation is produced. It is these final metabolic steps that form the Krebs cycle.

The series of reactions in this cycle was discovered a long time ago (1937), and Krebs himself called the sequence the citric acid cycle. Later it became known as the tricarboxylic acid cycle (TCA cycle), although not all of the compounds in it are tricarboxylic acids; indeed at the pH inside the mitochondrial matrix, where it takes place, the acids will be ionized, so it is more appropriate to talk about citrate than citric acid.

The cycle consists of a series of enzyme-catalysed reactions during which a two-carbon unit resulting from the breakdown of carbohydrate, fat or protein (amino acid) is added to a four-carbon compound to produce citrate (which has six carbons). During the course of the cycle, two carbons are successively removed from this as CO_2, regenerating the original four-carbon compound. Also during the cycle the 'carbon skeleton' gradually has hydrogen and electrons stripped away from it, and these are added to cofactors to give NADH and $FADH_2$.

It is not easy to explain the way in which this rather tortuous route was selected by evolution. One reason might be that in addition to its overall role of achieving a complete oxidation, i.e.

many of the compounds that appear as intermediates in the cycle have other roles too. They themselves may be breakdown products from food materials, or alternatively they may be used for building up other, more complex, molecules in the cell. The cycle is thus a 'metabolic hub', not simply a cycle; it is involved in both catabolism and anabolism.

Probably more important is another trick that cells have learned to do. In the course of some of the reactions of the cycle, H_2O is added to a compound, but then the hydrogens are removed in the next step (and used to reduce NAD^+), leaving the oxygen behind. In this way NADH is generated, which can be used for ATP synthesis in oxidative phosphorylation, but no O_2 is actually used.

The 'two-carbon unit' that feeds into the cycle is an important one because it has many functions in metabolism. It is called *acetyl-coenzyme A*, or *acetyl-CoA*, and plays a role in many pathways. It is basically an acetyl unit (CH_3CO), which corresponds to the two carbon atoms, joined to a complicated cofactor called coenzyme A (Figure 4.16, page 122). The join is through a sulphur atom and the acetate in acetyl-CoA is in quite an active state, ready to add itself on to other compounds. Probably this complicated cofactor not only provides this extra reactivity but also acts as a structure that is recognizable with great specificity by enzymes.

Figure 4.15 Breakdown of the three major classes of foods (or stored materials) eventually feeds molecules into the TCA cycle.

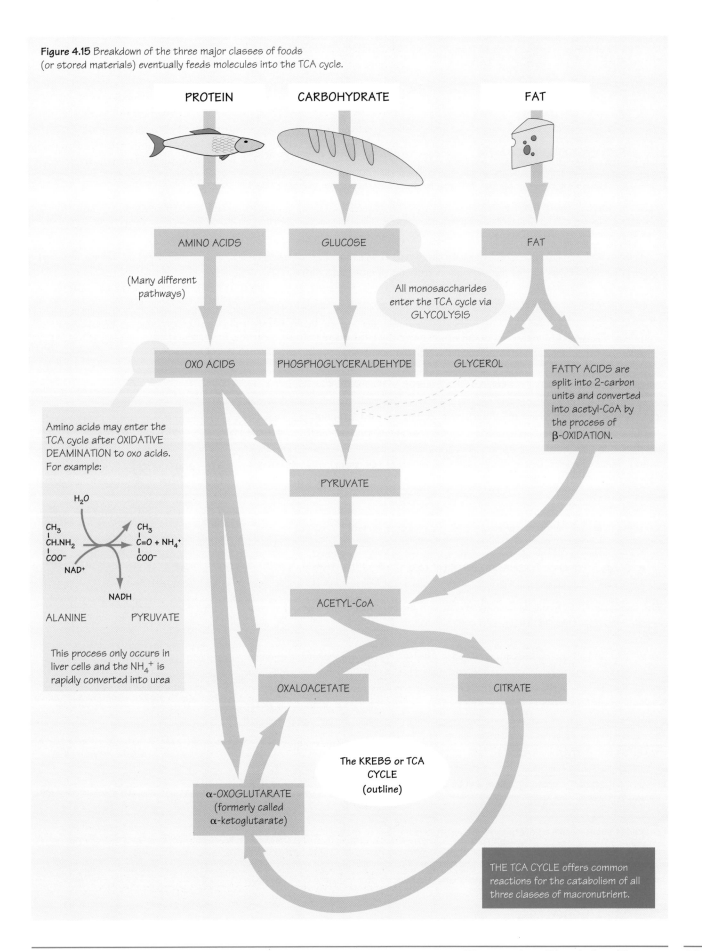

Mitchell's chemiosmotic theory

objectives
- To explain how the energy released when electrons travel down a potential gradient is used to create a gradient of protons
- To describe how the dissipation of this proton gradient via a membrane 'ATPase' enzyme drives the synthesis of ATP from ADP and phosphate (P_i)

Peter Mitchell proposed his CHEMIOSMOTIC THEORY many years ago (1961), but it was not generally accepted until 1979 when he was awarded the Nobel Prize. The theory is based upon the idea that ATP is produced by the utilization of energy held in a gradient of charge and concentration. The diagram opposite shows the proposed way in which this occurs in mitochondria.

Three stages in the overall process can be recognized:

1. There is an EXERGONIC SERIES of redox reactions between members of the ELECTRON TRANSPORT CHAIN on the inner mitochondrial membrane (IMM).

The electron transport chain consists of a series of dehydrogenase enzymes, with one oxidase enzyme at the end of the chain. These enzymes are arranged in a highly ordered fashion in the IMM: some of them traverse the whole membrane, others tend to sit on one side, while yet others are completely embedded within the membrane.

Each of the electron carriers has a cofactor that can exist in the oxidized or reduced state. The reduced coenzymes NADH and $FADH_2$ are the first members of this chain. The oxidized forms accept protons and electrons from metabolic intermediates during the catabolism of food and storage molecules, and pass them on to a quinone called UBIQUINONE. Ubiquinone, in turn, donates its electrons to the cytochromes — a series of increasingly powerful oxidizing agents (that is, compounds that will accept electrons). The last member of the chain, CYTOCHROME OXIDASE, passes electrons from cytochrome c to molecular oxygen.

2. These exergonic redox reactions can drive three proton pumps:

- NADH:COENZYME Q REDUCTASE
- COENZYME Q:CYTOCHROME c REDUCTASE
- CYTOCHROME c OXIDASE

which transport H^+ from the mitochondrial matrix to the intermembrane space, thus generating a PROTON GRADIENT across the IMM.

The electron and proton carriers are arranged in a highly ordered way within a membrane. This enables protons to be translocated to one side of the membrane when electrons flow through the carriers. In the case of mitochondria, electron flow results in the expulsion of protons into the intermembrane space. The medium outside the IMM thus becomes richer in protons (lower pH), as well as in positive charges, than the inside. In this way, the energy that has been released in passing electrons down a POTENTIAL GRADIENT is now stored as a CONCENTRATION GRADIENT.

3. Dissipation of the proton gradient can be coupled to the phosphorylation of ADP to give ATP

The IMM is impermeable to protons. However, at certain sites in the membrane, enzymes called ATPases harbour channels through which protons can pass. An ATPase catalyses the hydrolysis of ATP to ADP and P_i. When protons pass through the enzyme (that is, when the proton gradient dissipates), the ATPase 'works backwards' to produce ATP. It has been suggested that the proton flux alters the nucleotide-binding properties of the active site, thus promoting ATP synthesis.

CYTOCHROME c is a small iron-containing protein which can diffuse laterally across the IMM surface, transferring electrons to cytochrome c oxidase.

COENZYME Q (UBIQUINONE) is a small lipid-soluble molecule which shuttles electrons through the IMM from the first proton pump to the second proton pump.

NADH:COENZYME Q REDUCTASE catalyses the transfer of two electrons from NADH to coenzyme Q (ubiquinone). The energy released in this process drives the pumping of protons from the mitochondrial matrix to the intermembrane space. The electrons are transferred via a series of IRON–SULPHUR PROTEINS, whose structures are poorly understood.

The REDUCED COENZYMES NADH and $FADH_2$ introduce electrons into the electron transfer chain. These coenzymes are produced:
1. during the reactions of the tricarboxylic acid cycle,
2. from the β-oxidation of fatty acids,
3. from glycolysis via mitochondrial membrane shuttle systems.

ELECTRONS extracted from reduced coenzyme molecules are eventually accepted by molecular oxygen to produce water. This may represent 35% of the total oxygen consumption of a cell.

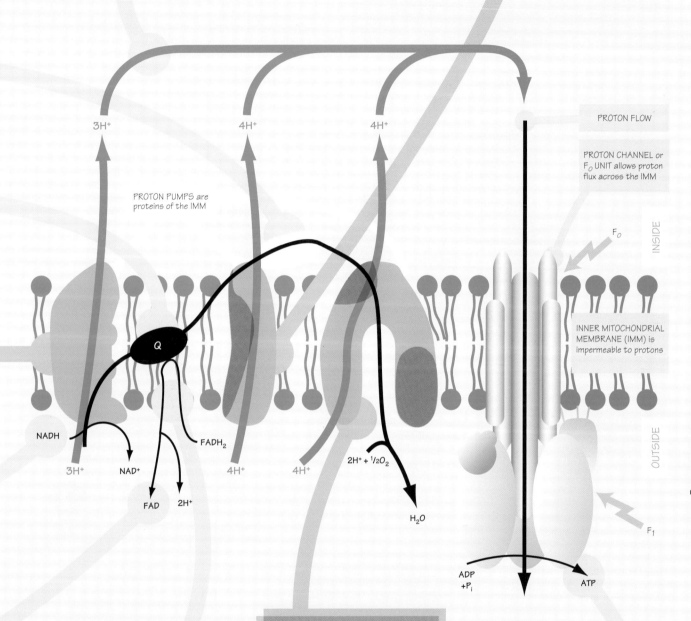

The Krebs tricarboxylic acid cycle

objectives

- To describe the principal steps of the Krebs cycle
- To explain that the effect of the cycle is to completely oxidize the materials that enter it
- To recognize that the reduced coenzymes produced by the cycle (NADH, FADH$_2$) are re-oxidized in the mitochondria with the production of ATP

Hans Krebs, working in Sheffield University in 1937, proposed the cyclic mechanism for the complete oxidation of pyruvate derived from the breakdown of carbohydrate. He called it the CITRIC ACID CYCLE, although it later became known as the TRI-CARBOXYLIC ACID CYCLE (or TCA cycle).

The TCA cycle is the major pathway of biological energy production in aerobic cells, and is the last steps in RESPIRATION, the outward signs of which are the consumption of oxygen and the liberation of carbon dioxide. It is this cycle that produces the bulk of the NADH and FADH$_2$ that is subsequently used in oxidative phosphorylation for ATP synthesis.

The cycle consists of a series of enzyme-catalysed reactions during which a two-carbon unit (ACETYL-CoA), resulting from the breakdown of metabolites, is added to a four-carbon compound (OXALOACETATE) to produce a six-carbon compound (CITRATE). Two of these carbons are successively removed as CO$_2$ during the course of the cycle, thereby regenerating oxaloacetate.

Sir Hans Krebs with a cycle.

In addition, at four points in the cycle, the 'carbon skeleton' passes on protons and electrons to the coenzymes NAD$^+$ and FAD, and subsequently to the electron-transport chain.

INHIBITORS that block the cycle are lethal because there is usually not an alternative pathway. For example, the enzyme ACONITASE is inhibited by FLUOROCITRATE, causing rapid accumulation of citrate, convulsions and death. Fluorocitrate is synthesized from FLUOROACETATE, found in the leaves of *Dichapetalum cymosum*, which is not itself poisonous. Thus the enzymic production of fluorocitrate is an example of LETHAL SYNTHESIS.

The principal reactions of the TCA cycle take place in the mitochondria.

SUMMARY

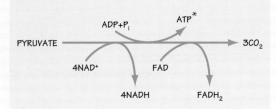

These reactions complete the regeneration of oxaloacetate, at the same time generating a further molecule of NADH.

SUCCINATE DEHYDROGENASE begins the regeneration of oxaloacetate by the dehydrogenation of succinate to fumarate. FAD is the hydrogen acceptor in this reaction because the free energy change is insufficient to reduce NAD$^+$. This enzyme is a NON-HAEM IRON PROTEIN and differs from the other enzymes of the TCA cycle in that it is part of the inner mitochondrial membrane. (It is, in fact, directly linked to the electron transport chain.) The enzyme also offers the classic example of COMPETITIVE INHIBITION, being inhibited by malonate, a respiratory poison.

$$\begin{array}{c} COO^- \\ | \\ CH_2 \\ | \\ COO^- \end{array}$$

Malonate

*SUBSTRATE-LEVEL PHOSPHORYLATION generates a molecule of ATP, by first coupling the hydrolysis of the thioester bond in succinyl-CoA to the phosphorylation of GDP:

$$GDP + P_i \rightleftharpoons GTP$$

and then transferring this phosphate group to ADP:

$$GTP + ADP \rightleftharpoons ATP + GDP$$

NUCLEOSIDE TRIPHOSPHATES such as ATP and GTP are very similar in their properties, such as ease of hydrolysis.

4.5 The Krebs tricarboxylic acid cycle

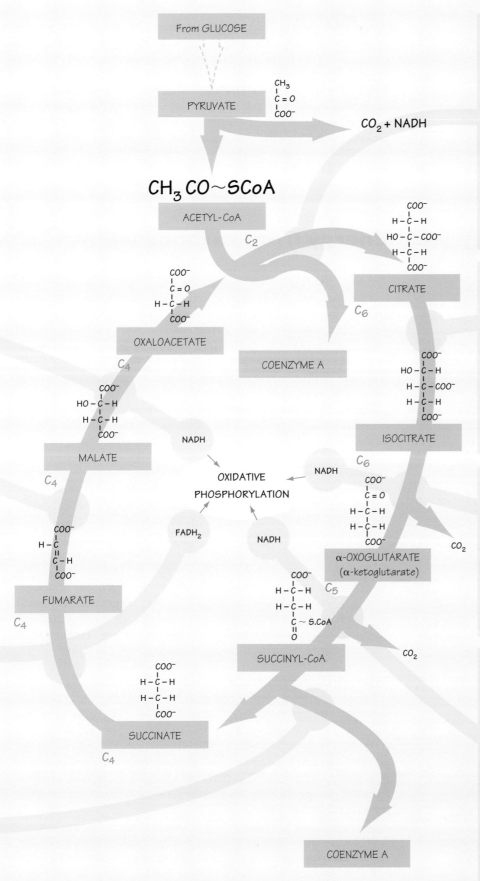

CITRATE SYNTHASE (originally called CONDENSING ENZYME) catalyses the reaction which introduces the acetyl group into the TCA cycle proper. Hydrolysis, which removes the CoA, pulls the overall reaction in the direction of citrate synthesis. The condensation is necessary since it is metabolically impossible to remove the CO_2 molecules by successive decarboxylations of a CH_3COO^- unless the CH_3- entity can be oxidized. This can only occur when the CH_3- entity becomes part of a larger molecule.

ACONITASE catalyses the isomerization of citrate to ISOCITRATE. This enzyme is inhibited by FLUOROCITRATE, formed from fluoroacetate as follows:

$$FCH_2CO_2^- + \text{Oxaloacetate} \longrightarrow \begin{array}{c} FCH\,COO^- \\ HO-C-COO^- \\ CH_2COO^- \end{array}$$

ISOCITRATE DEHYDROGENASE catalyses the first of four oxidation–reduction reactions in the TCA cycle. The dehydrogenation of the HO–C–H group, with the formation of NADH, generates an unstable intermediate which rapidly decarboxylates to α-oxoglutarate. Note that the CO_2 released comes from one of the $-COO^-$ groups in the acetyl 'acceptor', OXALOACETATE.

α-OXOGLUTARATE DEHYDROGENASE catalyses a reaction which is very similar to the conversion of pyruvate into acetyl-CoA:

a. a molecule of NADH is formed,
b. a high-energy thioester bond is produced, consuming a molecule of coenzyme A,
c. the reaction requires the coenzyme THIAMIN PYROPHOSPHATE (TPP),
d. a CO_2 molecule is released.

TPP deficiency causes BERI-BERI, characterized by elevated levels of pyruvate and α-oxoglutarate.

Chapter 4 • **Obtaining Energy**

Figure 4.16 *Coenzyme A is a coenzyme central in metabolism and participates, usually as acetyl-CoA, in many reactions. The complicated structure can be recognized with high specificity by enzymes, and is shown in outline form here. The arrow shows where an acetyl (CH_3CO-) unit may be joined.*

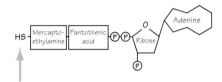

Spread 4.5 shows the TCA cycle. You can see how carbons are stripped away to form CO_2. The diagram also shows the structures of the intermediates and some of the enzymes that participate.

The TCA cycle is the final pathway of practically all metabolism in aerobic cells. We now ask how it is fed by the breakdown of other compounds that were originally nutrients — the carbohydrates, fats and proteins. These have their own individual pathways of catabolism that can feed into the cycle, but as we will see, most of the pathways produce acetyl-CoA.

4.5 Energy from carbohydrates

The major fuel for many, if not all, eukaryotic cells, is glucose. In animals this is most commonly supplied in the blood circulation or is stored actually inside cells as the polysaccharide glycogen. The origin of the glucose is usually disaccharides (sucrose, lactose) or polysaccharides (starch) in the diet. These are broken down by gut enzymes and the monosaccharides are absorbed and distributed. Plant cells, in the dark if they have chloroplasts or all the time if they have not, need energy to survive. Here again monosaccharide sugars are a major product of photosynthesis and are distributed round the plant as sucrose or are stored in starch granules. Some bacteria and fungi are also able to produce and store polysaccharides.

The pathway for degrading glucose is called *glycolysis* or the *glycolytic pathway*; 'lysis' means splitting. It is of fundamental importance in all eukaryotic cells, but in some cells it is also a major energy producer. Such cells include skeletal muscle cells when they are working anaerobically and mature red blood cells (which do not have mitochondria). The importance of this is that glycolysis can produce useable energy (i.e. ATP) under anaerobic conditions, albeit rather inefficiently, and is unique in this respect. Yeasts also do it when they are growing anaerobically. This is the economically important process which is called fermentation and which produces ethanol (Figure 4.17). Anaerobic glycolysis is sometimes called 'anaerobic respiration'. Here we shall reserve the term 'respiration' for processes that involve oxygen.

Glycolysis is a stepwise process whereby a six-carbon sugar (glucose) is split into two three-carbon compounds. Each of these is eventually converted into the two-carbon unit, acetyl-CoA. The process produces ATP but, rather strangely, it requires an initial input of ATP to get it started. In fact, the majority of the intermediate compounds in the glycolysis pathway have phosphate groups attached to them and these phosphates come from ATP. This seems to make them more reactive. Eventually, more ATP is generated than is actually used so that overall there is a net gain of ATP.

A problem with glycolysis (and one that living cells have overcome in various different ways) is that at one of the steps an oxidation takes place and this produces some NADH from NAD^+. The problem arises because there is not very much NADH (or NAD^+) in a cell, and this

Figure 4.17 Glycolysis needs a supply of NAD⁺ in order to continue; the NADH formed needs to be recycled.

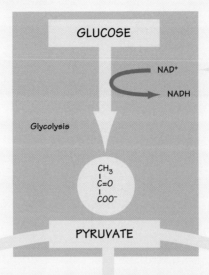

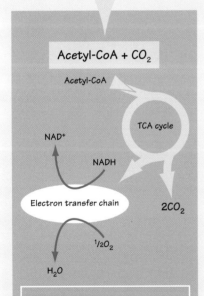

This OXIDATIVE DECARBOXYLATION of pyruvate occurs in the mitochondrial matrix and is the link reaction between glycolysis and the TCA cycle. It is irreversible and is catalysed by a large multienzyme complex called PYRUVATE DEHYDROGENASE. B-group vitamins are required. The TCA cycle may only continue in the presence of molecular oxygen when the NADH is rapidly re-oxidized to NAD⁺.

Alcoholic fermentation occurs in yeasts and in some other micro-organisms.

It is a two-stage process:

(a) a DECARBOXYLATION of pyruvate to ethanal.
(b) a REDUCTION of ethanal to ethanol.

Humans have long exploited this process to produce alcoholic beverages and to leaven (rise) bread.

Lactate fermentation occurs in many bacteria and is exploited by humans in the production of yogurt, cheeses, sauerkraut, soy sauce and even chocolate.

This process also occurs in muscle cells during strenuous exercise and high oxygen demand. The need to re-oxidize NADH produced in this way creates an OXYGEN DEBT.

Glycolysis

objectives
- To describe the pathway of glycolysis by which glucose is catabolized
- To explain that the C_6 molecule is phosphorylated and then split into two C_3 portions
- To identify those steps at which ATP is used and those where ATP is produced

GLYCOLYSIS deals with glucose which most cells of the body receive via the bloodstream. The pathway can also use glucose 6-phosphate which is produced by the breakdown of stored glycogen. Fructose in the diet may also be fed in at the fructose 6-phosphate stage.

The pathway consists of a series of reactions that take place in the CYTOSOL of the cell. Starting from glucose they result in the production of pyruvate ($C_6 \rightarrow 2 \times C_3$). Pyruvate normally feeds into the tricarboxylic acid cycle (see Spread 4.5), but, in anaerobic conditions, glycolysis can also continue and will generate ATP provided that the NADH produced at Step 5 can be reoxidized (see opposite). This is achieved in mammalian muscle by using it to reduce pyruvate to lactate (Figure 4.17).

'ACTIVATION' of glucose involves expenditure of ATP, but, by the formation of glucose 6-phosphate, this process (a) begins to drive glucose towards degradation and (b) prevents the loss of glucose from the cell since the plasma membrane is much less permeable to the charged glucose 6-phosphate. This reaction is catalysed by HEXOKINASE and the binding of glucose causes such a pronounced conformational change in the enzyme (induced fit) that water is excluded from the active site, promoting phosphoryl transfer and preventing undesirable ATPase activity.

ENERGY BALANCE SHEET

The first three steps use up 2ATP per glucose in creating the phospho-derivatives.

Steps 5 and 6 generate 2ATP per C_3 molecule (or 4ATP per glucose).

Steps 8 and 9 generate a further 2ATP per C_3 molecule (or 4ATP per glucose).

This gives a net gain of 2ATP for the whole pathway.

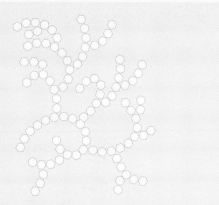

Glucose stored as glycogen may be used in liver and muscle to feed into glycolysis

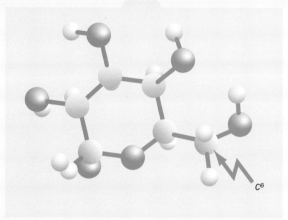

2,3-DIPHOSPHOGLYCERATE (glycerate 2,3-bisphosphate) is an intermediate in to the conversion of 1,3-DPG into 3-PGA:

1,3-DPG ⟶ 2,3-DPG ⟶ 3-PGA

It is also an important factor in the association of haemoglobin with oxygen (2,3-DPG promotes dissociation of oxyhaemoglobin) so that malfunction in the glycolytic pathway may cause disorders in oxygen transport.

A DEHYDRATION of 2-PGA to phosphoenolpyruvate produces an enol phosphate which has a high tendency towards phosphoryl transfer and hence 'drives' the formation of pyruvate and an ATP molecule.

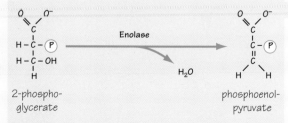

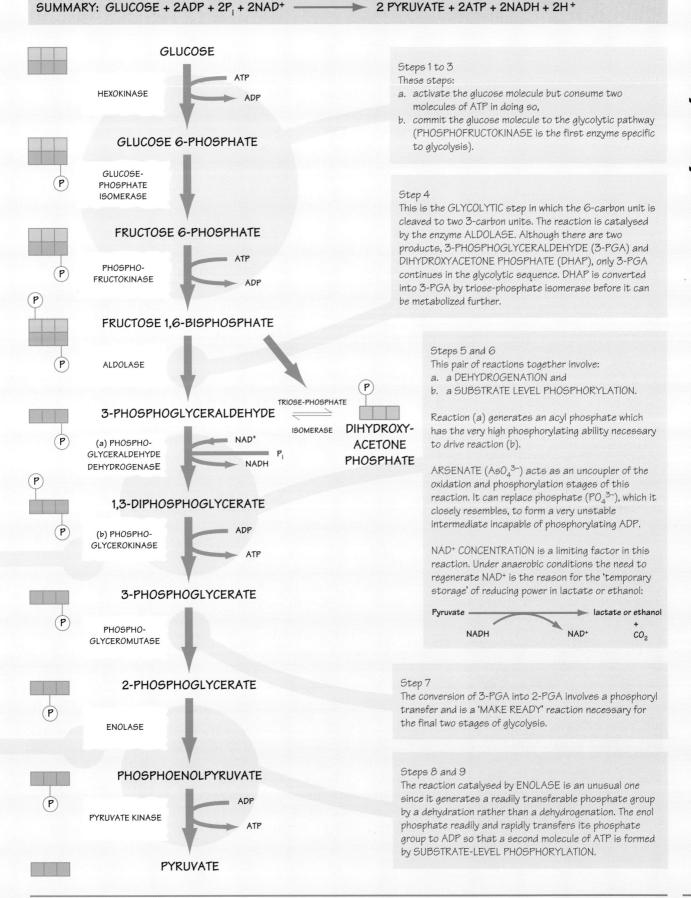

therefore needs to be recycled for glycolysis to continue. Under aerobic conditions the NADH can be recycled to NAD^+ via oxidative phosphorylation but in the absence of oxygen this is not possible. Cells from different organisms have evolved different ways of doing the recycling anaerobically (Figure 4.17, page 123).

The overall reactions of glycolysis may be summarized as follows:

$$\text{Glucose} + 2ADP + 2P_i + 2NAD^+ \longrightarrow 2 \text{ Pyruvate} + 2ATP + 2NADH + 2H_2O$$

However, as will be seen from Spread 4.6, this represents the sum of about 10 reactions, and on the 'input' side some ATP is required before the process can start.

The final product of glycolysis is really pyruvate (C_3) and this is converted into either lactate or ethanol (in yeast), anaerobically, or, under aerobic conditions, into acetyl-CoA (C_2). This latter conversion is a complicated one which involves a number of cofactors, which are formed from vitamins of the B complex, and is catalysed by an enzyme called pyruvate dehydrogenase. The C_1 that is removed in the process is released as CO_2 (see Figure 4.17).

4.6 Energy from fats

The fats used in energy production are the triacylglycerols (or triglycerides) which are either stored in the adipose tissue or are taken in with the diet. In the latter case they are hydrolysed by lipase in the small intestine, releasing fatty acids and glycerol. The glycerol can be modified slightly to produce an intermediate in the glycolytic pathway. The fatty acids resulting from this process (or fatty acids released from fat in the adipose cells) can be degraded by a process which 'chops off' successive two-carbon units as acetyl-CoA. This is called the β-oxidation pathway and is a sort of spiral (Spread 4.7). The acetyl-CoA can be fed into the TCA cycle. In the course of producing the two-carbon units, some more molecules of NADH and $FADH_2$ are produced and these can also give rise to ATP by oxidative phosphorylation in the mitochondria. Consequently, the breakdown of fat can produce a good deal of energy (ATP) in the mitochondria, and much more than from an equal weight of carbohydrate. However, this can only be done aerobically. In the absence of oxygen, no energy at all can be released by fatty acid breakdown (Figure 4.18).

In general, glucose is the preferred fuel for cells; fat is used less readily, although it has great potential for supplying ATP. In a 100 metre sprint, athletes will operate anaerobically, using up their muscle glycogen supply and incurring an oxygen debt. In marathons, in contrast, their bodies will be running on a mixture of glycogen and stored fat, both being used oxidatively (Figure 4.18). The glycogen stores would be insufficient for a marathon and the fat stores have to be mobilized.

Figure 4.18 In a marathon race the runners are operating on both carbohydrate and fat in an aerobic manner. There would not be sufficient glycogen stored for a long (anaerobic) run. Energy is more efficiently stored as fat, but this energy may only be released oxidatively. By contrast, in a 100 m sprint most of the energy comes from anaerobic glycolysis using stored muscle glycogen. (Courtesy of the Stockport Express, Advertiser and Times.)

4.7 Energy from proteins

Proteins in the diet are digested by the enzymes in the stomach and small intestine to release amino acids. These pass in the blood to other cells of the body and especially the liver. Some amino acids are needed for making new proteins, but if we eat excess protein over this requirement, then amino acids are catabolized to provide energy. Unlike carbohydrates and fats, proteins and amino acids are not stored, but some parts of some amino acids can be turned into glucose. Other parts of amino acids can be turned into acetyl-CoA, which may be sent round the TCA cycle, or, alternatively, if metabolic energy is not immediately required, may be converted into fat. Other amino acids can feed directly into the TCA cycle. It is because the 20 amino acids have such diverse chemical structures that a separate catabolic pathway exists for almost each one of them (see Figure 4.15, page 117).

In starvation, when all the body's glycogen and fat stores have been used up, then body protein may be broken down to supply energy. However, this is actually breaking down valuable structures in the body, such as muscle protein, and cannot go on indefinitely without permanent damage being done (Figure 4.19). It is like burning the furniture to keep warm.

Although the varied chemical structures of the 20 amino acids demand different catabolic pathways to allow them to be oxidized, the first step of the degradative process, in all cases, is the removal of the amino group. This is a *deamination* or *transamination*, and the toxic ammonia so produced is rapidly converted into something relatively harmless in mammals, namely *urea*, which is non-toxic and is excreted in the urine. This process takes place in the urea cycle (Figure 4.20, page 130) which was discovered by Krebs and his student Henseleit in 1931, well before the Krebs citric acid cycle was discovered. In birds and insects, the product is *uric acid*, but in water-dwelling animals, the ammonia is excreted as such, its toxic effects diluted by the water (Table 4.3).

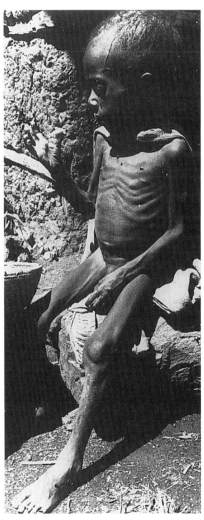

Figure 4.19 Child from a famine area in Ethiopia. Young children need essential amino acids for growth as well as calories (usually from carbohydrate or fat) to supply energy. Severe deficiency of protein (to supply essential amino acids) and energy foods leads to the condition, shown here, called marasmus. (Courtesy of the Catholic Fund for Overseas Development, London.)

Table 4.3
Disposal of excess dietary nitrogen depends partly on habitat

Ammonotelic organisms are typically aquatic (e.g. fish) and excrete the toxic and very soluble ammonia (NH_3) into water where it is diluted.

Ureotelic organisms (e.g. mammals) convert the toxic ammonia into the less toxic and less soluble urea that is excreted in urine.

Uricotelic organisms (e.g. insects, birds) convert ammonia into uric acid which is non-toxic and insoluble and crystallizes. Often crystals are excreted, so conserving water.

Plants and many micro-organisms can synthesize their own amino acids from ammonia or nitrates and carbon dioxide and do not need preformed amino acids. Some micro-organisms are, of course, vital to all life on Earth because they fix atmospheric nitrogen (see Chapter 5). Some plant seeds carry large stores of protein.

The oxidation of fatty acids

objectives

- To describe how long-chain fatty acids are catabolized by being broken down into C_2 units
- To explain that these C_2 units, as acetyl-CoA, feed into the TCA cycle to yield energy in the form of ATP
- To recognize that energy is only available from fats oxidatively. When the body needs energy under anaerobic conditions (e.g. during a sprint), carbohydrate must be used

FATS, both in the diet and stored in the adipose tissue, are triacylglycerols (or tri-glycerides). Triacylglycerols are hydrolysed in the small intestine by a lipase enzyme to yield FATTY ACIDS and GLYCEROL (see Spread 2.6).

The glycerol is modified slightly and can be fed directly into the glycolytic sequence. Here, however, we will concentrate on the breakdown of the fatty acids. These have first to be 'ACTIVATED', which requires the expenditure of one molecule of ATP which is hydrolysed to AMP. This process involves the addition of a molecule of coenzyme A to their carboxy groups to form 'acyl-coenzyme A' derivatives. As well as activating the fatty acids, making them ready for breakdown, adding coenzyme A also makes the fatty acid 'safe'. (Remember that free fatty acids can disorganize membranes by acting as detergents, and that fatty acid breakdown occurs in membrane-bound structures — the mitochondria.)

Activated fatty acids are degraded by a process which 'chops off' successive C_2 units as acetyl-CoA. This series of reactions constitutes a 'spiral' rather than a cycle, and is known as the β-OXIDATION PATHWAY. Each of the breakdown steps can produce one NADH and one $FADH_2$ — equivalent to 5 ATP molecules. Furthermore, each of the acetyl-CoA molecules produced can feed into the TCA cycle, eventually producing 12 ATP molecules.

Thus, in the example opposite, one molecule of palmitate (a C_{16} fatty acid) is broken down into eight molecules of acetyl-CoA, yielding 8 NADH and 8 $FADH_2$. This produces $(8 \times 12) + (8 \times 5) = 136$ ATP (minus two for the initial activation).

Since fatty acid oxidation yields so much ATP, why do cells oxidize glucose? There are two reasons:

1. Glucose is a very common dietary component.
2. Fatty acid oxidation is ENTIRELY AEROBIC whereas glucose → lactate offers an ANAEROBIC ENERGY SOURCE. This might be essential in, for example, rapidly contracting skeletal muscle.

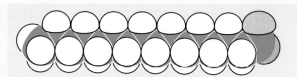

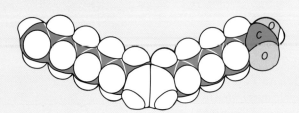

UNSATURATED FATTY ACIDS can also undergo β-oxidation. This is oleic acid (C_{18}; one double bond).

ACTIVATION of fatty acid consumes ATP and coenzyme A.

DEHYDROGENATION yields one molecule of reduced flavin coenzyme $FADH_2$.

HYDRATION of the C=C bond prepares the molecule for a second dehydrogenation.

DEHYDROGENATION yields one molecule of reduced nicotinamide coenzyme.

CLEAVAGE yields one molecule of acetyl-CoA and one molecule of 'activated' fatty acid which may return to the FAD-dependent dehydrogenation.

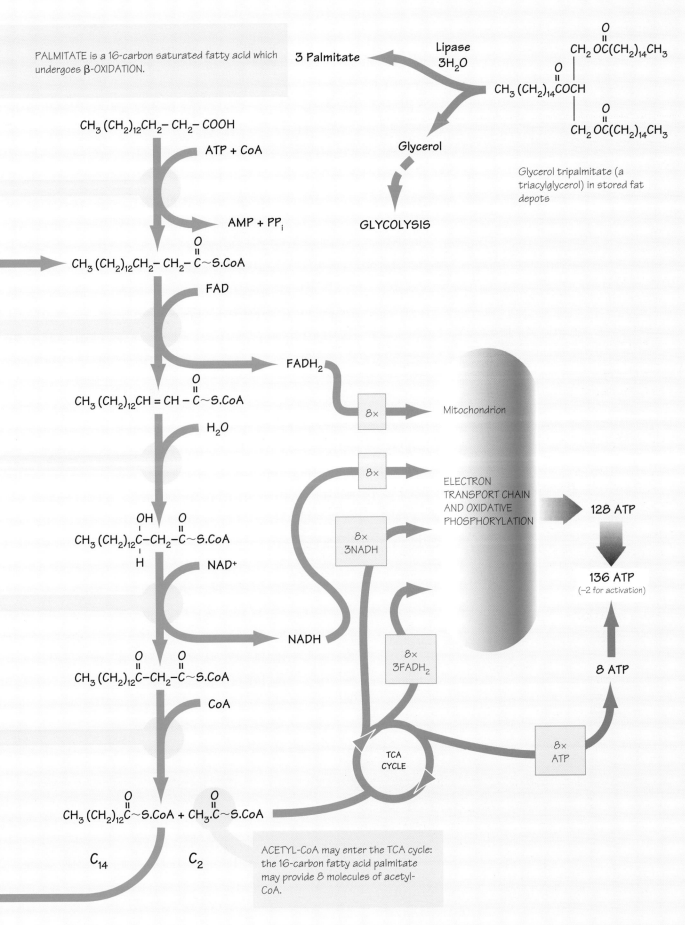

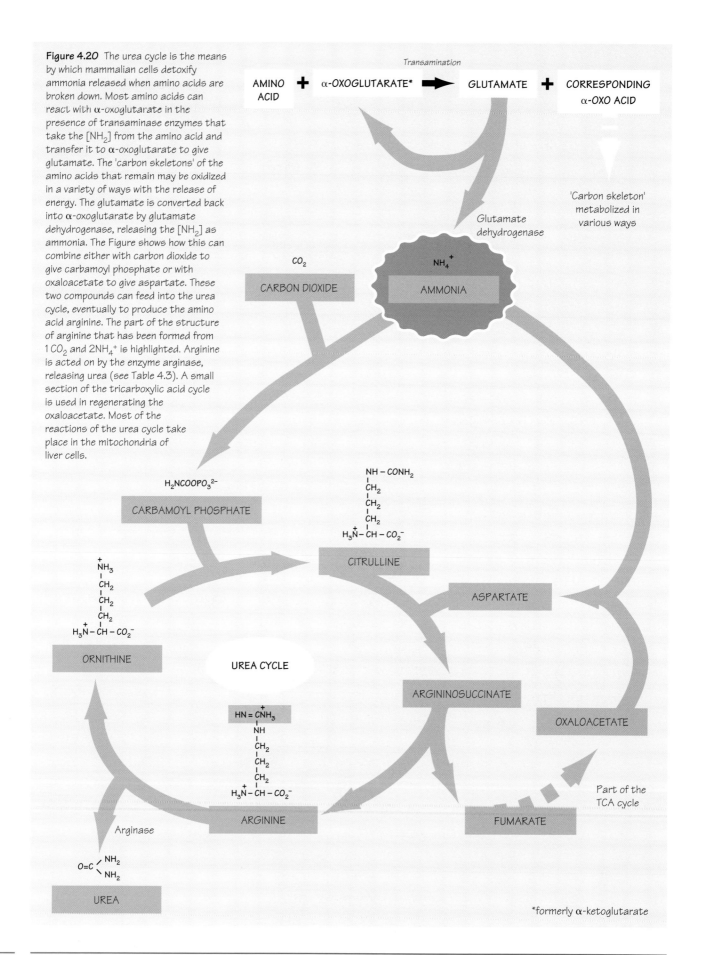

Figure 4.20 The urea cycle is the means by which mammalian cells detoxify ammonia released when amino acids are broken down. Most amino acids can react with α-oxoglutarate in the presence of transaminase enzymes that take the [NH_2] from the amino acid and transfer it to α-oxoglutarate to give glutamate. The 'carbon skeletons' of the amino acids that remain may be oxidized in a variety of ways with the release of energy. The glutamate is converted back into α-oxoglutarate by glutamate dehydrogenase, releasing the [NH_2] as ammonia. The Figure shows how this can combine either with carbon dioxide to give carbamoyl phosphate or with oxaloacetate to give aspartate. These two compounds can feed into the urea cycle, eventually to produce the amino acid arginine. The part of the structure of arginine that has been formed from 1 CO_2 and 2NH_4^+ is highlighted. Arginine is acted on by the enzyme arginase, releasing urea (see Table 4.3). A small section of the tricarboxylic acid cycle is used in regenerating the oxaloacetate. Most of the reactions of the urea cycle take place in the mitochondria of liver cells.

4.8 Conclusion

This chapter has dealt with a wide range of metabolic pathways, most of them catabolic routes whereby food or stored materials are broken down to provide energy for doing all the things that cells need to do. The pathways themselves are rather complex because a wide variety of nutrients needs to be dealt with, but overall there is a funnelling of material to acetyl-CoA and then the TCA cycle. The outcome of the whole process is production of ATP by oxidative phosphorylation.

Plants have photosynthesis as a mechanism for trapping light energy and using it to produce ATP and reducing power. Plants can make everything they need starting from CO_2, water and ammonia or nitrates. Animals and other heterotrophic organisms live on the organic materials produced by plant photosynthesis.

4.9 Further reading

Brown, B. (1991) Energy for life. Biological Sciences Review **3(3)**, 40–42

Gayford, C. (1993) ATP — the 'high-energy' compound that isn't. Biological Sciences Review **6(1)**, 12–14

Hall, D.O. and Rao, K.K. (1994) Photosynthesis (5th edn.). Cambridge University Press, Cambridge

Hinkle, P.C. and McCarty P.E. (1978) How cells make ATP. Scientific American **238 (Mar)**, 104–122

Leech, R. (1992) Harvesting the sun. Biological Sciences Review **3(5)**, 23–26

Miller, K.R. (1979) The photosynthetic membrane. Scientific American **241 (Oct)**, 100–111

Rothwell, N. (1988) Brown fat. Biological Sciences Review **1(4)**, 11–14

Weire, P.J. (1994) Photosynthesis. BASC Booklet; The Biochemical Society, London

Weitzman, D. (1988) Krebs cycling for the enthusiast. Biological Sciences Review **1(2)**, 6–8

West, I. (1990) Chemiosmotic coupling. Biological Sciences Review **3(1)**, 21–22

Youvan, D.C. and Marrs, B.L. (1987) Molecular mechanisms of photosynthesis. Scientific American **256 (June)**, 42–49

4.10 Examination questions

1. The diagram summarizes the light-dependent reaction in photosynthesis.

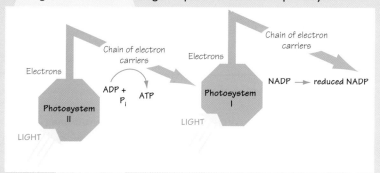

   (a) Where, in the chloroplast, does the light-dependent reaction take place? (1 mark)

   (b) During this reaction water molecules are broken down to yield oxygen, electrons and hydrogen ions (protons).
   (i) What is the name given to the process in which the water molecules are broken down? (1 mark)
   (ii) What happens to the electrons produced in this process? (1 mark)
   (iii) What happens to the hydrogen ions? (1 mark)

   [AEB (Biology) June 1992, Paper 1, No. 5]

2. A method used to isolate chloroplasts to investigate the light-dependent reaction of photosynthesis (the Hill reaction) is as follows:
   50 g of plant material is cut into small pieces and added to an ice-cold buffered solution of glucose. It is then blended and the resulting homogenate filtered through muslin into a beaker on ice. The filtrate is subsequently filtered again before it is centrifuged at a high speed. After centrifugation the supernatant is discarded and the chloroplast pellet resuspended in ice-cold buffer solution and stored in the dark.
   (a) (i) Why is the plant material cut into small pieces?
   (ii) Why is the buffer solution ice cold?
   (iii) Why is the solution buffered?
   (iv) Why is the blended mixture filtered through muslin?
   (v) Why is the filtrate centrifuged at a high speed?

   (b) (i) Why is the chloroplast suspension stored in the dark?
   (ii) A dye (such as methylene blue or DCPIP) is added to the suspension of chloroplasts prepared as above and the resultant mixture illuminated. The intensity of the coloured dye is measured using a colorimeter, with the results shown on the graph. Briefly explain these results.

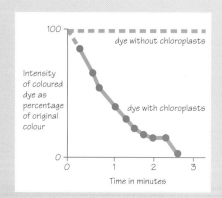

   [Northern Examinations and Assessment Board (formerly JMB) (Biology) June 1991, Paper 1A, No. 6]

3. Below is a diagram representing some of the changes which occur in the light-dependent stage of photosynthesis in plants.
   (a) What terms are given to the light-absorbing complexes A and B? (2 marks)

   (b) What are the chemical substances represented by C and D? (2 marks)

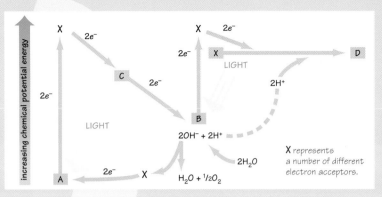

(c) (i) Besides oxygen, water and substance D, what other substance is produced in this process? (1 mark)

(ii) This substance and substance D are both used in the light-independent stage of photosynthesis together with carbon dioxide. Outline the steps in which these substances combine to form a sugar. (4 marks)

(d) (i) Excluding plants, which organisms can also photosynthesize? (1 mark)

(ii) What process can these organisms perform that is of direct nutritional benefit to plants? (1 mark)

(e) What two methods of nutrition may bacteria use to exploit the biomass generated by plants? (2 marks)

(f) What other major group of organisms also uses plant material in both of these ways? (1 mark)

(g) Almost all living organisms use the products of photosynthesis for respiration.
(i) What is the pathway called that oxidizes pyruvate (α-keto-propanate) to carbon dioxide? (1 mark)
(ii) Name two coenzymes which are involved in hydrogen transport. (2 marks)
(iii) Briefly describe how these two reduced coenzymes are re-oxidized. (4 marks)

[O & C (Biology) June 1992, Paper 1, No. 2]

4. (a) Give reasons why it is advantageous for glucose to be broken down in a cell by a series of enzyme-catalysed reactions rather than by one reaction. (3 marks)

(b) List three differences between glycolysis and the tricarboxylic acid (TCA or Krebs) cycle. (Each pair should relate to the same difference.) (3 marks)

(c) The graph opposite shows the relationship between the rate of oxygen uptake by respiring apple fruits and the external oxygen concentration.
(i) What is the rate of oxygen uptake in air? (1 mark)
(ii) Why does the curve flatten off (plateau) above 25% oxygen concentration? (1 mark)
(iii) What would be the expected rate of carbon dioxide evolution in 20% oxygen? (1 mark)
(iv) Suggest one reason why the rate of respiration of apple fruits is increased when the fruits are sliced or cut. (1 mark)

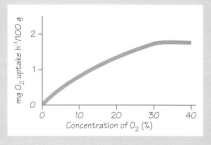

(d) The carbon dioxide output of a kilogram of apple fruits was 164 mg in one hour. Express this result in grammes of carbon dioxide produced per day. Show your working. (2 marks)

(e) Why should green plant tissues be placed in darkness when their respiration rates are being measured? (2 marks)

(f) Outline how (i) protein and (ii) fat may be used as substrates for respiration. (6 marks)

[O & C (Biology) June 1994, Paper 1, No. 9]

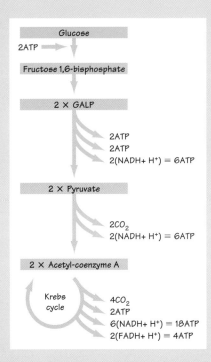

5. The diagram opposite summarizes stages in the complete aerobic breakdown of one molecule of glucose.
   (a) Comment on the use of ATP in the conversion of glucose to fructose 1,6-bisphosphate. (1 mark)
   (b) ATP can provide approximately 30.7 kJ mole^{-1} of usable energy. When glucose is completely oxidized to carbon dioxide and water in a calorimeter the total energy yield is 2880 kJ mole^{-1}.
       (i) Calculate the net energy capture by ATP during the complete aerobic breakdown of 1 mole of glucose. Show your working. (2 marks)
       (ii) Calculate the percentage efficiency of the energy capture. (1 mark)
       (iii) What happens to the energy which is not captured in ATP? (1 mark)
   (c) Explain why, when conditions are anaerobic, the net yield of ATP is only 2 moles of ATP per mole of glucose oxidized in actively working skeletal muscle. (3 marks)

[ULEAC (Biology) June 1994, Biology 1, No. 9]

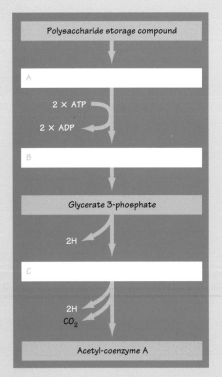

6. The diagram opposite shows stages in cellular respiration.
   (a) Write the name of one missing compound into each of the boxes A, B and C. (3 marks)
   (b) (i) What is the name given to the series of reactions from A to C? (1 mark)
       (ii) Where do these reactions occur in a eukaryotic cell? (1 mark)
   (c) What happens to the hydrogen atoms removed during this process? (1 mark)
   (d) Name one polysaccharide storage compound found in each of the following:
       (i) mammals. (1 mark)
       (ii) flowering plants. (1 mark)

[ULEAC (Biology) January 1994, Biology 1, No. 6]

7. Succinate dehydrogenase is an enzyme in the Krebs (TCA) cycle which converts succinate into fumarate by dehydrogenation.
   Two experiments were carried out investigating the changes in oxygen concentration in a suspension of isolated liver mitochondria, by placing the suspension in a reaction chamber containing an electrode which measured changes in oxygen concentration. In both experiments, A and B, the mitochondria were kept in a buffer solution containing sucrose and inorganic salts. In experiment B, succinate, a Krebs cycle intermediate, was included,

and malonate was added 6 minutes after the experiment had begun. Malonate is a competitive inhibitor of succinate dehydrogenase.

The results are shown in the table below.

	Oxygen concentration/percentage saturation	
Time (min)	Experiment A	Experiment B
0	100	100
2	97	89
4	94	78
6	92	66 (malonate added)
8	89	61
10	86	57
12	83	53

(a) Plot these data in a suitable form. (5 marks)

(b) Explain why the oxygen concentration decreased in both experiments. (3 marks)

(c) Compare the rates of oxygen utilization in experiments A and B during the first 6 minutes of the experiment and suggest an explanation for any differences you observe. (3 marks)

(d) Explain why, in experiment B, the rate of oxygen consumption changed after the addition of malonate. (3 marks)

[ULEAC (Biology) January 1994, Paper 1, No. 11]

8. (a) (i) Name two types of mammalian cell in which you would expect to find relatively large numbers of mitochondria. (2 marks)
   (ii) What is the function of the cristae in the mitochondria? (1 mark)

(b) The table below lists three cellular components (whole cell debris, mitochondria only and residual cytoplasm) and shows their ability to produce carbon dioxide and lactate when incubated with glucose in the presence of oxygen.

Cellular component	Product	
	CO_2	Lactate
Whole cell debris	✓	X
Mitochondria only	X	X
Residual cytoplasm	X	✓

Key
✓ = Present
X = Absent

   (i) Comment on the inability of mitochondria to produce carbon dioxide. (2 marks)
   (ii) Which cellular component contains the enzymes which catalyse the conversion of pyruvate into lactate? (1 mark)
   (iii) Cyanide is a respiratory poison but it has no effect on the production of lactate from glucose in the residual cytoplasm. Suggest a reason for this. (2 marks)

[ULEAC (Biology) January 1993, Paper 1 (adapted), No. 6]

Chapter 5

Using Metabolic Energy

5.1 Introduction

The major form of metabolic energy — the energy to drive muscle contraction and chemical synthesis and for active transport, and to maintain ionic gradients across membranes — is adenosine triphosphate, ATP. In the previous chapter we saw how ATP is produced by photophosphorylation in the chloroplasts of green plants and by oxidative phosphorylation in mitochondria of both plants and animals. This ATP should not be regarded as a *store of* energy; rather it is an *energy currency* that enables energy released from one set of processes (light energy capture, oxidation of food materials) to be used in other processes (mechanical work, chemical synthesis) (Figure 5.1). ATP is not stored in cells to any significant extent; it is made when required. If energy needs to be stored, then the ATP is used to make storage materials, such as the polysaccharides glycogen and starch, or fats, which can later be catabolized to release energy and re-form ATP.

'Growth' of organisms means increasing the number of cells. In general, cells stay about the same size; they increase a little in size but eventually divide to produce two cells. This is as true of a bacterium as it is of a human skin cell. Growth means producing new cell chemicals (new DNA, new protein, new cell wall components, and so on) from simpler precursors — a process which is called *biosynthesis*. This is the same as *anabolism*. Building up new compounds requires energy, as well as the raw materials — the building blocks from which these compounds are constructed.

Some organisms, such as many bacteria and green plants, can make everything in their cells from 'scratch'. Green plants use energy from sunlight and their raw materials are the simple molecules CO_2, H_2O and NH_4^+ or nitrate. They make the small organic molecules, such as monosaccharides, amino acids, fatty acids and nucleotides, and these are subsequently linked together to make the polymers: polysaccharides, proteins and nucleic acids, as well as triacylglycerols (triglycerides) and phospholipids. This linking process also requires energy (Figure 5.2) and this energy is usually provided in the form of ATP. Spread 5.1 gives an overview of how these biosynthetic processes are linked in a metabolic map. This type of nutrition is called the *autotrophic mode*.

In contrast, animal cells, and also many micro-organisms, only participate in the second phase of this process, but in a variety of ways. They need ready-formed precursor units which they then build up into the polymers. Human cells, for example, need at least half of the 20 amino acids in order to survive; they cannot make them themselves. These are called *essential amino acids*. Some of the vitamins and a few fatty acids (the *essential fatty acids*) also have to be supplied. These cells certainly cannot live on CO_2, H_2O and NH_4^+ alone. Their energy is obtained by breaking down organic molecules, such as carbohydrates, amino acids and fats; a selection of these is therefore needed in the diet. This is called the *heterotrophic mode* of nutrition.

Movement — one of the characteristic features of living cells and organisms — requires energy. Plants move their leaves to absorb the maximum amount of sunlight and animals use muscular tissue to

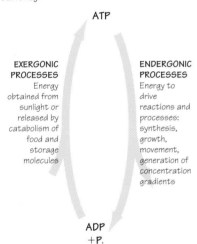

Figure 5.1 ATP acts as a universal energy currency.

locomote and breathe, and in the course of digestion in the intestine. All these movements use ATP as the energy currency. An exception is the case of bacterial movement. This is achieved by flagella, and the energy required is produced in the form of a gradient of protons. However, instead of manufacturing ATP, this gradient is used *directly* to power the flagella (Figure 5.3, page 140).

The other main use of energy by cells is in generating and maintaining gradients of ions and other substances across membranes. Maintaining different concentrations of substances on either side of a membrane requires energy (this is really the opposite of how ATP is produced during oxidative phosphorylation, if you think about it). In addition, food materials need to be taken into cells and this is an energy-requiring transport process too. Although this may sound rather trivial, in fact about 50 % of the energy expenditure of most cells is used in creating and maintaining such gradients. Failure to maintain these gradients causes death of cells. Likewise, when cells are deprived of nutrients or oxygen, death occurs because of this failure.

5.2 Biosynthesis

We now start to look at the processes by which energy is used in organisms by considering how autotrophs make their basic building blocks from simple materials. Chapter 4 explained how photosynthetic organisms are able to trap light energy as chemical energy, and we saw that the major products are ATP and the reduced coenzyme nicotinamide–adenine dinucleotide phosphate (NADPH). This reduced coenzyme can also be looked upon as a form of chemical energy because potentially its electrons could be used in an electron transfer system to make ATP. However, it is more useful as it is since the process of turning water and CO_2 into organic compounds needs a reduction step. (When glucose and other foods are broken down to provide energy, the process is an *oxidation*; clearly, building up such compounds will require a *reduction*.) In producing, say, carbohydrate from CO_2 and H_2O, atoms have to be joined together and this construction process requires energy (Figure 5.4, page 140). The process in green plants and photosynthetic bacteria by which CO_2 is taken from the atmosphere and incorporated into all the organic materials in the cells is often called 'fixing CO_2'. This term implies 'fixing' an atmospheric gas by chemical combination; we also speak of 'fixing nitrogen' whereby N_2 is trapped by micro-organisms in a similar way (see section 5.4).

If we want to understand life at the chemical level, we therefore have to understand the chemical processes whereby CO_2 is fixed and how this fixed carbon is converted in the chloroplast into other compounds, such as carbohydrate. The breakthrough in understanding these complicated processes taking place in the chloroplast was provided by the use of radioactive isotopes to 'trace' what happens to carbon after it has been fixed. The experiment was carried out by Calvin in the 1950s, and, as a result, we now know in great detail what happens during the so-called dark phase of photosynthesis. This expression *dark phase* describes the

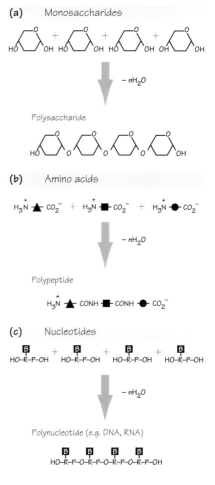

Figure 5.2 Linking monomer units to make polymers requires energy; each of the linking reactions requires the removal of a molecule of water in a 'condensation'. Only partial formulae are shown. In (b) the different types of amino acids are depicted as different shapes and in (c) the base–ribose–phosphate unit is shown as B–R–P.

Metabolic relationships in plant cells

objectives
- To explain how plants build up all their compounds from CO_2, H_2O and NH_4^+
- To trace the relationships, in plant cells, between the chloroplasts (Calvin cycle), mitochondria (TCA cycle), glyoxysomes (glyoxylate cycle) and fat body

Plants can make all the molecules they require from CO_2, H_2O and NH_4^+ (plus, of course, various minerals). CO_2 is fixed in the Calvin cycle (see Spread 5.2) to produce a range of compounds. Ammonia, as ammonium ions (NH_4^+), comes from the action of nitrogen-fixing micro-organisms or from fertilizers added to the soil (see Spread 5.3). It is incorporated into amino acids by amination of α-oxoglutarate taken from the TCA cycle. This mode of nutrition is called AUTOTROPHIC NUTRITION.

All the organic molecules that a plant cell needs are built up from CO_2, H_2O and NH_4^+. This includes the macromolecules DNA and RNA, proteins, and polysaccharides such as cellulose and starch, as well as lignin. Can you identify the routes by which these macromolecules are created? Animal cells, in contrast, start the synthesis of macromolecules from the monomer units (nucleotides, amino acids, monosaccharides), which must be obtained from the diet.

Plants use various intermediates from the TCA cycle to make new compounds for growth. If these intermediates were depleted without being replenished, the cycle would grind to a halt. This replenishment is carried out by a modified TCA cycle called the GLYOXYLATE CYCLE which operates in specialized organelles called GLYOXYSOMES. In plants the TCA cycle is often used *synthetically* to make new compounds; in animal cells it is usually used *degradatively* to release energy while catabolizing acetyl-CoA to CO_2, although it can be used synthetically too.

The glyoxylate cycle converts isocitrate into malate via a C_2 compound, glyoxylate. It involves the addition of an acetyl-CoA unit (C_2) so that the sequence of reactions can be used to build up compounds rather than break them down. Succinate is produced and is fed to the mitochondria where it supplements a normal TCA cycle, so avoiding depletion by intermediates being 'bled off'. Animal cells are not capable of carrying out the glyoxylate cycle.

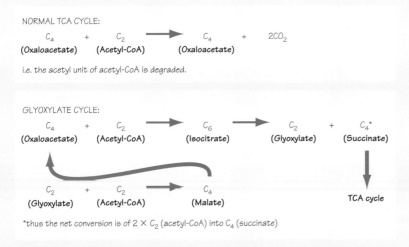

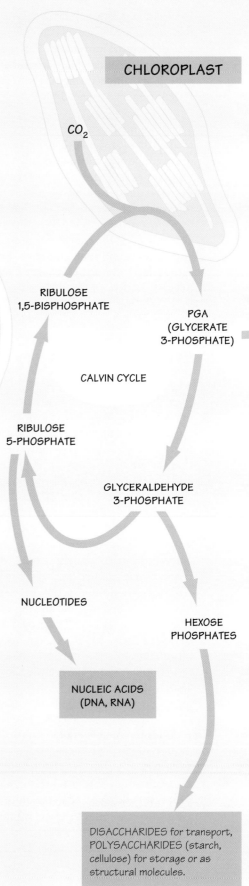

Plant cells, unlike animal cells, can achieve a net conversion of acetyl-CoA (derived from stored lipids in the fat body of seeds) into carbohydrates such as cellulose for growth.

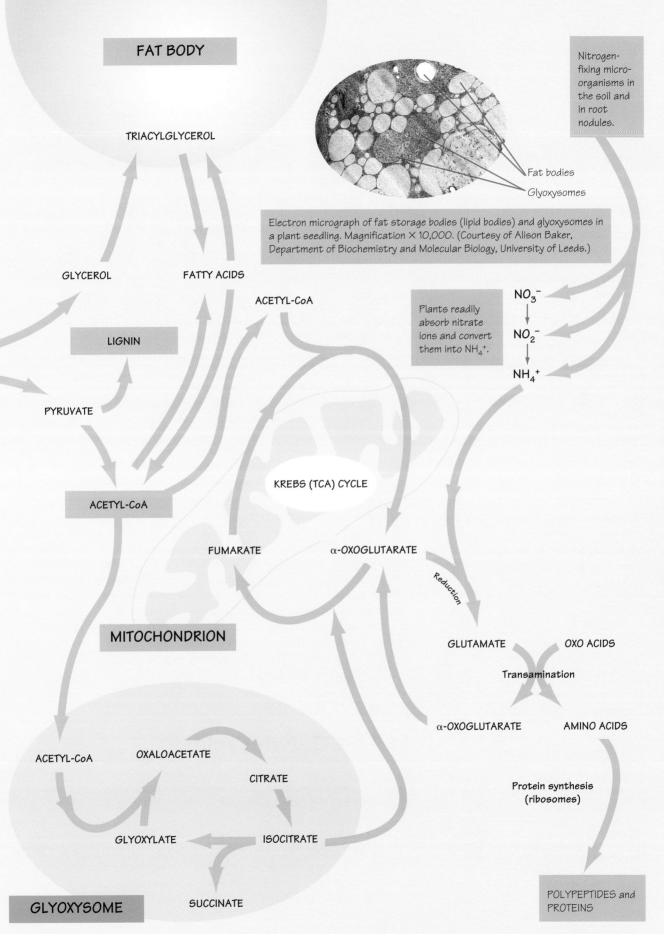

5.1 Metabolic relationships in plant cells

Electron micrograph of fat storage bodies (lipid bodies) and glyoxysomes in a plant seedling. Magnification × 10,000. (Courtesy of Alison Baker, Department of Biochemistry and Molecular Biology, University of Leeds.)

Chapter 5 • Using Metabolic Energy

Figure 5.3 Flagellum arrangement in *Escherichia coli*. Bacterial flagellar motion is driven by a gradient of protons generated by oxidative electron transport (compare this with the way a gradient of protons is generated across the inner mitochondrial membrane). However, ATP is not manufactured. Instead, the rotation is generated when the protons flow back through proteins at the base of the flagellum. The flagellum is made of different types of proteins and the three parts are the filament (the long flagellum), the 'hook', and the basal body whose rings fit into the plasma membrane. The corkscrew motion of the flagellum can change its direction of rotation, causing the bacterial cell to 'tumble' and so change direction.

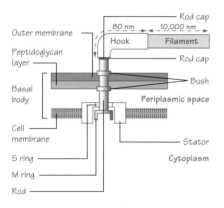

Figure 5.4 Carbohydrate synthesis requires energy.

Carbohydrate in the form of glucose is broken down to provide energy as follows:

$$C_6H_{12}O_6 + 6O_2 \longrightarrow 6CO_2 + 6H_2O$$

Glucose, Oxygen, Carbon dioxide, Water

If we write an equilibrium constant for this reaction we obtain:

$$K_{eq.} = \frac{[CO_2]^6 [H_2O]^6}{[Glucose][O_2]^6} = 10^{495}$$

It is not possible to bring this system to equilibrium in real terms. The very high equilibrium constant says that there is an extremely low chance of the back reaction occurring to any significant extent. Energy is released by the combustion of glucose; by the Law of Energy Conservation, energy has to be put into the system to reverse the reaction. In biological systems this 'reversal' is achieved by a succession of small, enzyme-catalysed steps which use ATP and NADPH.

reactions by which CO_2 is fixed and converted into carbohydrate. 'Dark' really means that if ATP and NADPH are supplied, then light is unnecessary. In fact, both the light phase and the dark phase go on simultaneously. Perhaps it would be more accurate to refer to the dark phase as the 'light-independent' phase.

Calvin used radioactively labelled CO_2 (as $NaHCO_3$ in solution), in which some of the carbon atoms were of the radioisotope ^{14}C rather than the normal ^{12}C. Any compound that picked up this carbon therefore became labelled and could be detected. The use of isotopes in biochemistry is very important and has been vital in elucidating metabolic pathways. The key feature is that, although the radioisotopes are detectable by virtue of their physical properties (e.g. radioactivity, emission of β-particles), chemically they are identical to the normal isotope and are treated no differently by the enzyme systems of the cell.

Another feature of the fixation of CO_2 is that the carbon taken into the cell very rapidly undergoes a series of enzyme-catalysed conversions. Calvin's contribution was to design experiments whereby some $^{14}CO_2$ could be incorporated, but then the reaction was stopped after only a few seconds. This enabled the first compounds to become labelled to be identified. Calvin designed an apparatus in which the unicellular green alga, *Chlorella*, could take up $^{14}CO_2$ in the presence of light (Figure 5.5). Then, after a few seconds, the contents of the apparatus could be quickly drained into boiling methanol which killed the cells and extracted the newly labelled compounds. These organic compounds were analysed by separating them using paper chromatography. Finally, placing the dried chromatography sheet against X-ray film in the dark allowed any radioactive 'spots' to be identified and compared with compounds of known chemical structure.

After as little as 30 seconds of photosynthesis a whole range of compounds, including monosaccharides and amino acids, became radioactively labelled. However, by using progressively shorter reaction times in his apparatus, Calvin was finally able to identify the very first compounds to become labelled. Eventually the sequence of enzyme-catalysed reactions could be mapped out, tracing the path of the newly incorporated radioactive carbon. This pathway is now called the *Calvin cycle*, and Melvin Calvin was awarded the Nobel Prize in 1961 for his painstaking studies (see Spread 5.2).

5.3 The Calvin cycle

There are a number of 'cycles' in metabolism, and they represent a way of dealing with small molecules, gradually building them up (as here) or breaking them down (as in the Krebs cycle), and ultimately regenerating the starting material. The Calvin cycle is very complicated in detail, involving many enzyme-catalysed reactions. However, if we divide it up into three separate processes, it becomes much easier to see what is actually being achieved. These three processes are:

(a) fixing CO_2 into an organic compound (an acceptor module);
(b) turning this fixed CO_2 into carbohydrate; and
(c) regenerating the acceptor molecule.

(a) The initial 'fixing' occurs when CO_2 reacts with a five-carbon compound, ribulose 1,5-bisphosphate. This reaction is catalysed by the enzyme ribulose-1,5-bisphosphate carboxylase ('Rubisco') and produces an extremely unstable six-carbon compound that breaks down immediately to produce two identical three-carbon units, glycerate 3-phosphate (or phosphoglycerate, PGA, as it is sometimes called).

(b) In the second set of reactions, glycerate 3-phosphate is gradually converted into a six-carbon carbohydrate (see Spread 5.2). This sequence of reactions involves a phosphorylation (hence the requirement for ATP) to drive the reactions, and also needs a reduction step (hence the requirement for NADPH). It is interesting to note that this is the only reduction step in the whole sequence. It involves the conversion of an acid (glycerate) into an aldehyde (glyceraldehyde), as shown in Figure 5.6.

(c) In the third set of reactions, some of the carbohydrate produced is tapped off as ribulose 5-phosphate. ATP is used to convert this into the double phosphate, ribulose 1,5-bisphosphate, which is, of course, the acceptor for CO_2, and hence the whole cycle may be repeated. Clearly one 'turn' of the cycle only fixes one carbon: six turns can therefore produce a C_6 monosaccharide.

All green plants operate this pathway (called the C_3 *pathway*). In addition, some plants, typically those that live in tropical climates where the light intensity is high, have some supplementary reactions that help them to operate more efficiently in the prevailing conditions. Carbon dioxide enters plants via the stomata on the leaves, but plants face a dilemma under hot dry conditions at high light intensity. In order to achieve maximal rates of photosynthesis they need to have their stomata open to let CO_2 in. However, this can lead to excessive loss of water vapour which is detrimental to the plant. One solution to this problem is to have a means of fixing the CO_2 more efficiently, i.e. at lower concentrations with the stomata closed. Rubisco is not very efficient at fixing CO_2, and some tropical plants (the so-called C_4 *plants*) are prepared to 'spend' a little extra ATP in order to fix it more efficiently (Figure 5.7, page 144).

An alternative to this solution is the system operated by the plants of the genus *Crassulaceae*. They divide the dark phase of photosynthesis into two parts spread over 24 hours. In the light they trap light energy, but with the stomata closed, and temporarily store this energy. In the dark (when it is cooler) they open their stomata to take in CO_2 and use the temporarily stored energy to fix the CO_2. This mode of operation is called *Crassulacean acid metabolism* or CAM (see Figure 5.8, page 144). Both C_4 and CAM plants have these extra reactions as an adjunct to the C_3 pathway.

The product of the Calvin cycle is carbohydrate and much of this may be transported around the plant (e.g. as the disaccharide, sucrose) or

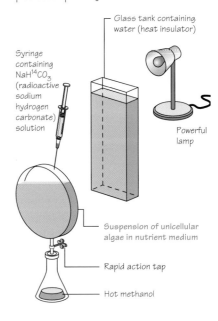

Figure 5.5 Simplified diagram to show Calvin's 'lollipop' experiment to find the first product of photosynthesis.

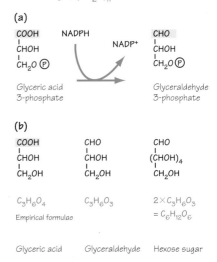

Figure 5.6 (a) The only step in photosynthesis at which a reduction takes place is the step in which the phosphate of glyceric acid is converted into the phosphate of glyceraldehyde. (b) If we look at the empirical formulae, ignoring the phosphate which is the same in each case, we see that the oxidation state of glyceraldehyde is the same as that of a hexose sugar, $(CH_2O)_n$.

Chapter 5 · **Using Metabolic Energy**

The Calvin cycle

objectives
- To explain how Calvin carried out pioneering experiments with radioactively labelled CO_2 in order to identify the 'first products' of photosynthesis
- To summarize how a cycle of reactions (the Calvin cycle) fixes CO_2 and incorporates it into hexose sugar with the simultaneous regeneration of the acceptor molecule, ribulose 1,5-bisphosphate

CARBON DIOXIDE FIXATION is the key step in the 'dark stage' of photosynthesis. CO_2 is 'fixed' into an organic compound by covalently bonding to the five-carbon acceptor molecule, ribulose 1,5-bisphosphate (RuBP).

```
    CH₂O—P                              CH₂O—P
    |                                   |
    C=O                                 HO—C—H
    |                                   |
    H—C—OH     + CO₂ + H₂O              COOH
    |                                       +
    H—C—OH                              COOH
    |                                   |
    CH₂O—P                              H—C—OH
                                        |
                                        CH₂O—P

Ribulose 1,5-bisphosphate       2mol of 3-phosphoglycerate*
```

*This is usually abbreviated 'PGA'. However, the correct chemical name is glycerate 3-phosphate.

The enzyme which catalyses the reaction, RuBP CARBOXYLASE ('Rubisco'), is the most abundant of all enzymes; there is probably about 40×10^{12} g of this protein on Earth (about 0.2 % of all the Earth's protein!).

The Calvin cycle may be summarized as:

```
        18 ATP
            \
             18 ADP + 18Pᵢ
6CO₂ + 6H₂O ———————————→ C₆H₁₂O₆
             /
        12 NADPH
            \
             12 NADP⁺
```

WHAT DID CALVIN DO?

In 1954 Melvin Calvin exposed unicellular green algae (*Chlorella*) to light in the presence of radioactively labelled carbon dioxide ($^{14}CO_2$). He then killed the cells after brief intervals of time (seconds) by dropping them into boiling methanol. The reasoning was that, after short time intervals, the first compounds to become radioactively labelled would be the 'first products' of photosynthesis.

Extracts of the killed algae were subjected to chromatography on a sheet of filter paper. Because several compounds became labelled, it was necessary to run the chromatogram in two directions at right angles (in different solvents) to spread out the 'spots'. The spots — each of which could be identified as a different compound — could then be made visible by placing the dried chromatogram against X-ray film for a few days (autoradiography). A pattern of spots was obtained on the developed film, each one corresponding to a single compound which was a product of the dark phase.

The 'first' compound identified by Calvin after about 3 seconds of photosynthesis was PGA (PHOSPHOGLYCERIC ACID or GLYCERATE 3-PHOSPHATE). Many other compounds (including hexose sugars) also became radioactively labelled, but only after somewhat longer periods of photosynthesis.

After a few seconds, algae are plunged into boiling methanol to stop all metabolic processes. A cell extract is prepared.

The cell extract is applied to paper and chromatographed in two directions at right angles.

The dried chromatogram is placed against photographic film.

The blackened areas on the film show the positions of any radioactive compounds.

5.2 The Calvin cycle

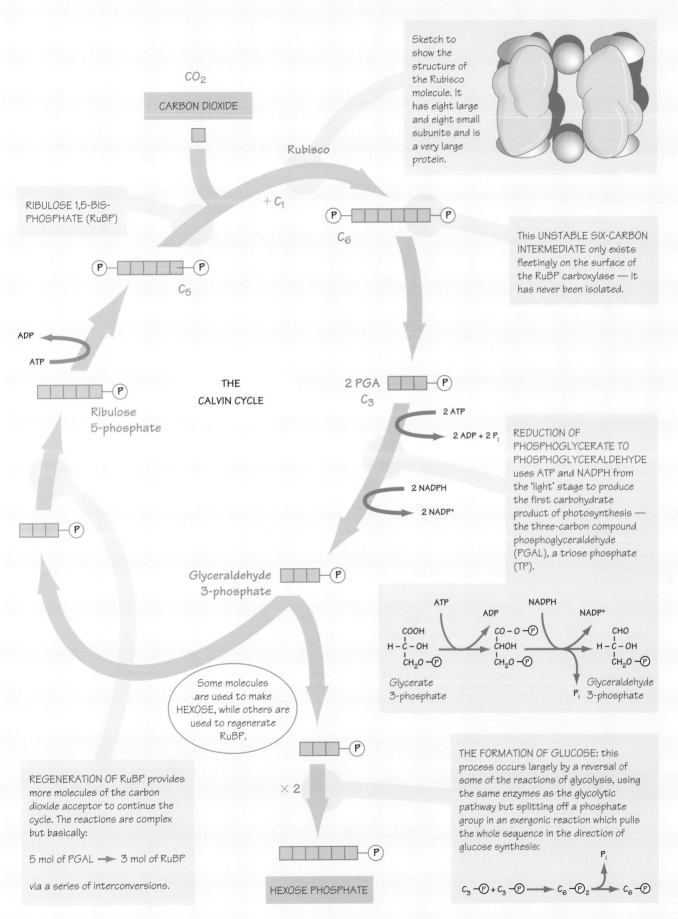

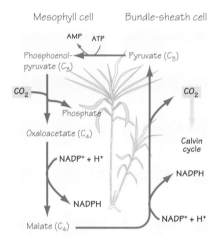

Figure 5.7 Carbon fixation in a C_4 plant. This provides a way of concentrating CO_2 in the bundle-sheath cells at the expense of ATP. The CO_2 is trapped by combination with phosphoenolpyruvate in the mesophyll cells to form oxaloacetate. Photosynthesis in C_4 plants (e.g. sugar cane, maize) consumes 5 ATP per carbon atom fixed (compare with 3 ATP in C_3 plants, which use the Calvin cycle alone). C_4 plants are so-called because they initially fix CO_2 into a C_4 compound, oxaloacetate.

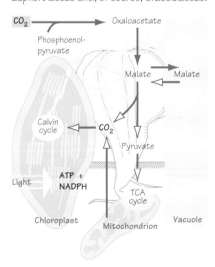

Figure 5.8 Crassulacean acid metabolism (CAM). This is a variant of the C_4 system. CO_2 is fixed in the dark (blue arrows) when stomata are open, resulting in the accumulation of malate in the vacuoles. During the day the stomata close to prevent water loss. The malate returns to the cytosol to be broken down into pyruvate and CO_2 which enters the chloroplast Calvin cycle. CAM plants include cactuses, bromeliads, some lilies and orchids, Euphorbiaceae and, of course, Crassulaceae.

stored as starch (when light and CO_2 are plentiful). Obviously plants have to survive, and therefore need to maintain their energy supply, during periods of darkness, and they do this essentially in the same way as animal cells, using stored carbohydrate or fat.

In addition to carbohydrate, however, plants have to manufacture all the organic compounds that make up their cells. Many of them, such as proteins and nucleic acids, contain nitrogen. Therefore some of the biosynthetic reactions of plants involve the incorporation of nitrogen and we should ask where this comes from. Plant roots take in ammonium (NH_4^+) or nitrates and nitrites from the soil and incorporate this nitrogen into their amino acids and nucleotides. Aside from that present in any nitrogenous fertilizer used by Farmer Giles, the natural source of this ammonia, nitrates and nitrites in the soil is nitrogen (N_2) in the atmosphere. Only certain prokaryotic cells, such as the cyanobacteria, can carry out the fixation of atmospheric nitrogen. All the rest of life on Earth — plants, and humans because we eat plants — depends on these nitrogen-fixing prokaryotes to provide them with nitrogen in a form that they can use (see Figure 1.7, page 12).

5.4 Nitrogen fixation

The element nitrogen is abundant. The Earth's atmosphere contains about 80 % N_2 but the process of fixing it is expensive in energy. The production of ammonia industrially by the Haber process requires a lot of heat energy, as well as a catalyst, to provide the necessary activation energy. Biologically about 1.4×10^8 tonnes of nitrogen are fixed annually by a few species of bacteria and cyanobacteria. It is still an energy-expensive process, but, remarkably, the bacterial enzyme catalysis can carry out the process in the soil at temperatures near to freezing, compared with the hundreds of degrees Celsius and high pressures required for the Haber process. Aside from this biological process and the production of artificial fertilizers, some nitrogen in the atmosphere is fixed by the action of physical processes such as lightning which can generate oxides of nitrogen.

Nitrogen-fixing organisms contain an enzyme complex called *nitrogenase* which catalyses the process of turning N_2 into ammonium ions. The reaction is a reduction but it also requires quite a lot of ATP and may be written:

$$N_2 + 8H^+ + 8e^- + 16ATP + 16H_2O \longrightarrow 2NH_3 + 16ADP + 16P_i$$

(although the NH_3 will be in aqueous solution as NH_4^+).

Nitrogenases are complicated, multi-subunit proteins that contain metal ions in their catalytic centres. The 'normal' nitrogenase contains both molybdenum and iron, and, until a few years ago, it was thought to be universal in nitrogen-fixing prokaryotes and the only enzyme that could do this. It was then found that some bacteria could fix nitrogen even when they were grown in medium devoid of molybdenum. The upshot of this was the discovery that some bacteria can produce a second type of nitrogenase that contains vanadium instead of molybdenum.

Nitrogen fixation is explained in Spread 5.3. The immediate supplier of the electrons in the equation above is the small iron-containing protein, ferredoxin. In photosynthetic nitrogen-fixing organisms, reduced ferredoxin is produced as a result of photosynthetic activity. In non-photosynthetic nitrogen-fixing organisms, reduced ferredoxin is produced by the electron transfer chain using electrons derived from the oxidation of food materials.

Once formed, the ammonium ions are released into the soil where they may be converted into nitrates and nitrites through the action of other micro-organisms (Spread 5.3). Nitrates, which are negatively charged, are more easily absorbed by plant roots than the positively charged ammonium ions.

In all cells, including the prokaryotic nitrogen-fixers, there are enzyme-catalysed pathways for incorporating fixed nitrogen (usually from NH_4^+) into amino acids and nucleic acid bases. The sequence that produces amino acids is shown in Spread 5.3 and is called the *glutamate cycle*.

Nucleotide bases required to form ATP, DNA and RNA are mostly formed using nitrogen donated by amino acids to make the nitrogen-containing purine and pyrimidine ring structures.

Getting rid of excess nitrogen

For plants, nitrogen is always at a premium. Plants generally exist in conditions of relative nitrogen shortage and this is why one of the major components of artificial fertilizers is nitrogen (in the form of ammonia or nitrates). Carbon, in contrast, is abundant, and the majority of plant bulk structural materials such as cellulose and lignin contain practically no nitrogen; it would be a waste of a precious resource to use nitrogen for structural materials.

The situation in animals is in some ways the opposite. Most animals need pre-formed amino acids in the diet for making their proteins, and they typically obtain these by digesting plant or animal proteins in the food. Many animals get too much nitrogen, especially if they are carnivorous. Much of their food is protein, and a high proportion of the amino acids released from this are catabolized in the body. There are amino acids in excess of those needed to manufacture new protein; their nitrogens are removed (as ammonia) and the remaining 'carbon skeletons' are broken down in a variety of ways to produce energy in the form of ATP. Some of the carbon skeletons are converted into pyruvate or acetate, while others form compounds that can be fed into the TCA cycle. Getting rid of the excess nitrogen in mammals requires the operation of the urea cycle (Figure 4.20, page 130).

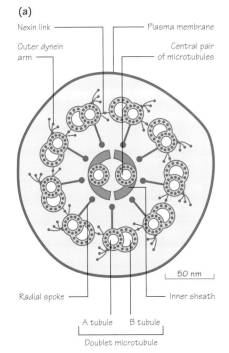

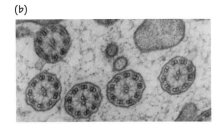

Figure 5.9 (a) The cilia and flagella of eukaryotic cells have a complex structure of nine doublet microtubules surrounding a central pair of singlet microtubules. The whole structure is covered by an extension of the plasma membrane. The microtubules are made of stacked subunits of the protein tubulin, but in addition many other proteins are present, including dynein and nexin. Dynein arms are thought to bend the flagellum. (b) Electron micrograph of cilia in tracheal lining. Magnification × 55,600.

The nitrogen cycle

objectives

- To explain the relationship between atmospheric nitrogen and 'fixed' nitrogen in living organisms
- To recognize that all life depends on the activities of nitrogen-fixing organisms in the soil and in root nodules
- To recognize that nitrogen fixation is an expensive process in energy terms
- To explain the uses of nitrogen in different life forms

Rice growing in Bali. An aim of plant genetic engineers is to place nitrogen-fixing genes in rice plants.

All living organisms are dependent on the activities of soil and root nodule nitrogen-fixing bacteria. Although nitrogen is abundant in the atmosphere, it is difficult chemically to 'fix' it, i.e. to combine it into inorganic and organic compounds. Nevertheless, all organisms need nitrogen:

NITROGEN is in
- amino acids — proteins
- nucleotides — nucleic acids
- vitamins, hormones, coenzymes, haem

Some micro-organisms make a living (i.e. obtain their metabolic energy) by oxidizing ammonia to nitrite, or nitrite to nitrate. Micro-organisms also release nitrogen tied up in dead organisms in the process called putrefaction.

Root nodules containing nitrogen-fixing bacteria.

DENITRIFICATION: represents a loss of 'available' nitrogen from the ecosystem, and may occur under anaerobic conditions such as might exist in a waterlogged soil. For example:

$$2NO_3^- \xrightarrow{\text{Pseudomonas denitrificans}} N_2 + 3O_2$$

$$S + 2NO_3^- \xrightarrow{\text{Thiobacillus denitrificans}} N_2 + SO_4^{2-} + O_2$$

this oxygen is required to metabolize carbohydrate

LIGHTNING MAY FIX ATMOSPHERIC NITROGEN: high-energy discharges may combine nitrogen and oxygen, and the products may dissolve in rainwater and be precipitated as weakly acidic solutions of nitrous or nitric acid:

$$N_2 + 2O_2 \longrightarrow 2NO_2 \xrightarrow{H_2O} 2HNO_3$$

$$N_2 + O_2 \longrightarrow 2NO \xrightarrow{H_2O} 2HNO_2$$

NITRIFICATION: a sequence of oxidations which convert ammonium compounds into nitrite and release energy for the chemoautotrophic bacteria which carry them out. For example:

$$NH_4^+ \xrightarrow{\text{Nitrosomonas}} NO_2^- + \text{ENERGY}$$

$$NO_2^- \xrightarrow{\text{Nitrobacter}} NO_3^- + \text{ENERGY}$$

These reactions generate nitrate, the form in which nitrogen is most easily absorbed by plant roots.

PUTREFACTION: releases nitrogen from combined organic forms. For example:

Protein $\xrightarrow{\text{Proteases}}$ Amino acids

Amino acids $\xrightarrow{\text{Deaminases}}$ Ammonium ions + oxo acids

Urea $\xrightarrow{\text{Urease}}$ Ammonium ions + carbon dioxide

These processes are carried out by saprophytic bacteria and fungi (= decomposers) sometimes called AMMONIFIERS since ammonium ions are the end product.

5.3 The nitrogen cycle

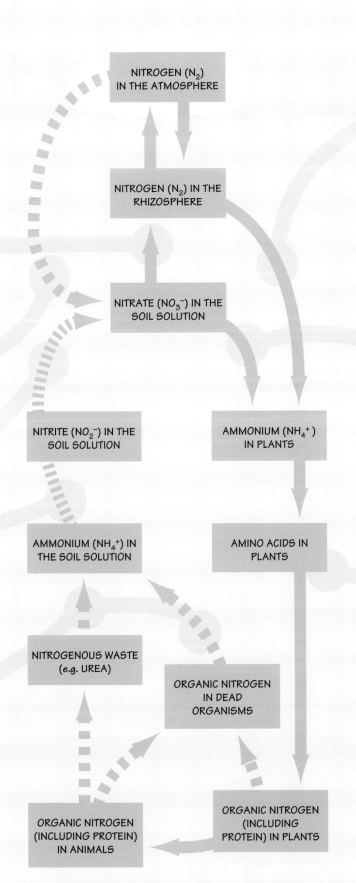

NITROGEN FIXATION: this process reduces nitrogen from the rhizosphere (soil atmosphere) to ammonium ions:

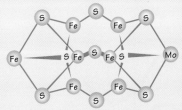

$$N_2 + 6H^+ + 6e^- \xrightarrow{16ATP \to 16ADP + 16P_i} 2NH_3$$

This is a multi-stage process catalysed by an enzyme complex called NITROGENASE, and requires both iron and molybdenum for enzyme function.

Structure of the FeMo cluster in nitrogenase (composition $MoFe_7S_8$)

There are some free-living nitrogen-fixing bacteria, such as *Azotobacter*, but the great majority of nitrogen fixation is carried out by symbiotic bacteria such as *Rhizobium* and cyanobacteria such as *Nostoc*.

NITRATE REDUCTION: occurs in two stages, each catalysed by a different enzyme and operating in different parts of the cell:

$$NO_3^- \xrightarrow{NADH \to NAD^+} NO_2^- \xrightarrow{FERREDOXIN_{RED} \to FERREDOXIN_{OX}} NH_4^+$$

nitrate reductase in cytoplasm   nitrite reductase in plastids

NITRATE REDUCTASE requires molybdenum for activity and is abundant in fast-growing crops, but much less so in acid-rich, nitrogen-deficient soils.

NITRITE REDUCTASE is about 10 times more abundant than nitrate reductase so that nitrite never accumulates (fortunately for human consumers, as nitrites may be converted into carcinogens in the gut).

INCORPORATION OF AMMONIUM occurs by the synthesis of the amino acid glutamate from the oxo acid α-oxoglutarate:

$$\alpha\text{-Oxoglutarate} + NH_4^+ \xrightarrow[\text{glutamate dehydrogenase}]{NADPH \to NADP^+} \text{Glutamate} + H_2O$$

The glutamate may then take part in many transamination reactions to produce any of the other 19 amino acids.

Chapter 5 • **Using Metabolic Energy**   147

Figure 5.10 (a) Simple diagram to show the movement of a beating flagellum. Successive waves of motion pass towards the tip of the flagellum (whiplash movement), and the cell is driven in the opposite direction. (b) A single cilium moves like an oar in a rowing boat, curving during the recovery stroke to produce less resistance. (c) In ciliates such as *Paramecium* large numbers of cilia move in a co-ordinated fashion called metachronal rhythm.

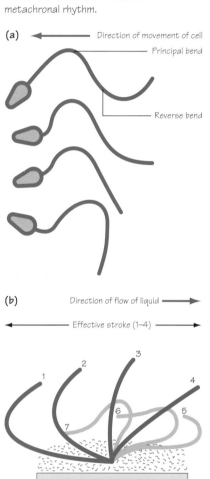

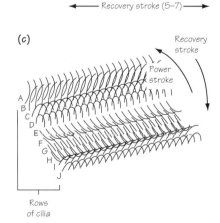

5.5 Biosynthesis of polymers

The biosynthesis of the macromolecules of the cell (nucleic acids, proteins, polysaccharides), as well as fatty acids and fats, is a vast area of metabolism, most of which is now quite well understood. The intermediates are known and the participating enzymes are characterized. Although the biosynthesis of polymers occurs basically by the same chemical processes — namely condensation reactions by which the monomer units are joined together — the processes themselves differ in detail. We are not going to describe the reactions in great depth here, but will concentrate instead on some basic principles. Clearly, the formation of polymers requires a supply of the monomer units, plus a source of energy (e.g. ATP), together with the appropriate enzymes. In the case of polysaccharides it is the specificity of the enzymes that decides which polymer is produced; for example, cellulose, glycogen and starch are all forms of 'polyglucose', but are quite different in their properties. Their biosynthesis is outlined in Spread 5.4.

In the case of proteins and nucleic acids, however, where the *order* of the monomer units is important, in addition to building blocks and energy, we also need *information* specifying this order. How these polymers are made is described in the next chapter where we consider the process by which this information in obtained.

Fatty acids are put together by an energy-requiring process that is more or less a reverse of their catabolic pathway, although it is not identical. An outline of the processes whereby fats are biosynthesized is given in Spread 5.4.

5.6 Locomotion

Organisms use a variety of mechanical systems to move themselves or to transport materials (Table 5.1). The flagella of bacterial cells have already been mentioned (see Figure 5.3) and these rotate to propel the cells. Many eukaryotic organisms (protozoa) also locomote powered by flagella or cilia, but these have a radically different structure from bacterial flagella. Eukaryotic flagella and cilia both have the same basic structure but differ in length (flagella 100–200 μm; cilia 2–10 μm) and have a diameter of about 0.4 μm. They have an internal skeleton consisting of two central microtubules and nine peripheral microtubules running their whole length (Figure 5.9, page 145). The structure undergoes movement as a result of a co-ordinated movement of the microtubules, which is driven by ATP hydrolysis. Successive waves of motion pass towards the tip of the flagellum, generating a whiplash movement which drives the cell in the opposite direction (Figure 5.10). Individual organisms possess only a few flagella per cell but may have thousands of the shorter cilia. *Paramecium*, for example, has more than 10,000 cilia per cell. Each cilium beats like an oar, but during the recovery stroke it is curved and gives less resistance. All the cilia beat in a co-ordinated fashion called *metachronal rhythm* (Figure 5.10).

Table 5.1
Some examples of mechanical systems for movement

Skeletal muscle	— Moves whole organisms, or their parts, e.g. arms and legs
Smooth muscle	— Moves food through intestine
Actin fibres	— Moves cells, changes cell shape, allows cell migration
Microtubules	— Responsible for chromosome segregation at cell division
Cilia	— Moves particles on surface of cells
Flagella	— Moves whole cell by whiplash movement

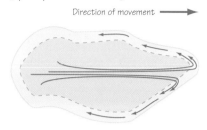

Figure 5.11 Protozoa such as Amoeba move by protoplasmic streaming.

In contrast to this, protozoa such as *Amoeba* have neither cilia nor flagella, but show a characteristic type of movement which is based on *cytoplasmic streaming* (Figure 5.11). A directed flow of cytoplasm extends part of the cell's surface out to form a pseudopodium The tip of this structure then anchors itself on the substratum and the rest of the cell is pulled towards it. The protoplasmic streaming is driven by bundles of actin fibres formed from subunits of the protein actin which are situated immediately beneath the plasma membrane. The actin filament contraction is powered by ATP hydrolysis. The toadstool toxin, phalloidin, can bind to the actin filaments and prevent movement.

Many other cells have actin filaments that are used for locomotion. Figure 5.12 shows human fibroblasts from the dermis of the skin. These can move about these cells and exist in a state of 'tension' because of their intracellular actin fibres. Actin also forms the basis of the multicellular structures responsible for movement that we call *muscle*. Vertebrates have three types of muscle, *smooth*, *cardiac* and *striated* (skeletal), that differ in their structures (Figure 5.13), as well as their physiological properties (Table 5.2, page 152). Insect flight muscle has been extensively studied and is capable of remarkable rates of contraction (Figure 5.14, page 152).

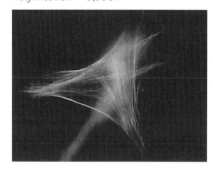

Figure 5.12 Fibroblast from human dermis stained to show the fibres of actin. Magnification × 9,500.

Striated muscle

The characteristic fibres of striated or skeletal muscle are formed by the fusing together of a number of cells to produce a large, multinucleate 'cell' called a *syncytium*. Each muscle fibre is several centimetres long and is 50–100 μm in diameter. The mechanical power results from the pulling effect of the shortening of these fibres when ATP is hydrolysed. Muscle fibres are surrounded and intimately connected with fibrous connective tissue which extends beyond the muscle to form tendons. These transmit the contractile force to the skeleton (Figure 5.15, page 152). In addition there is a nerve supply that controls and co-ordinates the contractile process. Muscles that work mostly aerobically have large numbers of mitochondria in close association with the muscle fibres (see Figure 5.14).

The muscle fibres are packed together with bundles of fibrous proteins to form *myofibrils*, and each myofibril is, in turn, composed of units called *sarcomeres* (Figure 5.16, page 152). The structure of these units is

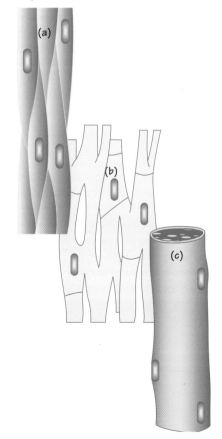

Figure 5.13 Different types of muscle: (a) smooth muscle; (b) cardiac muscle; and (c) striated (skeletal) muscle. Magnification × 9,500.

Biosynthesis of polysaccharides and fats

objectives

- To explain that the biosynthesis of polysaccharides and fats requires metabolic energy as small molecules are being joined together; reducing power (NADPH) may also be required to change the oxidation state
- To recognize that the energy is supplied by ATP, but that simple hydrolysis of ATP to ADP and phosphate is not a satisfactory way of using ATP's energy

BIOSYNTHESIS OF SUGAR POLYMERS always involves a nucleotide diphosphate sugar. This is formed as follows:

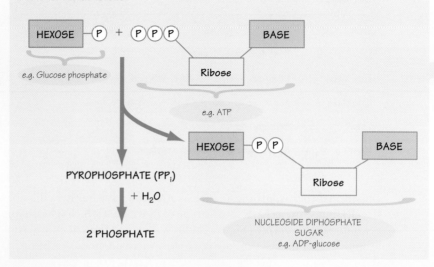

Joining monomer units or small molecules to form macromolecules requires metabolic energy. This energy is supplied in the form of ATP or frequently another nucleoside triphosphate. This energy is used in various ways, but enzyme reactions couple the release of free energy with biosynthetic processes. The straightforward hydrolysis of ATP (to ADP + P_i) never occurs during these BIOSYNTHETIC reactions.

In the biosynthesis of DISACCHARIDES and POLYSACCHARIDES such as STARCH and GLYCOGEN, a complicated series of reactions occurs that involves nucleoside diphosphate sugars. At least two so-called 'high energy' phosphate bonds are used to make one bond between two monosaccharide molecules. Remember that this is a CONDENSATION reaction whereby one molecule of water is removed as the bond is formed.

Fatty acids are synthesized by joining together two-carbon units (essentially a reversal of the breakdown pathway). However, additional steps are needed, and an input of energy is required. In addition, as the fatty acid chains are built up, reductions have to take place (again a reversal of the breakdown pathway). This requires the coenzyme, NADPH.

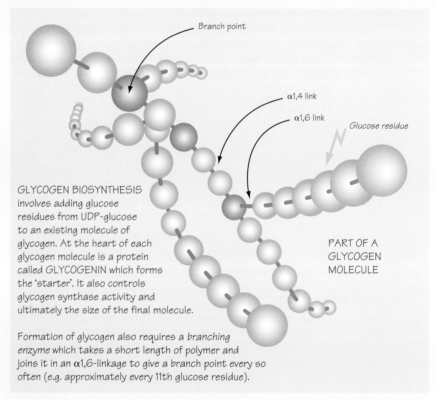

GLYCOGEN BIOSYNTHESIS involves adding glucose residues from UDP-glucose to an existing molecule of glycogen. At the heart of each glycogen molecule is a protein called GLYCOGENIN which forms the 'starter'. It also controls glycogen synthase activity and ultimately the size of the final molecule.

Formation of glycogen also requires a branching enzyme which takes a short length of polymer and joins it in an α1,6-linkage to give a branch point every so often (e.g. approximately every 11th glucose residue).

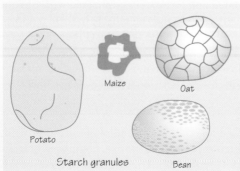

STARCH is the storage polysaccharide in plant cells that is analogous to glycogen in human liver cells. It is made in plant chloroplasts and is deposited as granules. As with glycogen a primer is required. Different plants use different nucleoside diphosphate sugars, e.g. ADP-glucose, CDP-glucose or GDP-glucose. Plants also make cellulose, using UDP-glucose.

Usually compounds other than ATP are used, such as CTP or UTP. This reaction has a strong tendency to proceed in the direction shown because hydrolysis of the pyrophosphate formed gives two molecules of phosphate in an EXERGONIC reaction.

The nucleoside diphosphate sugars can then donate their sugar unit to another sugar (to make a disaccharide) or to a primer to make a larger polysaccharide such as starch or glycogen:

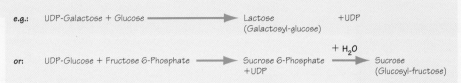

e.g.: UDP-Galactose + Glucose ⟶ Lactose (Galactosyl-glucose) + UDP

or: UDP-Glucose + Fructose 6-Phosphate ⟶ Sucrose 6-Phosphate + UDP —+H_2O→ Sucrose (Glucosyl-fructose)

BIOSYNTHESIS OF FATTY ACIDS, TRIACYLGLYCEROLS AND PHOSPHOLIPIDS

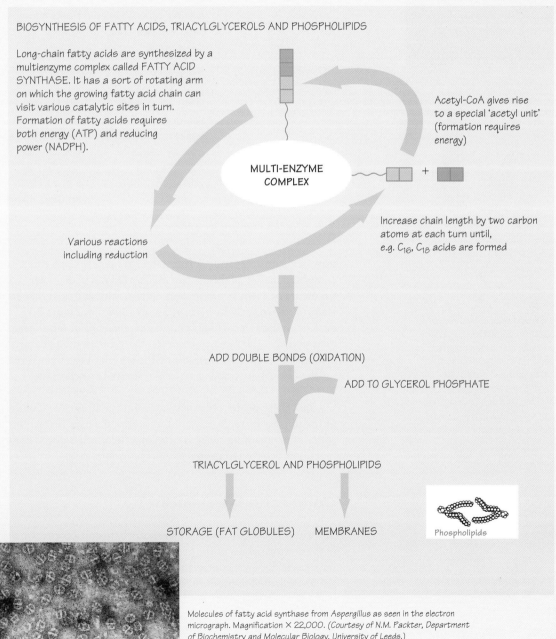

Long-chain fatty acids are synthesized by a multienzyme complex called FATTY ACID SYNTHASE. It has a sort of rotating arm on which the growing fatty acid chain can visit various catalytic sites in turn. Formation of fatty acids requires both energy (ATP) and reducing power (NADPH).

Acetyl-CoA gives rise to a special 'acetyl unit' (formation requires energy)

MULTI-ENZYME COMPLEX

Various reactions including reduction

Increase chain length by two carbon atoms at each turn until, e.g. C_{16}, C_{18} acids are formed

ADD DOUBLE BONDS (OXIDATION)

ADD TO GLYCEROL PHOSPHATE

TRIACYLGLYCEROL AND PHOSPHOLIPIDS

STORAGE (FAT GLOBULES)   MEMBRANES

Phospholipids

Molecules of fatty acid synthase from Aspergillus as seen in the electron micrograph. Magnification × 22,000. (Courtesy of N.M. Packter, Department of Biochemistry and Molecular Biology, University of Leeds.)

5.4 Biosynthesis of polysaccharides and fats

Figure 5.14 Insect flight muscle has remarkable powers of contraction. Shown here is an electron micrograph of a section of Tsetse fly muscle cut at right angles to the fibres. The dark areas are mitochondria which supply ATP to the muscle fibres. Magnification × 10,000. (Courtesy of M. Anderson, University of Birmingham, U.K.)

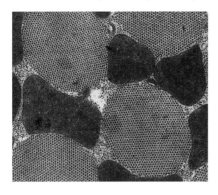

Figure 5.15 Muscle contraction for the movement of limbs in a vertebrate. Pairs of muscles work in opposition giving precise control of movement.

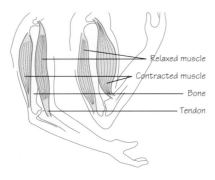

Figure 5.16 Structure of a vertebrate voluntary muscle showing the relationship between myofibrils, fibres (see Figure 5.13c) and fibre bundles. The muscle fibres themselves are multi-nucleate cells.

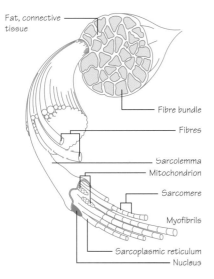

Table 5.2
Structural features and properties of muscle types

Muscle	Structural features	Physiological properties
Involuntary (smooth)	Sheets or bundles, unstriated, each cell distinct, spindle-shaped	Under control of autonomic nervous system; capable of sustained, rhythmical contraction without noticeable fatigue
Cardiac (heart)	Three-dimensional network of cylindrical branching short columnar fibres; abundant mitochondria; each fibre surrounded by plasma membrane called the sarcolemma	Impulses that cause contraction arise within the heart muscle; all fibres contract in unison to produce pumping action of heart
Voluntary (skeletal; striated)	Elongated, multinucleate*, muscle fibres lying parallel to each other, striated (striped); groups of fibres bundled and surrounded by connective tissue; each multinucleate fibre surrounded by sarcolemma	Supplied by motor nerves from brain and spinal cord; each fibre is served by a nerve fibre ending in a *motor end plate*; when stimulated, fibres contract rapidly and powerfully, with a short refractory period, but fatigue relatively early

*This multinucleated structure is called a syncytium.

now well understood. Within the sarcomeres are thick filaments composed mainly of the protein myosin (15 nm in diameter) and thinner filaments of actin (7 nm in diameter) arranged in a highly structured fashion to produce the various zones shown in Figure 5.17. There are dark-staining lines which limit the extent of each sarcomere, called Z lines or discs. The repetition of sarcomeres, with their characteristically arranged protein filaments, along the length of the fibre is responsible for the distinctive stranded pattern of striated muscle tissue.

Overall, these protein filaments account for about 80 % of the volume of the muscle fibre. Nuclei and mitochondria are forced to the outer edge. The remaining space is cytosol, called *sarcosol*, and endoplasmic reticulum, referred to as *sarcoplasmic reticulum*.

Contractile mechanism

Myosin is known as a 'molecular motor'; this means that small changes in the conformation of the protein can somehow lead to contraction of the muscle fibre. Of course, in a muscle fibre many millions of molecules are co-ordinated to produce the contraction actually observed.

Each myosin molecule is made up of two polypeptide chains and the ends of these chains are folded up into globular structures. A myosin (thick) filament is made up of a number of these molecules arranged in a characteristic way. Actin (thin) filaments are made up of chains of actin molecules joined together rather like a string of beads, but the whole filament is made up of two intertwined chains (Figure 5.18).

Non-covalent cross-bridges form between the globular 'heads' of the myosin and the actin molecules of the thin filament. Contraction of the muscle fibre occurs when the two filaments slide against each other. Cross-bridges break and then new cross-bridges form, continuing the movement of one filament against the other. The two ends of the myosin filaments are effectively pulling in opposite directions, ensuring that the actin filaments, with their fixed ends attached to the Z bands, are forced together, and thus the muscle becomes shorter and thicker (Figure 5.19).

The detailed biochemical processes that occur in this sliding filament mechanism are not properly understood, but it is known that the globular heads of the myosin molecules can catalyse the hydrolysis of ATP:

$$ATP + H_2O \longrightarrow ADP + P_i + energy$$

The 'energy' released in this reaction somehow produces the change in protein shape that results in the contraction of the sarcomere. Recently very detailed studies of the three-dimensional structure of the myosin molecule have been carried out using X-ray crystallography. The myosin molecule has a long α-helix running down its length, and a pocket where ATP binds and may be hydrolysed has been identified. It is proposed that the binding of ATP causes this pocket to close, which, in turn, causes the tip of the myosin molecule to move by about 5 nm as the protein flexes. This would represent the 'power stroke'. Hydrolysis of ATP to ADP and phosphate reverses this, perhaps after these products are released, allowing another power stroke to be initiated (Figure 5.20).

Contraction of muscle fibres is controlled by many other proteins present in the fibre and also by the release and re-absorption of Ca^{2+}, and indeed the sarcoplasmic reticulum surrounding the fibrils acts as a Ca^{2+} store. Nerve stimulation releases Ca^{2+} into the sarcosol which stimulates the myosin heads to hydrolyse ATP. After stimulation the Ca^{2+} is re-absorbed by the sarcoplasmic reticulum and contraction ceases.

Muscle ATP stores

Clearly, muscle contraction requires a great deal of ATP, and, from what we have learnt previously about ATP not being stored to any significant extent, this means that muscle needs efficient ways of regenerating ATP from ADP and phosphate. In fact, there are several ways of achieving this depending upon whether the muscle is operating anaerobically or aerobically. In addition, when muscles have to start to contract they need an almost 'instant' supply of large amounts of ATP. This is achieved using a store of so-called *phosphagens*. In vertebrate muscle the major phosphagen is *creatine phosphate*. This is explained in Figure 5.21.

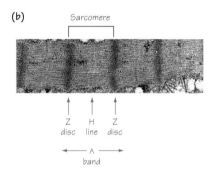

Figure 5.17 (a) Arrangement of actin and myosin filaments in a sarcomere in relation to the Z discs. (b) Electron micrograph of mouse back muscle showing actual appearance. The dark spots are glycogen granules. Magnification × 4000.

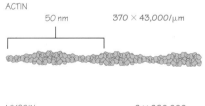

Figure 5.18 Actin and myosin are muscle proteins. Actin fibres are formed from small subunits and have a helical cable structure. Myosin molecules twist together in pairs with protruding heads; a number of these can join together (see Figure 5.17).

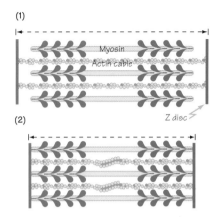

Figure 5.19 The sliding filament model for muscle contraction. Notice how the two Z discs move closer together, creating a contraction force.

Figure 5.20 The mechanism by which a myosin molecule 'walks' along an actin filament [based on I. Rayment et al. (1993), Science, 261, 50]. **(1)** At the start of the cycle the myosin head has no ATP and is tightly bound to an actin filament. **(2)** When ATP binds it causes a small change in conformation which reduces the affinity of the myosin head for actin and allows it to move along the filament. **(3)** The myosin undergoes a further change of shape causing it to be displaced along the actin filament by about 5 nm. ATP is hydrolysed. **(4)** Phosphate is now released and the myosin head binds tightly to the actin. This triggers the power stroke during which the head regains its original shape and loses its bound ADP. **(5)** The myosin head is now tightly bound to the actin as in **(1)**, ready for another cycle.

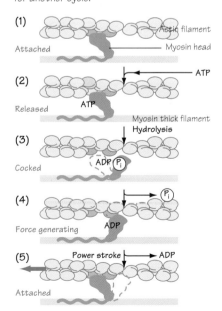

As will be realized from the above, a great deal is known about muscle although much still remains to be discovered. The methods of biochemistry, structural studies using X-ray crystallography, physiology, cell biology and thermodynamics are all required to enable us to elucidate this very fundamental process. It is a very exciting area and an increasing amount of knowledge will allow us to understand why some people are good at sprinting and why others are good at marathons, and, perhaps more importantly, what is going wrong in people with muscle-wasting diseases such as muscular dystrophy.

5.7 Transport across biological membranes

The final section in this chapter deals with the transport of materials across membranes. Maintaining differential gradients of substances across membranes is a very important activity for cells and therefore for organisms that are made up of cells. To take just one example, the maintenance of different concentrations of Na^+ and K^+ inside and outside of a nerve cell is vital for the conduction of nerve impulses, which, in turn, can make muscles function.

Materials can cross biological membranes by one of two processes. The first of these is *diffusion* which requires no energy input; the second is *active transport* which costs energy to drive it. Diffusion may be *simple* (or *passive*), or it may be *facilitated*. The characteristics of these transport mechanisms are summarized in Table 5.3.

Table 5.3
Characteristics of transport mechanisms

	Simple diffusion	Facilitated diffusion	Active transport
Protein required	No	Yes	Yes
Energy source	Concentration gradient	Concentration gradient	ATP hydrolysis
Direction	With gradient	With gradient	Against gradient
Specificity	Non-specific	Specific	Specific
Saturable	No	Yes	Yes

Passive diffusion

Solutes will tend to move from a region of high concentration to one of low concentration, so that eventually the concentration gradient is evened out. (If you put sugar in your tea, even if you do not stir, the sugar molecules eventually distribute themselves through the whole cup of liquid.) Since the interior of a cell has a strikingly different composition from the external environment, it will be expected that substances and ions would diffuse in or out and the cell would gain or lose materials by this process. However, the lipid bilayer presents a hydrophobic barrier. Hydrophobic substances may be able to diffuse through, but hydrophilic substances are prevented from diffusing. Thus cells are able to take in hydrophobic gases such as O_2 and long-chain fatty acids by this mechanism. The Laws of Diffusion state that the greater the concentration on one side of the membrane, the faster the rate of diffusion, and, of course, diffusion may only take place down a concentration gradient.

Small polar molecules such as water and ethanol are also able to diffuse across membranes fairly freely. These small molecules only form a few hydrogen bonds with water molecules, and very little energy is required for them to come out of the aqueous phase and go through the lipid phase of the membrane.

In contrast, larger polar molecules such as sugars and amino acids form many hydrogen bonds with water molecules and do not easily escape to go through the membrane, although they may do so via a facilitated mechanism (see below). Similarly, charged ions (Na^+, K^+) are strongly hydrophilic and are highly hydrated with a shell of water molecules, and cannot cross the lipid bilayer except where special transport mechanisms exist.

Both facilitated diffusion and active transport require the presence of integral membrane proteins which form channels through the membrane or act as carriers to transport these substances through the membrane lipid (see Spread 1.4, page 22). These integral membrane proteins form pores which have a hydrophilic interior, allowing small molecules and ions to pass through the membrane (Figure 5.22).

Carrier proteins in the membrane can bind specifically to solute molecules to be transported. The complex of proteins and small molecules results in transport across the membrane and release of the small molecule on the other side. If transport is 'down' the prevailing concentration gradient then the process is said to be *facilitated*. If transport is 'up' a concentration gradient then an energy input will be required and the process is referred to as *active transport*.

Facilitated diffusion

As stated above, facilitated diffusion can allow the passage of relatively large polar molecules across a membrane, but in a direction dictated by that of the concentration gradient. (Macromolecules cannot cross membranes by any of these mechanisms, and special means such as endocytosis and pinocytosis have evolved for this purpose.)

In vertebrates, the blood glucose concentration is tightly controlled and kept fairly constant by the action of insulin and other hormones. As a consequence, the cytosol of cells, which are using energy, always has a lower glucose concentration than the blood. Within the plasma membrane of cells there are specific glucose transporter proteins which facilitate the movement of glucose down the concentration gradient and into the cell interior. Such transporters are called *permeases*, implying

Figure 5.21 Creatine phosphate acts as a store of energy in voluntary muscle. ATP is not stored to any great extent and the ratio of ADP/ATP in muscle does not change very much. When a muscle suddenly needs to contract vigorously it needs an immediate supply of ATP. However, in order to obtain this by glycolysis it is necessary to break down some glycogen, which takes time. Furthermore, to initiate glycolysis requires some ATP input before any 'new' ATP is produced. Creatine phosphate supplies sufficient ATP for a few contractions when it is necessary to move in a hurry.

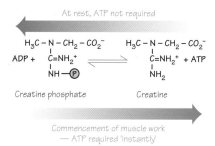

Figure 5.22 Hypothetical arrangement of a membrane protein. This is a transporter molecule that has ATP-binding domains. **(a)** Possible arrangement of the polypeptide in the membrane where sections of α-helix are shown as cylinders, **(b)** shows how these may be folded up to create the transport protein. The sections that span the membrane will be highly hydrophobic. [Redrawn with permission from B.A. Alberts et al. (1995), Molecular Biology of the Cell, 3rd edn., Garland, New York.]

(a)
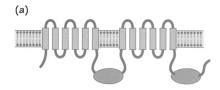

(b)
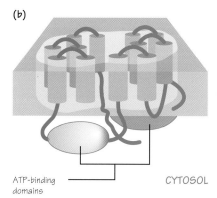

Figure 5.23 Molecules and ions may pass through membranes by simple diffusion by dissolving in the membrane; through a protein channel; or via a specific carrier. The carrier may simply facilitate transport or it may require energy (ATP), as in active transport (see Figure 5.24). (a) In the case of facilitated diffusion the molecules or ions travel down a concentration gradient, but since there are a limited number of carriers it is possible for these to become saturated. (b) The carrier proteins are thought to alternate between two conformations so that the region where the solute binds is sequentially accessible on one side of the lipid bilayer and then on the other.

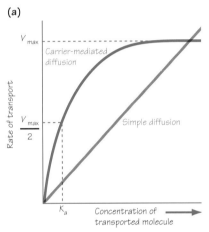

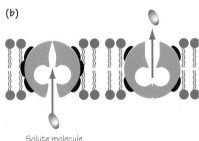

that they are enzymes. They certainly have an 'active site' for binding glucose — but the 'substrate' and the 'product' are identical.

The glucose permease of red blood cells has been studied in great depth. Increasing the concentration of glucose does increase the rate of glucose uptake but only up to a point. This is unlike simple diffusion. The maximum rate is reached when the permease molecules are all saturated with glucose; they cannot transport any faster (Figure 5.23). The concentration of glucose that gives half the maximum rate of uptake is called the *affinity constant* (K_a) of the permease for the solute. If you look back to Spread 3.2 (page 78), you will see that this constant is very similar to the Michaelis constant, K_m, for an enzyme.

Active transport

Active transport is the movement of materials against a concentration gradient, and this is often necessary in cells. For example, a typical mammalian cell has intracellular concentrations of Na^+ and K^+ of 10 and 140 mmol·dm^{-3}, respectively, whereas the external concentrations are about 5 and 145 mmol·dm^{-3}. These differences must be maintained, but there is constant slow leakage of ions across the membrane which must be made good. This requires the transport of Na^+ out and K^+ in, and these transfers go against the gradient.

The balance of ions across the membrane is achieved as a result of the activity of a membrane protein enzyme called Na^+/K^+-adenosine triphosphatase or Na^+/K^+-ATPase. This enzyme in the membrane catalyses the transfer of two K^+ ions into the cell for every three Na^+ expelled, and ATP is hydrolysed to supply the energy required for this process (Figure 5.24). Since materials are co-transported in opposite directions, this enzyme is called an *antiporter*, and because three positive charges are exported for every two positive charges imported, a potential difference is developed across the plasma membrane of between –60 and –70 mV (inside negative). This potential difference is called the *membrane potential*. Cells may use 30–50 % of their energy to maintain these concentration differences between the cytosol and the exterior. Remember, too, that inside the cell there are other membrane systems, such as the mitochondrial membrane, where active transport takes place.

The opposite of an *antiporter* is a *symporter* where the active transport of one solute is linked to the facilitated uptake of a second substance. For example, some bacterial cells absorb lactose from the environment using a Lac/H^+-symporter. The bacteria actively transfer protons to the exterior resulting in the outside H^+ concentration exceeding that on the inside. These protons may therefore be re-absorbed by facilitated diffusion. However, the H^+ ions are returned to the inside of the cell by the Lac/H^+-symport system, which only operates if it can co-transport a lactose molecule at the same time as a H^+ (Figure 5.25). This means that lactose can be taken up against its concentration gradient.

Similar symport mechanisms operate in eukaryotic cells, although the co-solute is more usually Na^+ rather than H^+. For example, gut epithelial cells absorb glucose from the digested food by an active mechanism. The gut contents are normally rich in Na^+ and the epithelial cells have Na⁺/glucose-symports on the region of their plasma membrane facing the gut lumen. They use the facilitated uptake of Na⁺ to drive the active uptake of glucose from the gut lumen into the cells. At the other 'face' of the epithelial cells, different mechanisms operate so that the glucose passes into the blood and hence to the liver.

In cases of the disease cholera, caused by the bacterium *Vibrio cholerae*, the action of cholera toxin results in extensive diarrhoea and water loss so that the patient's ionic balance is upset and little glucose is absorbed. In order to re-hydrate and re-feed the patient it is important to give not just glucose but a mixture of glucose and Na⁺ (as NaCl).

5.8 Further reading

Bassham, J.A. (1962) The path of carbon in photosynthesis. Scientific American **206 (June)**, 88–100 *[although this is rather old, it is a classic article on the subject.]*

Berg, H.C. (1975) How bacteria swim. Scientific American **233 (Aug)**, 36–44

Björkman, O. and Berry, J. (1973) High efficiency photosynthesis. Scientific American **229 (Oct)**, 80–93

Gayford, C.G. (1985) Energy and cells. Macmillan Education, London

Hall, D.O. and Rao, K.K. (1994) Photosynthesis (5th edn.). Cambridge University Press, Cambridge

Hirschhorn, N. and Greenough, W.B. (1992) Progress in oral rehydration therapy. Scientific American **264 (May)**, 16–22

Jones, D. (1990) The fast whites and the slower reds. Biological Sciences Review **2**, 2–6

Postgate, J. (1989) Fixing nitrogen. Biological Sciences Review **1**, 2–6

Storr, C. and McMillan, B. (1995) Human Biology. Wadsworth Publishing/International Thomson, London

Stossel, T.P. (1994) The machinery of cell crawling. Scientific American **271 (Sept)**, 40–47

Tobin, A. and Whitehouse, D. (1993) More than one way to photosynthesise. Biological Sciences Review **5**, 30–33

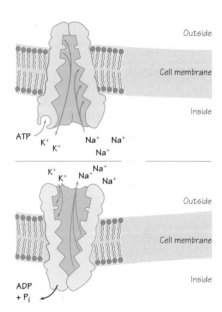

Figure 5.24 Hypothetical structure of the sodium–potassium pump (Na⁺/K⁺-ATPase). This is an active transport system that requires energy (ATP) to move the ions against their concentration gradients. The membrane protein has a cavity and can change conformation. It also has an ATP-binding domain and the protein is likely to have several α-helical sections repeatedly crossing the lipid bilayer. The Figure shows how the release of two K⁺ ions is followed by the binding of three Na⁺ ions and ATP on the inner membrane. The protein changes shape when it opens to the outside, releasing three Na⁺ ions and picking up two K⁺ ions. Changes in the affinity for Na⁺ and K⁺ now cause the channel to open to the inside, to increase its affinity for Na⁺, and the cycle begins again. [Redrawn with permission from A.G. Lowe, Biological Sciences Review (1994), **7**(1), 29.]

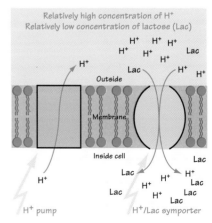

Figure 5.25 Possible mechanism for the lactose transport system.

5.9 Examination questions

1. (a) Compete the flow diagram, relating to photosynthesis, by inserting the appropriate names in the boxes. (6 marks)

(b) (i) ATP plays a key role in the operation of this cycle. Indicate, on the diagram, one step involving ATP. (1 mark)

(ii) ATP in the chloroplast is formed by proton pumping. What is meant by the term *proton pumping*? (3 marks)

(c) Indicate how the structure of a chloroplast is related to its functions. (5 marks)

(d) When the intensity of light falling on photosynthetic tissue is gradually increased from zero, there is at first an uptake of oxygen, followed by a period of no net gas exchange and then by increasing oxygen release. Explain these observations. (5 marks)

(e) Which one of the following percentages indicates the efficiency of photosynthesis in converting light energy into chemical energy: 1; 5; 10; 15; 20? (1 mark)

(f) In the U.K., measures are being taken to reduce overproduction of food in agriculture. One way would be to use agricultural products for non-food purposes. State three non-food commercial uses of compounds produced by plants as a result of photosynthesis. (3 marks)

[O & C (Biology) June 1993 Paper 1, No. 1]

Flow diagram boxes:
- Name of acceptor molecule .... ← Carbon dioxide
- H_2O → H^+ OH^- → Oxygen
- Name of enzyme ....
- Glycerate 3-phosphate
- NADPH+H^+
- Name of cycle regenerating the acceptor molecule ....
- NADP$^+$
- Name of molecule produced ....
- 2 molecules condense to produce ....
- Name of storage compound produced ....

2. An investigation was carried out into nitrogen fixation in soil. Samples of two different soils (A and B) were incubated at 30°C for 29 days in either aerobic or anaerobic atmospheres containing the same percentage of nitrogen.

Half of the samples, in both the aerobic and the anaerobic atmospheres, were watered with 1% glucose solution. The other half were watered with equal amounts of distilled water.

At the end of the incubation period the amounts of fixed nitrogen were measured, and rates of nitrogen fixation were calculated. The results are shown in the table below:

Soil sample	Watering solution	Rate of nitrogen fixation in mg × 10^{-4} nitrogen per g dry soil per day	
		Aerobic	Anaerobic
A	Water	0.00	1.63
	Glucose	1.78	3.76
B	Water	1.48	4.14
	Glucose	1.96	8.17

Adapted from Chang and Knowles, Can. J. Microbiol., **11**, 29–38.

(a) (i) Explain why some samples in the experiment were watered with water only. (2 marks)

(ii) Explain why the soil samples were incubated. (2 marks)

(b) Suggest a method by which the rate of nitrogen fixation might be measured. (3 marks)

(c) (i) Describe the effect of the added glucose on nitrogen fixation in the two soil samples. (2 marks)

(ii) Suggest an explanation for this effect. (3 marks)

(d) (i) Compare the effects of aerobic and anaerobic atmospheres on nitrogen fixation in these two soils. (3 marks)

(ii) The activity of certain enzymes is known to be affected by the presence of oxygen. Suggest an explanation for the different rates of nitrogen fixation in aerobic and anaerobic conditions. (3 marks)

(e) From the results of this investigation, suggest two practical ways in which a farmer could increase the rate of nitrogen fixation in the soil. (2 marks)

[ULEAC (Biology) June 1991, Paper 1, No. 14]

3. (a) (i) Using only the following compounds, construct a single flow diagram to show the incorporation of carbon dioxide into carbohydrates, proteins and fatty acids: carbon dioxide, phosphoglyceraldehyde, acetyl-CoA, ribulose bisphosphate (diphosphate), fatty acids, amino acids, phosphoglyceric acid. (6 marks)

(ii) State two ways in which this flow diagram would differ from a similar flow diagram outlining the process of glycolysis. (2 marks)

(b) The graph opposite indicates the effect of temperature on the rates of photosynthesis and respiration in leaves.

(i) At what temperature was the net gas exchange zero? (1 mark)

(ii) Suggest a possible reason why the rate of respiration in the leaf is less affected by the higher temperature than is the rate of photosynthesis. (1 mark)

(iii) What would be the effect of a rise in temperature on the rate of photosynthesis if the intensity of light falling on the leaf was very low? Give a reason for your answer. (3 marks)

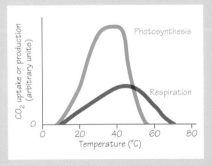

(c) Shade in, on the graph, the area which indicates the apparent rates of photosynthesis. (2 marks)

(d) Explain how water molecules are (i) utilized in the process of photosynthesis, and (ii) produced during aerobic respiration. (4 marks)

[O & C (Biology) June 1987, Paper 1, No. 4]

4. The equation represents a reaction which occurs in the liver and other body cells:

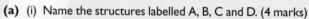

(a) Give the name of: (i) the type of compound labelled molecule A; (ii) the molecular group common to all four molecules in the equation; (iii) the molecular group which molecule A is donating to molecule B in this reaction; (iv) the type of reaction of which this is an example. (4 marks)

(b) Give one reason why this reaction is important in the body. (1 mark)

(c) Suggest one possible fate of molecule C. (1 mark)

[AEB (Biology) June 1991, Paper 1, No. 18]

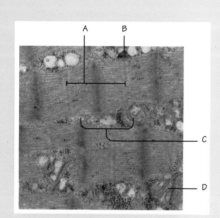

5. The illustration is an electron micrograph of a longitudinal section of a striated muscle fibre.

(a) (i) Name the structures labelled A, B, C and D. (4 marks)

(ii) State two ways in which the histology of striated muscle differs from that of cardiac muscle. (2 marks)

(b) Indicate briefly the roles of actin, myosin and ATP in muscle contraction. (9 marks)

[O & C (Biology) June 1993, Paper 1, No. 10]

6.1 Introduction

It has been emphasized repeatedly that life depends on macromolecules: *proteins* to act as enzyme catalysts and in many other roles, and *polysaccharides* to act as storage and structural materials, for example. In this chapter we deal with the functions of the third group of biological macromolecules — the nucleic acids, DNA and RNA. These molecules may be described as informational macromolecules: they store information in their sequences of four bases. This information is digital in the same sense that computer information, a series of '1's and '0's, is digital. Some of the background of the chemistry and function of DNA and RNA was explained in Chapter 2. Now we are going to look at the importance of this information in detail.

We can think of this information as *genetic information*: it represents the set of instructions passed on from generation to generation when a cell divides or as an organism produces progeny. This is an enormous area of biochemistry (or molecular biology): in the 40 years since the discovery of the double helix by Watson and Crick (Figure 6.1), knowledge has increased exponentially. Furthermore, scientists have learned how to manipulate this genetic information and so we are now in the era of genetic engineering, gene therapy and transgenic organisms (see Chapter 7). There is no sign of the flood of knowledge decreasing — just the opposite — and, indeed, both commercial exploitation and exciting medical possibilities are driving research on with ever increasing intensity.

There are three areas that we will look at in this chapter, and each of these can be put in the form of a question. (1) How is the DNA present in every cell precisely duplicated whenever a cell divides? (2) How are the instructions coded in the DNA used (translated) to make the molecules that will form a new cell? (3) How do we account for the phenomena of mutations and evolution? (It may be helpful at this stage if you look back at page 55 and Spread 2.4 to review the chemical nature of DNA and RNA.)

6.2 How is DNA replicated at cell division?

When a cell divides, its complement of genetic information must be copied with high fidelity otherwise the daughter cells will receive DNA with mistakes in the base sequence and may not function properly. It is vital that the genetic message be passed from generation to generation with high accuracy. If we remember that the human genome has a sequence of about 3×10^9 bases it can be seen that formidable accuracy is required. (If this book contains about 10^5 words each of five letters, the human genome is equivalent to 6000 *different* such books; imagine copying all of these without making any errors!) How, then, is the copying done and how is the high accuracy achieved? The answer has to involve enzymes, of course. Usually the copying or duplication process is referred to as *replication*.

Chapter 6
DNA: Dealing with Information

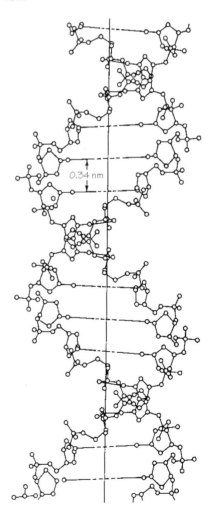

Figure 6.1 A computer-generated image of how the bases in the DNA double helix complement each other in the centre of the helix.

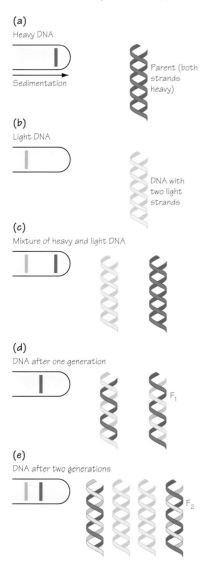

Figure 6.2 *The Meselson and Stahl experiment (1958) showed that DNA replication is semi-conservative. When Escherichia coli cells are provided with ammonia, they use nitrogen from the ammonia to make the bases (A, C, G, T) in DNA. If they are supplied with $^{15}NH_4^+$ instead of $^{14}NH_4^+$, the DNA produced is a little heavier, and 'light' and 'heavy' DNA may be separated by centrifugation through a gradient of caesium chloride. E. coli were grown for several generations in $^{15}NH_4^+$ to ensure that their DNA was all heavy [result shown in **(a)**], in comparison with what it would have been with 'normal' DNA **(b)**. When the cells with heavy DNA were switched to medium containing $^{14}NH_4^+$, after one generation all the DNA had an intermediate density **(d)**, between that of light and heavy. This showed that all the DNA was made up of one heavy and one light strand. The result after two generations in $^{14}NH_4^+$ is shown in **(e)**.*

DNA replication

Recall that the structure of DNA, as deduced by Watson and Crick, is a double helix in which the bases complement each other at the centre of the structure like the steps of a spiral staircase (Figure 6.1). Wherever there is an adenine (A) it is always paired with a thymine (T), and wherever there is a cytosine (C) it is paired with a guanine (G). The sequence of one strand of the DNA is thus *complementary* to the sequence of the other strand. (In computer terms, the sequence 1011001 is the complement of 0100110.) Although the two sequences are not *identical*, if you know one of them you can easily write the other. This suggests a method for replicating a DNA double helical model and, indeed, in Watson and Crick's report on their proposed structure for DNA, they had already noticed this possibility. "It has not escaped our notice", they wrote in 1953, "that the specific base pairing suggests a possible copying mechanism for the genetic material". On the face of it, it is really very simple; all you need to do is separate the two strands of DNA, and then on each of them build up a new, complementary strand by bringing in the individual base units to match those in the existing strand, and then linking them up. This gives two DNA double helices identical to the original one. In essence, this is exactly what does happen, although the enzyme-catalysed steps involved are rather complicated (see Spread 6.1). In 1958, Meselson and Stahl, in a very famous experiment using the ultracentrifuge, proved that this is what happens (Figure 6.2). The method of replication is called *semi-conservative* because, in every new DNA double helix, one of the two strands is conserved unchanged from the original double helix.

Three rather fundamental points need mentioning here. The first is the question of *energy*. Building up a new molecule of the polymer DNA from its monomer units obviously requires energy, but where does this come from? The free base units that drift into pair with bases on one of the existing strands of DNA are *deoxynucleoside triphosphates* (Figure 6.3). If we want to add an 'A' to the growing chain, we bring in dATP (not ATP because this is *deoxy*ribonucleic acid), which, like ATP, is a so-called high-energy compound. In an enzyme-catalysed reaction, two of the phosphates are hydrolysed, and the remaining adenine–deoxyribose–phosphate unit is joined to the growing chain. In other words, the energy released by hydrolysing two phosphate bonds is spent in adding one new base unit. The same applies if we wish to add C, G or T; we bring in dCTP, dGTP or dTTP respectively, all of which have free energy that is easily released like that of ATP. This is a direct illustration of the fact that energy is needed for the biosynthesis of macromolecules.

The second fundamental point is that, as you may remember, the DNA strands or molecules have 'direction'. Because of the way the phosphates link up to the deoxyribose sugar units, the chains go –phosphate-5′-deoxyribose-3′-phosphate–, where the 3′ and 5′ refer to the positions of the -OH groups on the sugar ring. In the DNA double helix, the two DNA strands are *antiparallel* (see Figure 6.4), written in shorthand form:

3′ ——————————— 5′

5′ ——————————— 3′

Now the triphosphate groups of dATP, dCTP, dGTP and dTTP are on the 5′ position of the sugar (see Figure 6.3). Consequently when a new base–sugar–phosphate unit is added to a growing DNA chain, it can only add to the 3′ position, and hence the chains always grow in the 5′→3′ direction. This causes a problem because if the double helix is opened out to form a replication fork (see Spread 6.1), then one new chain can form continuously, but the other cannot. It is interesting to see how Biology has solved this problem.

The third fundamental point about DNA replication is that it needs to be nearly perfect in its accuracy. Errors in putting in new bases in the sequence would corrupt the genetic message — and yet chemical reactions cannot be 100% perfect. When an 'A' should be added there is always a chance, albeit a small one, of picking up some other base unit. In order to protect against this happening, or rather to correct the error when it does occur, the enzyme responsible for DNA replication, *DNA polymerase*, also has a proofreading function. Putting in a wrong base distorts the double helix a little (see Spread 6.1). This distortion is detected by the enzyme, and the new unit is excised and the correct unit put in. This costs some energy, but is obviously worth doing, and an impressive level of accuracy — an error rate of 1 in 10^8 — is achieved. (Many organic chemistry reactions are regarded as pretty good if the yield of product is 70%; what we are talking about here is a yield of 99.999999%.)

However, this is not quite the end of the story, because a little inaccuracy is in a way desirable in order to produce *mutations*. The sequence of bases in DNA is a code that tells the cellular machinery how to make proteins; specifically, what order to join up amino acids into strings to make proteins. Put them together in one sequence and you get haemoglobin; put them in a different sequence and you get trypsin. Occasional changes in the DNA sequence can lead to the production of new proteins with different sequences. Most of these changes or mutations may be bad or even fatal to an organism, but some will be good, and new species will eventually result. In fact, mutations do not only arise during replication (they may also be induced by chemicals or by physical influences such as X-rays or ultraviolet light), but occur they do, and they represent the raw material of evolution. Some of the consequences of mutations will be discussed below.

DNA packaging

Most bacteria contain a single 'chromosome' — one circular double-helical DNA molecule (although they may also contain some smaller DNA circles called plasmids; see page 193). The *Escherichia coli* genome contains

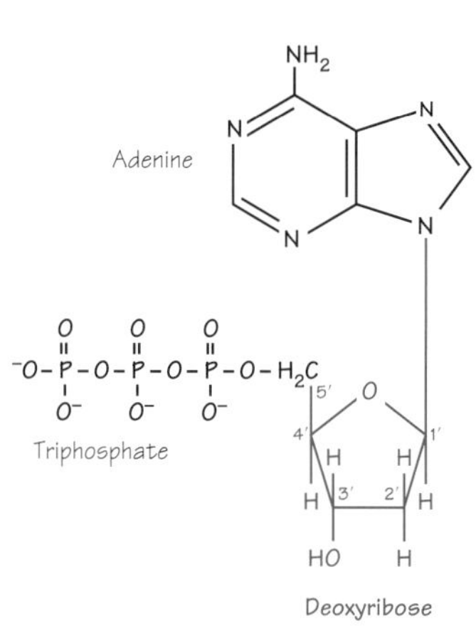

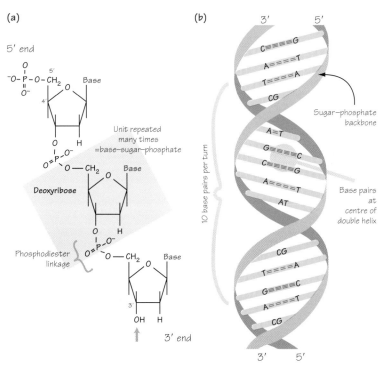

Figure 6.3 Structure of dATP (deoxyadenosine triphosphate). In nucleotides the carbon atoms in the sugar ring are labelled 1′, 2′, 3′, etc., so that the sugar in dATP is 2′-deoxyribose. How does dATP differ from ATP (refer to Spread 4.1)?

Figure 6.4 (a) The chemical structure of DNA shows that the molecule has a 'direction', i.e. the ends are different from one another. An incoming dATP (see Figure 6.3) can only link on to the 3′ end (OH, arrowed) so that when ATP is being synthesized the chain can only 'grow' in one direction. (b) In the DNA double helix the two molecules (strands) run in 'opposite' directions and are said to be antiparallel. This causes a problem when DNA is being replicated because the chain can only grow

Replication of DNA

objectives
- To describe how double-stranded DNA is replicated with high fidelity
- To explain that energy (ATP) is required to drive the necessary complex topological changes, in addition to the energy required for chemical synthesis (condensation)

HIGH FIDELITY COPYING OF DNA

When DNA is copied the error rate is better than one mistake in a billion bases. The DNA polymerase enzyme complex has a proofreading function for removing wrong bases and replacing them.

THE REPLICATION OF DNA begins at a replicating fork and then proceeds in opposite 5'→3' directions. On one strand of DNA, two or more DNA replication complexes can work at once in opposite directions to produce a 'replication bubble':

DNA HELICASE (rep protein) unwinds the double helix — this is a difficult procedure mechanically and is powered by the hydrolysis of ATP. About two ATP molecules are consumed for each base pair that is separated.

SS-BINDING PROTEIN stabilizes the single-stranded DNA so that the unwound DNA strand stays unwound and can serve as a template.

The LEADING STRAND of DNA is synthesized continuously in the 5'→3' direction (that is, using the unwound 3'→5' DNA strand as a template).

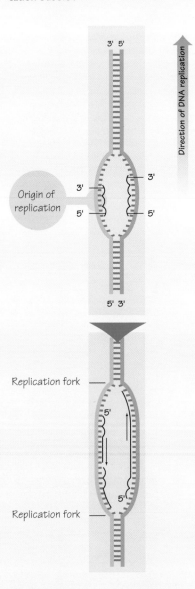

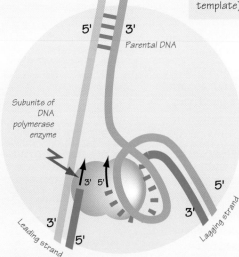

DNA polymerase probably works on both strands; this simple diagram shows the likely arrangement at the replication fork.

DNA POLYMERASE III is a template-dependent enzyme capable of synthesizing DNA in the 5'→3' direction so long as a 3'-OH group is available. This is the enzyme which synthesizes most of the new DNA — it requires the presence of a number of other protein subunits for full activity. The multisubunit complex is called DNA POLYMERASE III HOLOENZYME.

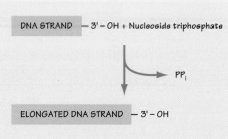

Life chemistry and molecular biology

6.1 Replication of DNA

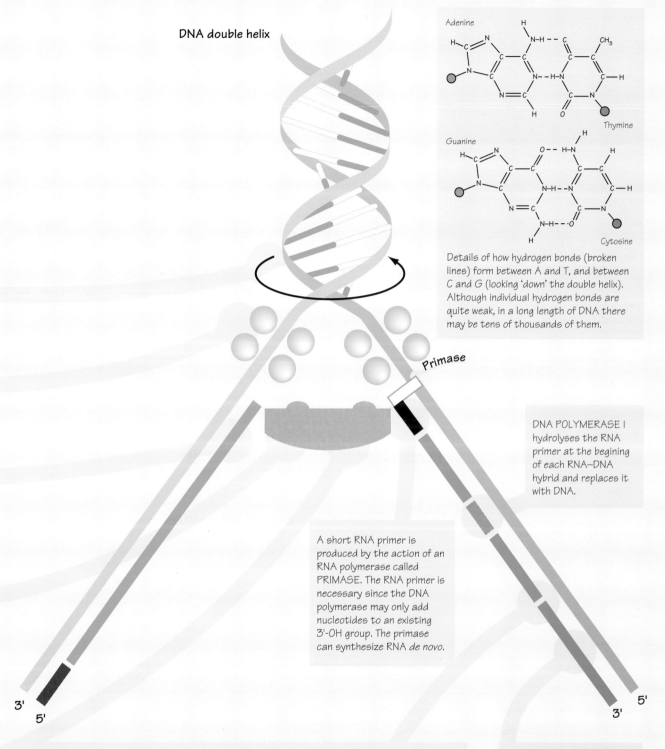

DNA double helix

Details of how hydrogen bonds (broken lines) form between A and T, and between C and G (looking 'down' the double helix). Although individual hydrogen bonds are quite weak, in a long length of DNA there may be tens of thousands of them.

Primase

DNA POLYMERASE I hydrolyses the RNA primer at the begining of each RNA–DNA hybrid and replaces it with DNA.

A short RNA primer is produced by the action of an RNA polymerase called PRIMASE. The RNA primer is necessary since the DNA polymerase may only add nucleotides to an existing 3'-OH group. The primase can synthesize RNA de novo.

DNA LIGASE joins the ends of adjacent DNA fragments, but only if the fragments are part of double-stranded DNA. This enzyme is thus responsible for the condensation of the Okazaki fragments into functional DNA.

The LAGGING STRAND is synthesized as a series of 5'→3' fragments (Okazaki fragments, named after the Japanese scientist who discovered them) using the unwound 5'→3' DNA strand, but in the direction 3'→5', i.e. towards the replicating fork.

Chapter 6 • **DNA: Dealing with Information**

4.5×10^6 nucleotides and is about 1.5 mm in length. The genome of *Haemophilus influenzae*, another bacterium, is 1.8×10^6 nucleotides long and its complete sequence was elucidated in 1995. These molecules of DNA are several hundred times the length of the bacterial cell (e.g. about 1 μm) and are coiled up in some way. This problem of packaging is an interesting one because the DNA must be made compact and yet it must also be accessible so that the code of the sequence of bases can be read when necessary. In bacteria, the DNA (negatively charged because of the phosphate groups) is associated with small, basic (positively charged) protein molecules, which allows the DNA to 'supercoil' into a compact volume.

In the case of eukaryotic cells, the problem is even more acute because the DNA in, say, any human cell is about two metres long and yet is packaged so that it fits into a nucleus 2–3 μm in diameter. Here, again, the code on the DNA must be accessible at the appropriate times in the cell cycle. It must be admitted that we do not understand very much about how this is achieved. The relationship between DNA, packaging and the chromosomes is described in Spread 6.2.

6.3 The genetic code

The sequence of bases (A, C, G, T) in DNA carries a code which specifies how amino acids should be joined together in strings to make proteins. It is now time to ask what this code is and how it is 'translated' to get the instructions to make proteins.

DNA is the genetic material

In the first half of this century people were unsure about how the genetic message was carried from generation to generation. It was known that proteins were rather complicated structures and it was commonly believed that DNA or RNA — just four bases, sugar and phosphate — were much too simple to carry the information to make proteins. [Computers, of course, had not been invented, and consequently most people did not think in terms of storing vast amounts of information digitally as strings of '1's and '0's as is now done (Figure 6.5).] However, if proteins carry the genetic message, it has to be asked what codes for the proteins that carry the message, and so on? It is a difficult question.

The answer started to be revealed in the 1930s when several experiments were carried out using bacteria and viruses, which showed that nucleic acid, at least in these organisms, carried the genetic information. The experiments with *Pneumococcus* carried out by Griffith (Figure 6.6) showed that the 'transforming principle', which would carry a genetic characteristic from one type of bacterial cell to a different type, was pure DNA. The famous Hershey and Chase experiment (Figure 6.7, page 170) showed that when a bacterial virus (or *bacteriophage*) infects a bacterial cell, it is the viruses' nucleic acid that carries the genetic information. In these two instances it was DNA, but with a different virus, tobacco mosaic virus, it was shown to be RNA that carried the information.

Figure 6.5 Digital information. (a) The bar code on supermarket items is 'read' using a laser, the width of the lines being translated into a binary code. The code is digital, i.e. either 'white' or 'black'. (b) Digital information as a string of '1's and '0's has its complement. The two strings of information are not identical, but if you know one you can predict the other with 100% accuracy. (c) How the digital information is stored in DNA. Instead of '1' and '0' we now have A, C, G and T, but the same rule about complementarity applies.

(a)

0 245790 000322

(b) 1011 100100110

0100 011011001

(c) C G C T A G A T T G

G C G A T C T A A C

Gradually it became accepted that nucleic acids could carry the genetic message, and, later, evidence accumulated that in eukaryotic cells DNA was always the carrier. For example, chemicals that affected DNA caused mutations, but things that affected protein did not. Today, we know so much about DNA and its structure and function that there is absolutely no doubt that DNA is the genetic material. Every day scientists in laboratories all over the world transfer bits of DNA from one cell to another and in doing so transfer genetic information.

Discovering the code

The discovery of the DNA double helix by Watson and Crick in 1953 effectively fired the starting pistol for a race to discover the nature of the genetic code: how amino acid order in a polypeptide was specified, how the 'letters' in the code were read and how polypeptide synthesis occurred. The 1950s and early 1960s were extremely active times for biochemical researchers trying to find answers to these questions.

One important discovery was the sorting out of the roles of DNA and RNA. The DNA in the nucleus carried the information, but protein synthesis took place in the cytosol. This meant that the message was first *transcribed* from DNA into RNA (messenger RNA; mRNA), and it was this message that was *translated* into proteins. Since both RNA and DNA have a base sequence this is not difficult to grasp. It did, however, allow experiments to be carried out to find out what the code was. Since there are only four bases and 20 amino acids, codes of one base, or even two bases, are insufficient. The code for one amino acid has to be at least a three-base or three-letter code (and it could be more). A three-letter code actually gives 64 possibilities, which is more than enough, and some codes could be redundant, or used for the same amino acid, which is called *degeneracy*. There should also presumably be codes for START and STOP since the DNA in the cell is an unbroken sequence coding for many polypeptides.

Knowledge about RNA enabled some simple experiments to be done that started to reveal the genetic code. A preparation of cytosol (i.e. excluding the nuclei) was made from cells, and, when mRNA was added to this, protein was produced. This was easy to follow because, if radioactive amino acids were added, then radioactive protein resulted. At the time it was not possible to get hold of, or make, mRNA of known base sequence, but it was possible to make RNA with the sequence AAAA...., or UUUU...., etc. When these were used, the code started to reveal itself. Thus when RNA of the sequence UUUU... was added to this system, under rather artificial conditions, a polypeptide of the sequence Phe-Phe-Phe-Phe... was produced (Figure 6.8, page 171). Thus the code for phenylalanine must be UUU (or UUUU, etc.); this was the first code word to be broken. (UUU obviously corresponds to TTT in DNA.)

After this a great many experiments along these lines were carried out, some of them rather complicated, that gradually revealed all 64 *codons* of the *triplet code* that is so well known today. The code is indeed degenerate: there is one START signal, AUG, that confusingly also codes for the amino acid methionine, and three different STOP, or chain

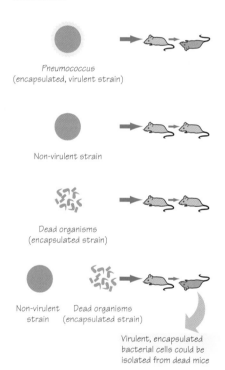

Figure 6.6 The experiment with *Pneumococcus* bacteria carried out by Griffith in 1928. Two strains of the bacterium are known, virulent and non-virulent. The virulent strain has a polysaccharide capsule and, when injected into mice, causes death from pneumonia. When living, non-virulent bacteria and heat-killed virulent bacteria (which were harmless to the mice) were injected together, the mice died. It was possible to isolate living, virulent organisms from the dead mice. The non-virulent bacteria had acquired a new genetic character — the ability to synthesize the polysaccharide capsule which causes virulence. It was shown subsequently that this transfer of genetic information required that the non-virulent bacteria strain took up DNA (and only DNA) from the heat-killed virulent bacterial cells, thus showing that DNA carries the genetic information.

Packaging of DNA

objectives
- To explain why DNA molecules in cells need to be carefully packaged
- To describe how DNA is folded in the nucleoid of prokaryotes
- To outline how the DNA of eukaryotic cells is packaged at a series of levels in order to fit it into the cell nucleus

The genetic information of organisms is stored as the sequences of bases in DNA. In prokaryotes, the DNA occurs as circular molecules, whereas the chromosomes of eukaryotic cells contain linear DNA molecules.

It is important that the DNA molecules must be packaged very tightly in the cell. Bacterial chromosomal DNA circles are typically 300–400 μm in diameter, whereas bacterial cells are only 1–2 μm in size. A human diploid cell has 46 chromosomes, each containing a single DNA molecule. The combined length is over 2 metres, contained within a nucleus only 5 μm in diameter!

Despite their lengths, DNA molecules are only 2 nm in diameter. Long thin structures are easily snapped, and so careful packaging is required to protect the DNA from shearing forces. The DNA must also remain accessible to the proteins and enzymes involved in replication and transcription.

The nucleoid of the bacterium *Escherichia coli* occupies a space within the cell relatively free of ribosomes. Its chromosome contains a single circular DNA double helix, which is folded into a series of 50–100 loops. Each loop consists of coils of the double-stranded DNA twisted around each other. This is called SUPERCOILING; such an arrangement greatly reduces the overall length of the DNA, thereby enabling it to be packaged in the cell. The loops are stabilized by the attachment of small basic proteins. These proteins interact with the negative charges on the phosphate groups of the DNA.

The coiling of DNA in eukaryotic chromosomes is even more complex than in prokaryotes. It occurs by a series of foldings, each of which produces a fatter but shorter structure. This is called CHROMATIN. The folding requires the presence of basic proteins called HISTONES. The specific arrangements of foldings are not fully understood, but the DNA is known to be wound around groups of histones to generate a string of bead-like structures called NUCLEOSOMES. The nucleosomes are then condensed into a tightly wound spiral (see diagram opposite) to form a fibre about 30 nm in diameter which is then elaborately folded into further loops and coils to form the highly compacted chromatin.

DNA containing genes which the cell is actively using is relatively loosely folded and is called EUCHROMATIN. In contrast, DNA which is not being used is folded into a very tight arrangement called HETEROCHROMATIN. The most highly folded chromosomes — the mitotic and meiotic chromosomes — are produced prior to cell division.

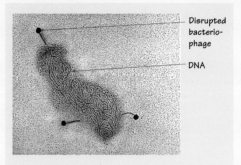

Some idea of the extent of packaging of DNA can be gauged from the lengths of these bacteriophage DNA molecules, which have been released by disrupting the viral particles.
(Courtesy of V. Virrankoski-Castrodeza, University of Leeds.)

PACKAGING IN PROKARYOTES

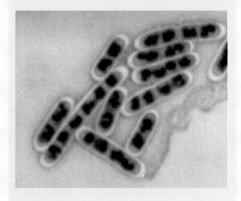

The nucleoids of these stained Bacilli show clearly as darker areas.
(Courtesy of C.F. Robinow, The University of Western Ontario, Canada.)

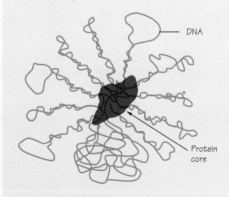

Schematic of an isolated bacterial chromosome. For the sake of clarity the double-stranded DNA is represented as a single line, only a few of the 100 or so loops of supercoiled DNA are shown and the proteins which stabilize the assembly have been omitted.
[Redrawn with permission from Kellenberger, E. (1990) The Bacterial Chromosome (Drlica, K. and Riley, M., eds.), The American Society for Microbiology.]

6.2 Packaging of DNA

PACKAGING IN EUKARYOTES

Nuclei in rat seminiferous tubules (left; magnification × 800) and young pea cotyledons (right; magnification × 600). (Courtesy of D.G. Robinson, Pflanzenphysiologisches Institut der Universität Göttingen, Germany.) The DNA is complexed to form darkly staining material called chromatin. The less darkly stained euchromatin is genetically active, while the most darkly stained heterochromatin is genetically inert.

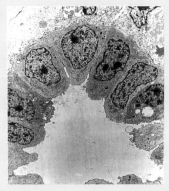
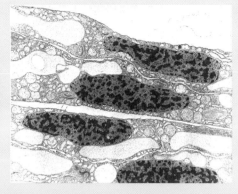

SCHEMATIC MODEL OF CHROMOSOME PACKAGING

DNA double helix — 2 nm

forms

'Beads-on-a-string' nucleosomes — 11 nm

to give

Chromatin fibre of spirally packed nucleosomes — 30 nm

which folds up to form loops

Section of chromosome — 300 nm

Condensed section of metaphase chromosome — 700 nm

Entire metaphase chromosome — 1400 nm

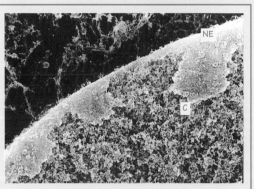

The fibrous nature of chromatin (C) can be observed through the damaged nuclear envelope (NE) of this cell. Magnification × 10,300. (Courtesy of T.A. Allen, Paterson Institute, Manchester.)

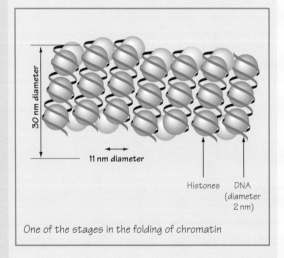

30 nm diameter; 11 nm diameter; Histones; DNA (diameter 2 nm)

One of the stages in the folding of chromatin

Chapter 6 · DNA: Dealing with Information

Figure 6.7 *The Hershey and Chase experiment (1952). It is known that when a bacteriophage infects a bacterial cell its protein coat remains on the outside of the bacterial cell while the DNA is 'injected' into the bacterium.* **(a)** *By growing Escherichia coli in medium containing either radioactive phosphate or radioactive sulphur, either the bacteriophage DNA or its protein could be made radioactive (DNA contains no sulphur and protein contains no phosphate).* **(b)** *When bacterial cells were infected with bacteriophage containing radioactivity in the DNA, and then the bacteriophage and bacterial cells broken apart by shaking them up vigorously in a kitchen blender, it was found that the radioactivity (i.e. the DNA) was now inside the bacteria. These bacterial cells, of course, have new genetic information — the information to make new virus particles. When the experiment was carried out with the bacteriophage protein made radioactive, no radioactivity was found in the bacterial cells, showing that it is only bacteriophage DNA that enters the bacterial cell.* **(c)** *Electron micrograph of bacteriophages. Magnification* × *850,000.*

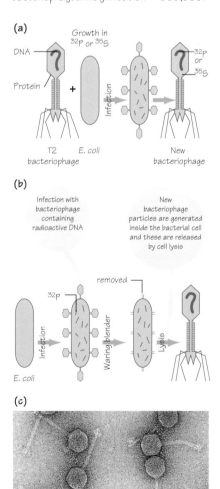

termination, signals. The start, AUG, signal is a little confusing, but it only signals START when the sequence of bases before it ('upstream') is correct; otherwise it signals that methionine should be incorporated.

Incidentally, the experiment using UUU... (or 'poly U') as message should not have worked, because there is no START signal. This is why slightly artificial conditions had to be used (e.g. high Mg^{2+} concentration). Nevertheless, the results obtained were still valid. The genetic code is shown in Figure 6.9 (page 176); significantly, this code is essentially universal in all life on Earth, with a couple of very small exceptions. People believe that all life originated from one common form that had evolved this coding mechanism.

What is a gene?

Geneticists talk about *genes* as the units of inheritance. Our purpose here is to ask, in biochemical terms, what is a gene? A series of experiments, quite a long time ago, established the 'one gene–one enzyme' hypothesis, and this concept is essentially correct. A gene is a length of DNA with a code in the bases that specifies one polypeptide. (The change from 'enzyme' to 'polypeptide' recognizes that some enzymes have more than one subunit polypeptide, and that not all polypeptides are enzymes. A case in point is haemoglobin, which has two α and two β polypeptides; these are encoded by two separate genes.)

In biochemical terms we can see a gene as a length of DNA in which the base sequence is 'read' in a triplet code. We start off on the right foot with the AUG (in mRNA) meaning START, and, because the code has no punctuation, and nor does it overlap, we just read along in triplets. The START keeps us in the right reading frame, and each triplet is translated into the appropriate amino acid in sequence, until we reach STOP. Exactly how the mechanics of all this work will be revealed later; it is sufficient for now to understand the relationship between gene and polypeptide.

In addition to this coding sequence, most, if not all, genes have additional sequences of DNA associated with them. These are 'upstream', i.e. before the gene in the direction of reading, sometimes quite a long way upstream. These are regions that switch genes on or off; some of them are called *promoters*. The idea of genes being turned on or off ('on' means producing mRNA and hence proteins) was first formed when it was found that in bacteria some proteins can be *induced* and others *repressed* (see Spread 6.3) by various environmental factors. In eukaryotic organisms this control is even more sophisticated. Although all the somatic cells of the body have a full complement of genes (or DNA), most of these genes are turned off most of the time. Only red blood cells make haemoglobin, only islets of Langerhans cells make insulin, yet all cells make all the enzymes of glycolysis. Furthermore, embryonic cells make totally different proteins compared with adult cells. Gene expression is at least partly controlled by hormones but there is a great deal that remains to be discovered about the control of gene expression. It would be fascinating to know which genes control cell proliferation (and could cause cancer) and which genes control the process of cell suicide, called *apoptosis* or programmed cell death, that might be triggered to cause a tumour cell to self-destruct.

The manipulation of genes and DNA has now reached a very sophisticated level (see Chapter 7). Bits of genes can be stitched together, genes can be transferred between different cells, synthetic bits of DNA can be manufactured. This so-called *genetic engineering* is having a profound effect on how research is carried out to find out about how organisms work, and it also has enormous commercial possibilities. Over the next few years we are going to hear about the first successful gene therapy trials, and gene therapy is likely to be an important medical treatment in the next century.

Mutations

Now that we have a fairly good idea of the chemical nature of a gene (remember that Biochemistry is the study of the chemistry of living things and biological phenomena), and how the genetic code works, we can ask more precisely what *mutations* are. Mutations have been known for a long time to geneticists as subtle changes in organisms that may be disadvantageous, but equally may lead to the development of variations and eventually new species. In human terms, many diseases are genetic diseases — sickle cell anaemia, brittle bone disease, cystic fibrosis — that are the result of mutations having occurred.

What we may observe in a mutant organism is a change in a gene product, i.e. a protein, that either does not work properly or is not produced at all. The classic example is the mutation occurring in sickle cell disease (see Spread 6.4). In this condition, the haemoglobin is changed such that it precipitates and forms fibres at low oxygen tensions (Figure 6.10, page 176). This occurs as the erythrocytes pass through the tiny capillaries in the tissues and the consequence is that the capillaries become blocked and red cells lyse. The blockages prevent oxygen being supplied to the tissues properly and the destruction of red cells causes anaemia. This anaemia, of course, exacerbates the oxygen delivery problem.

Analysis of the haemoglobin isolated from people with sickle cell disease shows that sickle cell haemoglobin differs in only one respect from normal adult haemoglobin (HbA): it has a valine residue in the β-chain at position number 6 where HbA has a glutamic acid. This happens to be on the outside of the molecule and the unfortunate consequence is that when the haemoglobin becomes deoxygenated and changes its overall shape a little (Spread 6.4), this hydrophobic valine acts like a 'sticky patch' so that the deoxyhaemoglobin molecules stick together (Figure 6.11, page 177) with the disastrous consequences mentioned above. How this mutation was discovered is explained in Spread 6.4 (page 174).

Now let us try to understand the mutation at the level of the gene. In the sickle cell haemoglobin, Glu is replaced by Val. If we look at the genetic code (Figure 6.9, page 176) we see that the code for Glu is GAA or GAG, and that for Val is GUA, GUC, GUG or GUU. In fact, out of these possibilities, the change is from GAG to GUG, or, in the DNA, to GTG — a single base change. This is called a *point mutation* and many such

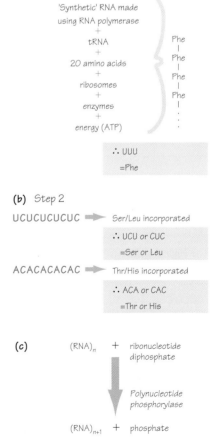

Figure 6.8 Cracking the genetic code. (a) It was possible to make an artificial mRNA with a repeating sequence UUU... using the enzyme RNA polymerase which does not require a DNA template. When this was added to a protein-synthesizing system a polypeptide of sequence Phe-Phe-Phe... was generated, showing that the code for Phe (in mRNA) is UUU. This experiment should not have worked because there is no START codon (AUG). In fact an abnormally high concentration of Mg^{2+} ions was added to get the synthesizing system started. Part (b) shows the next step, using a more complicated sequence. It was not possible at this time to create mRNA molecules of known sequence other than the very simple repeats shown here. However, a process of elimination soon started to reveal the code. (c) The enzyme polynucleotide phosphorylase enables RNA molecules of known repeating sequence (e.g. UUUU...) to be synthesized. This enzyme does not require a DNA template and indeed its function is normally to break down RNA (i.e. the opposite to that shown here). However, if high concentrations of the nucleoside diphosphate (e.g. UDP) are supplied then RNA is synthesized. Supplying ADP would produce AAAA... and so on.

Regulation of gene activity

objectives
- To explain why some enzymes are induced while others are repressed
- To describe how induction of enzymes works

LACTOSE IS AN INDUCER:
THE DNA of every cell contains the genes coding for all the proteins that the cell can manufacture. However, not all the genes are expressed at any given time and indeed most are not expressed. To take an extreme example, only immature red blood cells make haemoglobin; no other cell in the body expresses the genes for the haemoglobin protein subunits.

The first indication of how genes are controlled, i.e. switched on and switched off, came from work with bacteria.

JACOB and MONOD proposed a hypothesis to explain the control of gene expression in bacteria. They experimented on the gut bacterium, *Escherichia coli*, by measuring the activity of enzymes concerned with the metabolism of lactose, namely:

1. β-galactosidase — hydrolyses lactase to glucose plus galactose,
2. β-galactoside permease,
3. thiogalactoside transacetylase.

} control the uptake and metabolism of lactose

Jacob and Monod found that:

(i) when no lactose is available in the growth medium, less than 10 molecules of each enzyme protein are present per cell;
(ii) when lactose is available in the medium, 3000–5000 molecules of each enzyme are present per cell and that ALL THREE PROTEINS ARE SYNTHESIZED AT THE SAME RATE.

This makes sense economically, and bacteria live in a very competitive environment so that any economy may have a survival advantage. Thus when there is no lactose present in the medium, it would be a waste of amino acids and energy to make the enzyme β-galactosidase. It is much better to manufacture the enzyme only when it is needed. Hence β-galactosidase is an INDUCED enzyme. In fact the three proteins of the pathway for lactose utilization are controlled simultaneously.

Some MUTANTS of *E. coli* make large amounts of all three proteins even in the absence of lactose; these are CONSTITUTIVE mutants, and Jacob and Monod proposed that the mutation had occurred in a REGULATOR GENE.

The regulator controls the synthesis of a SIGNAL MOLECULE (a protein) called the *lac* REPRESSOR, which 'switches' off the action of the structural genes that code for the synthesis of the three protein molecules.

Other *E. coli* mutants are unable to receive this signal from the repressor; these were found to have a mutation very close to the structural gene for β-galactosidase. The mutated region is called the OPERATOR GENE.

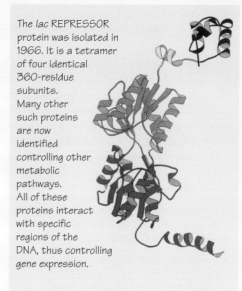

The *lac* REPRESSOR protein was isolated in 1966. It is a tetramer of four identical 360-residue subunits. Many other such proteins are now identified controlling other metabolic pathways. All of these proteins interact with specific regions of the DNA, thus controlling gene expression.

HOW GENE EXPRESSION IS CONTROLLED IN EUKARYOTIC CELLS
The situation in eukaryotic cells is much more complicated than in prokaryotic cells, and much less is known about how eukaryotic genes are regulated. Recently it has been found that some genes are controlled by steroid hormones (e.g. androgens, oestrogens, glucocorticoids) as well as by vitamins A and D. When these hormones/vitamins reach the interior of the cell they combine with receptor proteins which have two sites on them:

(i) a site that can bind the hormone/vitamin, and
(ii) a site that can interact with DNA in the nucleus when a hormone/vitamin molecule is bound.

This causes genes to be switched on or switched off. These receptor proteins have a so-called 'zinc finger' structure, producing loops of polypeptide that can attach to specific places in the grooves of the DNA.

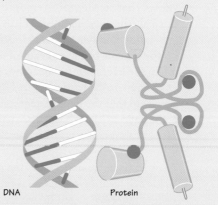

Schematic diagram showing how a zinc finger protein might interact with the DNA double helix.

Life chemistry and molecular biology

6.3 Regulation of gene activity

HOW INDUCTION OF ENZYMES WORKS IN BACTERIA

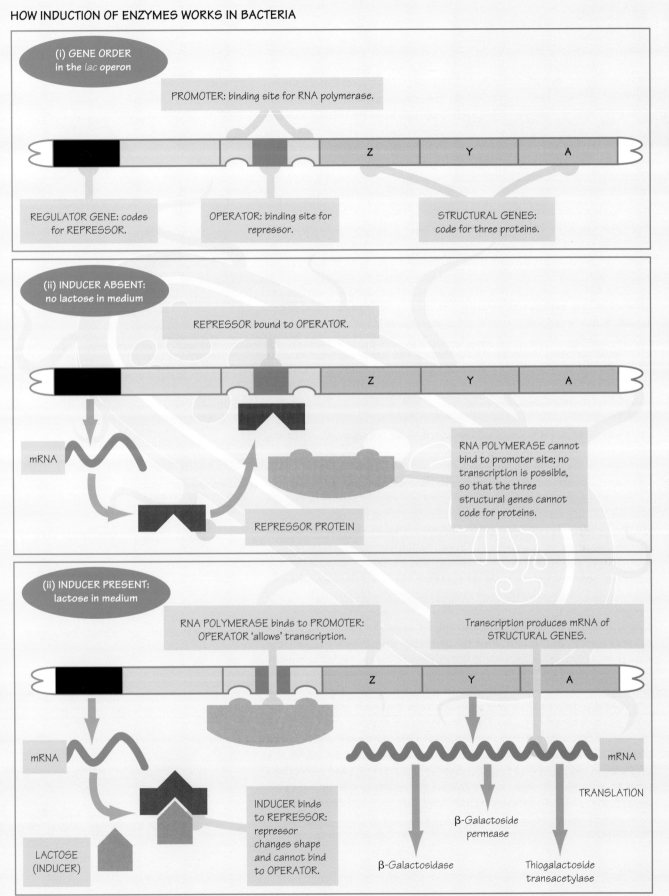

Mutations produce sickle cell disease

objectives
- To explain how a point mutation produces an abnormal protein
- To understand how the mutation for sickle cell disease was detected by chromatography and electrophoresis
- To appreciate the resistance to malaria produced by the sickle cell trait

THE DISCOVERY OF SICKLE CELL HAEMOGLOBIN:
SICKLE CELL ANAEMIA is known as a 'molecular disease', that is to say, one in which there is a clearly identified molecular deficiency responsible for the observed clinical symptoms. We know now that the condition is caused by a point mutation resulting in a form of haemoglobin (HbS) that differs from the normal adult form (HbA), and has specific properties that produce the disease symptoms. However, it should be remembered that, when sickle cell anaemia was first described, the structure of proteins had not been unravelled; nor was it known that DNA carries the genetic message.

In 1904 a Chicago doctor, James Herrick, described symptoms of anaemia in a 20-year-old black student (the symptoms are listed in the Table opposite). In particular, Herrick noted that the shape of the patient's red blood cells was irregular with many of them having 'sickle'- or crescent-shaped forms. Many other identical cases were described subsequently. The disease was subsequently found to be inheritable and was characteristic of individuals of black African origin. There was no treatment available at that time, and there is no cure today other than treatment for relief of the symptoms.

Many years later, in 1949, American chemist and Nobel prizewinner Linus Pauling carried out electrophoresis of sickle cell haemoglobin in comparison with normal adult haemoglobin. Electrophoresis involves subjecting proteins in solution (usually on an inert support such as filter paper) to an electric field. Proteins have a charge on them due to their amino acid side chains and move to the positive or negative electrodes. The greater the charge, the greater the rate of movement or 'mobility'. At pH 8.6 sickle cell haemoglobin moved more slowly towards the positive electrode than normal haemoglobin. Thus sickle cell haemoglobin is more negatively charged than HbA at this pH. Individuals who are heterozygous for the condition (i.e. have inherited the sickle cell gene from one parent but have received a normal gene from the other) had two haemoglobins (HbA and HbS).

Shortly after this, Vernon Ingram in Cambridge tracked down the defect in sickle cell haemoglobin. At that time it was not possible to determine the sequence of large proteins such as haemoglobin, so Ingram digested the haemoglobins with the digestive enzyme TRYPSIN. This produced many small peptides and these were separated from one another. This was quite difficult to achieve, but eventually it was found that if the peptides were separated in one dimension on a piece of filter paper by electrophoresis (i.e. by differences in charge) and then at 90° to this by chromatography, a pattern of spots could be seen after staining. The patterns for HbA and HbS differed in only one spot and eventually this was tracked down to a difference in one amino acid residue: valine (no charge) in HbS had been substituted for glutamic acid (one negative charge) in HbA. Much later on, when the genetic code and DNA sequences became known, this could be related to a single base change (mutation) in the DNA coding for the β-chain of sickle cell haemoglobin.

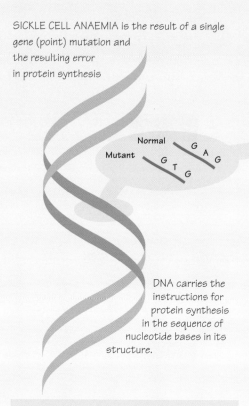

SICKLE CELL ANAEMIA is the result of a single gene (point) mutation and the resulting error in protein synthesis

DNA carries the instructions for protein synthesis in the sequence of nucleotide bases in its structure.

DETECTION OF THE MUTATION:

ELECTROPHORESIS (separation by charge) at pH 8.6. Sickle cell haemoglobin moves more slowly towards the anode. Can you explain why?

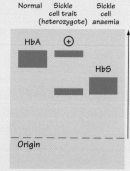

PAPER CHROMATOGRAPHY ELECTROPHORESIS
'Fingerprints' of tryptic peptides of haemoglobins (electrophoresis in one dimension; chromatography in the other dimension on paper; ninhydrin staining).

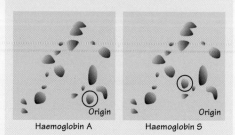

Only one tryptic peptide is different, corresponding to this mutation.

6.4 Mutations produce sickle cell disease

A MUTATION is a change in the DNA. In this case a BASE SUBSTITUTION has taken place — the base A (adenine) has replaced the base T (thymine) so that the triplet codeword CAT has replaced the normal triplet codeword CTT.

In mRNA, GUA instead of GAA gives valine (Val) instead of glutamic acid (Glu).

The haemoglobin molecule has four polypeptide chains: 2 × α and 2 × β. In sickle cell haemoglobin the mutation is at position 6 of the β chain.

Haemoglobin A	Val	His	Leu	Thr	Pro	Glu	Glu	Lys
Haemoglobin S	Val	His	Leu	Thr	Pro	Val	Glu	Lys
β chain	1	2	3	4	5	6	7	8

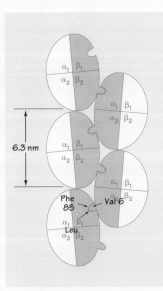

The valine side chain is non-polar whereas the glutamate side chain is highly polar: the presence of valine makes haemoglobin S much less soluble than haemoglobin, causing the molecules to stack to form fibres which distort the red blood cells. This produces the following symptoms.

- Rapid destruction of red cells: anaemia, shortage of breath, lethargy and bone marrow overactivity, repeated infections.
- Clumping of cells causing interference with circulation including right-sided heart failure.
- Collection of red cells in spleen causing spleen enlargement and fibrosis.

SYMPTOMS OF SICKLE CELL DISEASE PATIENTS

Cough, fever, weak and dizzy, headache, palpitations, shortness of breath. Crises in which these become worse. Sores on legs which are slow to heal.

Mucous membranes pale, tinge of yellow in whites of eyes. Lymph nodes enlarged. Heart enlarged, murmur present. Fatal often before age 30 due to infection, renal failure, cardiac failure or thrombosis.

	Sickle cell	Normal range
Red cell count	2.6×10^6 cm^{-3}	$(4.6–6.2) \times 10^6$ cm^{-3}
Haemoglobin content	8 g/100 cm^3	14–18 g/100 cm^3

Blood histology: sickle-shaped red cells (see Figure 6.10), many smaller than normal red cells; presence of many nucleated red cells (reticulocytes) compared with normal.

[Note: heterozygotes show these symptoms to a much lesser extent.]

HETEROZYGOTES (sickle cell trait) have both HbA and HbS and are not seriously affected by the disease.

In areas in which MALARIA is common:

HOMOZYGOUS NORMAL: HbA HbA
no sickle cell symptoms, but
PRONE TO MALARIA

HOMOZYGOUS ABNORMAL: HbS HbS
early death from
SEVERE SICKLE CELL SYMPTOMS

BUT

HETEROZYGOUS: HbA HbS
mild sickle cell symptoms but
PROTECTED FROM MALARIA
(Plasmodium cannot infect sickle cells)

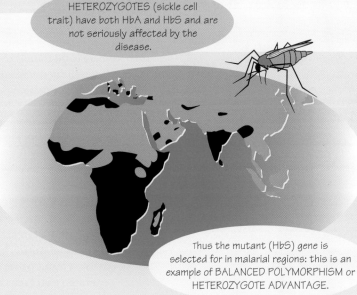

Thus the mutant (HbS) gene is selected for in malarial regions: this is an example of BALANCED POLYMORPHISM or HETEROZYGOTE ADVANTAGE.

A map indicating the region of the world where malaria caused by *Plasmodium falciparum* was prevalent before 1930, together with the distribution of the sickle-cell gene.

Figure 6.9 The genetic code shown in an unconventional way as a 'rising sun' diagram. Triplet codes (in mRNA) are read starting at the centre: AUG means Met or 'START'. The code is degenerate: only Met and Trp have a single codon each, and Ser and Leu have six codons each. Chemically similar amino acids often have similar codons. Thus a single base substitution (mutation) converts any of the serine codons into a threonine codon. Codons that have U in their second position invariably code for a hydrophobic amino acid.

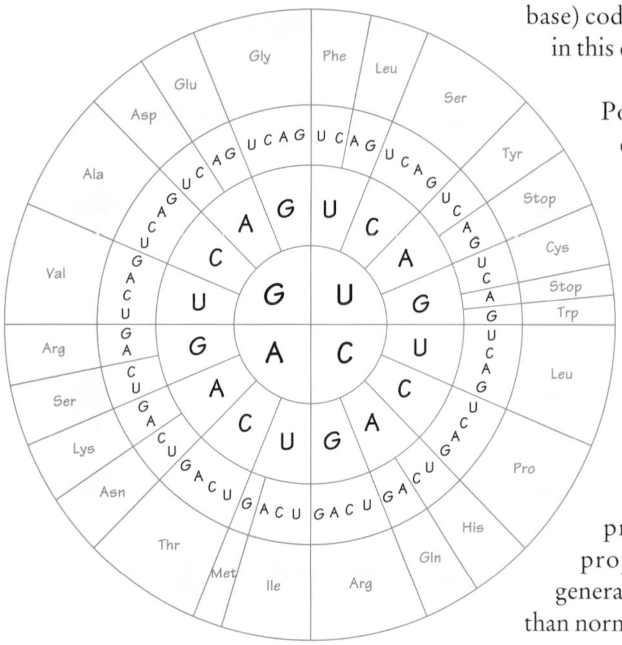

Figure 6.10 Erythrocytes in sickle cell anaemia. Magnification × 1000.

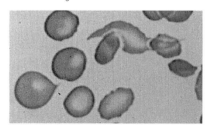

mutations are known. Although changing an A to a T may seem rather trivial in 3×10^9 bases, the consequences for an individual with sickle cell disease are pretty serious. Individuals that have received the mutant gene from only one parent, the heterozygotes, can produce normal β-chain as well and are not so seriously affected. Homozygotes, who have the faulty gene from both parents, have a serious medical condition. With knowledge of the structure of the gene and the protein we have a complete understanding of sickle cell anaemia at the molecular level.

Point mutations are common, although you will see by looking at the genetic code (Figure 6.9) that many mutations will have no effect at the protein level because the code is degenerate. Since GGN (where N is any base) codes for glycine, mutations in the third code letter are irrelevant in this case, for example.

Point mutations in START or STOP codons may have serious consequences. A mutation in AUG at the start of the mRNA may mean that the message will not be translated, and no protein will be produced. If this is a vital protein, and the condition is homozygous, then not only is no protein produced but the organism may not survive even to the end of the embryonic stage. Such fatal mutations tend to disappear from a population because the individuals carrying them do not reach reproductive age to pass them on. However, the heterozygotes are likely to survive to do this.

Mutations in STOP codons may lead to a larger than normal protein being produced, which may or may not function properly. Another possibility is that a STOP codon may be generated in mid-chain as a result of a mutation In this case, a shorter than normal protein will be generated which may not function properly.

In much the same way, adding or subtracting one base from a sequence may have dire consequences because then all the other codons will be out of phase (known as a *frame-shift mutation*) and be read as rubbish.

In addition to point mutations, much bigger mutations are known where large chunks of DNA are lost, put in reverse or placed somewhere else in the genome with more or less disastrous consequences (Figure 6.12). All of these types of mutations are known, and since determining the sequence of DNA is fairly easy these days, if the gene can be identified then the mutation can be characterized. The challenge, in the next century, will be to identify such faulty genes and replace them (see Chapter 7) using gene therapy.

Mutants as tools

In the past, mutants, often of micro-organisms because they are easy to generate and isolate, have been used as tools to find out about metabolic pathways and other biological phenomena. If an organism can be generated that lacks one of the enzymes of a metabolic pathway the effect is the same as having a specific inhibitor for that enzyme-catalysed step (see Figure 6.13). In such an organism the metabolic intermediate before the blocked step will accumulate and may be identified. Using a selection of mutants blocked in different steps, a pathway can be elucidated.

This same technique, in a more sophisticated form, is at present a very important molecular biology technique for finding out about how molecules such as proteins work. If the sequence of the gene coding for a protein is known, then a technique known as *site-directed mutagenesis* may be used to change bases in the DNA, and hence amino acid residues in the protein, at will. One can go through a polypeptide sequence, changing residues one by one. If amino acids vital for the biological activity of the polypeptide are changed, then it will lose its activity. This is a very powerful technique and has been called *protein engineering*. This technique can potentially produce new proteins with new properties.

6.4 Protein biosynthesis

Now we come to the complicated process by which proteins are made; i.e. how the information carried in the sequence of bases in the DNA is used to direct the synthesis of polypeptides. It will be recalled that in eukaryotic cells the DNA is in the nucleus (and stays there) and that protein biosynthesis takes place in the cytosol at the ribosomes. Protein produced may stay in the cytosol but it may also be moved into the mitochondria or other cellular organelles, or be exported from the cell. There have to be mechanisms for doing this. We will deal with this process in three sections: (1) how mRNA is produced, (2) how mRNA directs polypeptide synthesis, and (3) how proteins are targeted to specific locations and modified after they have been manufactured.

Messenger RNA (mRNA) production (transcription)

As was mentioned above, by no means all of the DNA (i.e. all the genes) is translated at once. In fact, only a very small number of genes are active at any one time. We therefore have to envisage the process whereby sections of DNA (i.e. genes) are somehow exposed and activated so that their genetic message is transcribed into mRNA. We say 'transcribed' because the genetic information in DNA is 're-written' in the mRNA sequence and the process is referred to as *transcription*.

The 'copying' process (transcription) is carried out by enzymes called *RNA polymerases*. They require a length of DNA that is somehow identified as being ready for transcription, and, because this synthesis of RNA is an energy-requiring process, the four ribonucleotide triphosphates, ATP, CTP, GTP and UTP. The relevant stretch of DNA is unwound (and may be referred to as the *template*; it contains the information) and the ribonucleotides assemble in the sequence dictated by the sequence of bases in the DNA. Only one of the strands of the DNA double helix is 'read' and this is called the *transcribing strand*. There are sequence signals at the start of the stretch of DNA to be

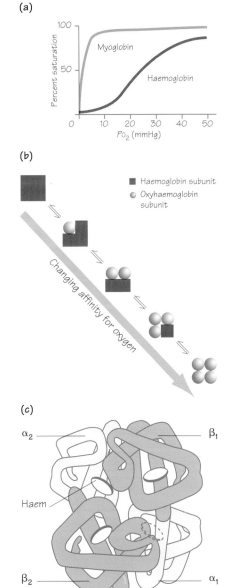

Figure 6.11 (a) The oxygen-binding curve of haemoglobin, which is a tetrameric protein, is sigmoidal, while that of myoglobin, a monomer, is hyperbolic. (b) In haemoglobin the subunit interactions mean that the binding affinity for oxygen changes as oxygen molecules successively bind. Haemoglobin is said to be allosteric. (c) Simplified diagram to show how the four subunits of haemoglobin ($\alpha_2\beta_2$) fit together very snugly. When one changes its shape a little upon binding an oxygen molecule, the other subunits change a little too, increasing the affinity with which they bind oxygen (cooperative effect).

Figure 6.12 The effects of mutations. The normal sequence (triplet code) is shown in (a), but if one base is substituted [a point mutation; (b)] we may get a protein with different properties. If the change generates a STOP codon then we may get a shorter polypeptide or no polypeptide at all. Frameshifts [(c) and (d)] create nonsense. Of course if they occur near to the end of a polypeptide they may have less effect, but are likely to wreck the reading of the STOP codon, producing a longer polypeptide. Other mutations involve transpositions or inversions of larger chunks of code.

(a) THE BIG DOG CAN RUN

(b) THE BIG DOE CAN RUN
Point mutation

(c) TEB IGD OGC ANR UN-
Frameshift (deletion)

(d) THE BIG DOO GCA NRU
Frameshift (insertion)

Figure 6.13 Mutations can help in the elucidation of metabolic pathways. In the Figure, a mutation has caused enzyme c to be inactive or absent; consequently compound C accumulates. (Normally metabolic intermediates are only present in tiny quantities.) Compound C may then be isolated and identified as an intermediate in the pathway. Repeating this experiment with another mutant, say in enzyme d, will allow another intermediate (D) to be identified. This method works well in bacteria where it is easy to make a range of different mutants at will by using chemical mutagens or ultraviolet radiation.

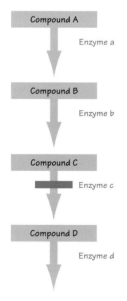

transcribed that are recognized by the RNA polymerase so that it knows what to do.

Although RNA polymerase works with high fidelity in the copying process, it is nowhere near as accurate as DNA polymerase. The occasional mistake can be tolerated, although not too many of them. Such mistakes do not have the same genetic consequences as those occurring during DNA duplication. The error is not passed on to the next generation

The use of radioactive tracers has shown that the mRNA synthesized in the nucleus rapidly moves into the cytosol ready for protein biosynthesis. There is thus a flow of information from nucleus to cytoplasm.

A given mRNA molecule may be long or short depending on the length of the gene. However, in eukaryotic cells, genes are rather more complicated than those in prokaryotes. Prokaryotic genes are continuous — a sequence of bases which, when translated, has a concordance with the sequence of amino acid residues in the polypeptide produced. Most eukaryotic cell genes, in contrast, seem to come in shortish sections. Regions of DNA that code for polypeptides are divided by sections of DNA that are non-coding (or nonsense) to be followed by further sections that code (Figure 6.14). The DNA itself is, of course, continuous; it is the information that is split up into chunks. When the mRNA is made, the initial linear transcript is *spliced*, that is to say, the non-coding bits are removed (by enzymes) and the coding portions are joined together to give a transcript that codes for the polypeptide. This is called *RNA processing* and it is the spliced transcript that leaves the nucleus to go into the cytoplasm.

Quite why eukaryotic genes should come in bits is not certain. The gene for collagen, for example, exists in over 50 sections. It may be that sections code for specific 'protein motifs' — little sections with a certain structure. Proteins may evolve by these sections joining together in different ways at the DNA level. Nor do we really know the function of the non-coding bits: do they carry information or is it really nonsense? These bits are called intervening sequences or *introns*. The bits of DNA that carry the polypeptide sequence information are called *exons*.

In eukaryotic cells messenger RNA undergoes a further step in processing before it is translated. A 'cap' of a few 'unusual' bases is added at the leading end, and a 'poly(A)' tail — a tail of up to 200 adenine residues — is added at the other. Possibly the cap ensures correct starting of the translation process, but the function of the poly(A) tail is uncertain; one idea is that it may control the stability of mRNA in the cytosol.

Protein synthesis (translation)

The process of protein synthesis requires the linking together in a specific sequence of several hundred or even thousand amino acids. The chemical reaction is that of forming several hundred peptide bonds (Chapter 2) which are condensation reactions, and hence energy will be required as well as information. The process is called *translation* because the information in the message (mRNA) is converted into a sequence of amino acid residues. We can also think of the language of the DNA (base sequence) being translated into the language of the amino acids; the 'translators' are small molecules of RNA, called *transfer RNA* (tRNA). The importance of these (and there is at least one for each amino acid) is that at one 'end' they recognize a specific triplet code on the mRNA, and at the other 'end', with the help of specific enzymes, they recognize a given amino acid. Thus they act as adaptors in bringing the amino acids together on the mRNA. Spread 6.5 shows how they work.

One other RNA-containing component is required in protein synthesis, and this is the ribosome, of which there are many thousands in a typical cell. These little organelles or particles, made up of about half protein and half RNA (*ribosomal RNA*; rRNA), are the 'anvils' on which the links in the polypeptide chain are forged. The structure of the ribosomes and their role in the sequential addition of amino acids to a growing polypeptide chain is explained in Spread 6.5.

Getting the proteins to the right location (targeting)

The typical cell manufactures a whole range of proteins. Some, such as the enzymes of glycolysis, remain in the cytosol where they do their work, while others, such as those of the Krebs cycle, need to be sent into the matrix of the mitochondria. Some proteins go into the endoplasmic reticulum where they are modified, or 'finished off', perhaps by adding carbohydrate to them to make glycoproteins. Others, such as trypsinogen and insulin, are intended for export from the cell. Many of these translocations require the proteins to cross membranes, which is not something proteins easily do. Moreover, since all proteins are synthesized on ribosomes that are thought to be identical, we might ask how some proteins 'know' that they should stay in the cytosol while others 'know' they should go through two separate membranes to get to the matrix of the mitochondria. Or how does the cell know and how does it direct them?

We do not know the answers to all of these questions, but we have a fairly good idea in general of how it is done. When polypeptides are synthesized they have sections of the chain which act as 'post codes' or targeting signals. For example, polypeptides that are destined to pass into the endoplasmic reticulum have a *leader sequence* usually consisting of 20–30 hydrophobic amino acids. This somehow guides the polypeptide, as it is being formed on the ribosome, to thread its way into the endoplasmic reticulum cavity. A set of proteins also ensures that the ribosome stays associated with the endoplasmic reticulum membrane. When the polypeptide has been fed through (or perhaps while it is being threaded through) the membrane, this leader or signal sequence is cut off by an enzyme. This sequence is not needed in the functional protein. Moreover, once this cleavage has taken place the polypeptide cannot escape from the endoplasmic reticulum (Figure 6.15, page 182).

Other types of signal make sure that other polypeptides find their way into the mitochondria and other organelles. Some proteins have inbuilt signal sequences that let them thread their way into a membrane but

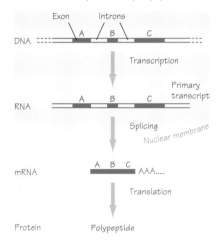

Figure 6.14 Eukaryotic genes come in sections in a continuous length of DNA. Inside the nucleus the primary transcript is a direct RNA copy of the DNA, but this is then 'edited' or 'spliced' to remove the introns. A cap and a poly(A) tail are added to give mRNA which can be translated on the ribosome to produce a polypeptide.

Protein synthesis

objectives
- To outline the function of the three types of RNA involved in protein synthesis (tRNA, rRNA, mRNA)
- To describe the process of protein synthesis
- To explain what is meant by the terms *transcription* and *translation*

PROTEIN SYNTHESIS is a very complicated and highly orchestrated process. The information for making new proteins is carried in the nucleus as sequences of bases (genes) in the DNA of the chromosomes. Protein synthesis takes place in the cytosol rather than the nucleus An intermediary, messenger RNA (mRNA), takes the 'message' (again the sequence of bases) from the nuclear DNA into the cytosol. Here, the RIBOSOMES carry out the process whereby amino acids are joined together in the right sequence to make a given type of protein (e.g. collagen, trypsin, insulin).

The ribosomes are complicated structures made up of certain proteins and a class of RNA known as ribosomal RNA (rRNA). rRNA does not code for proteins but, along with ribosomal proteins, forms a sort of 'work-bench' on which polypeptide chains are assembled. Several ribosomes may be working their way along a given mRNA molecule at any one time, forming a polyribosome, or 'polysome'.

A third type of RNA called transfer RNA (tRNA) acts as an adaptor between the genetic code sequence of bases and the amino acids they encode. There is at least one type of tRNA for each type of amino acid. One 'end' of the tRNA molecule can combine with a specific amino acid while the other 'end' consists of a sequence of three bases, known as the ANTICODON. The anticodon is complementary to (and therefore binds to) the CODON — a sequence of three bases in the mRNA which specifies that particular amino acid.

Some proteins are manufactured on ribosomes that are floating about in the cytosol. However, proteins that are designated for export from the cell are synthesized on ribosomes that are attached to the endoplasmic reticulum (see Figure 6.15) giving it a 'rough' appearance in the electron microscope. In this case the newly formed polypeptide passes into the cavity of the endoplasmic reticulum.

In bacteria (prokaryotes) the mechanism of protein synthesis is similar although not identical. For example, the DNA is not in a nucleus and the ribosomes are smaller.

Some protein synthesis also takes place within the mitochondria and chloroplasts of eukaryotes, since these organelles have their own DNA. In these cases the ribosomes are very similar to those found in prokaryotes.

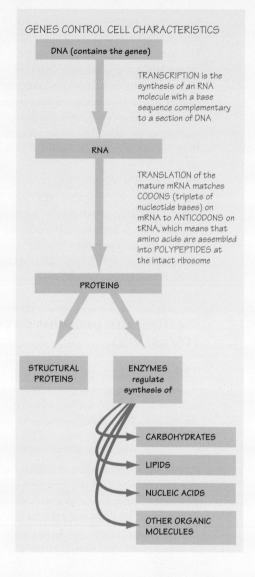

GENES CONTROL CELL CHARACTERISTICS

DNA (contains the genes) → RNA

TRANSCRIPTION is the synthesis of an RNA molecule with a base sequence complementary to a section of DNA

TRANSLATION of the mature mRNA matches CODONS (triplets of nucleotide bases) on mRNA to ANTICODONS on tRNA, which means that amino acids are assembled into POLYPEPTIDES at the intact ribosome

RNA → PROTEINS → STRUCTURAL PROTEINS / ENZYMES regulate synthesis of → CARBOHYDRATES, LIPIDS, NUCLEIC ACIDS, OTHER ORGANIC MOLECULES

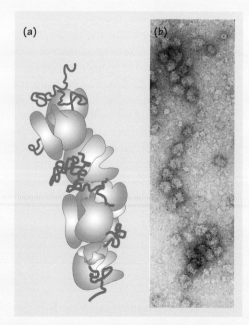

(a) Diagramatic view of part of polysome. [Redrawn with permission from Zimmerman, R.A. (1995) Nature **376**, 391.] (b) Electron micrograph of polysomes. Magnification × 130,000.

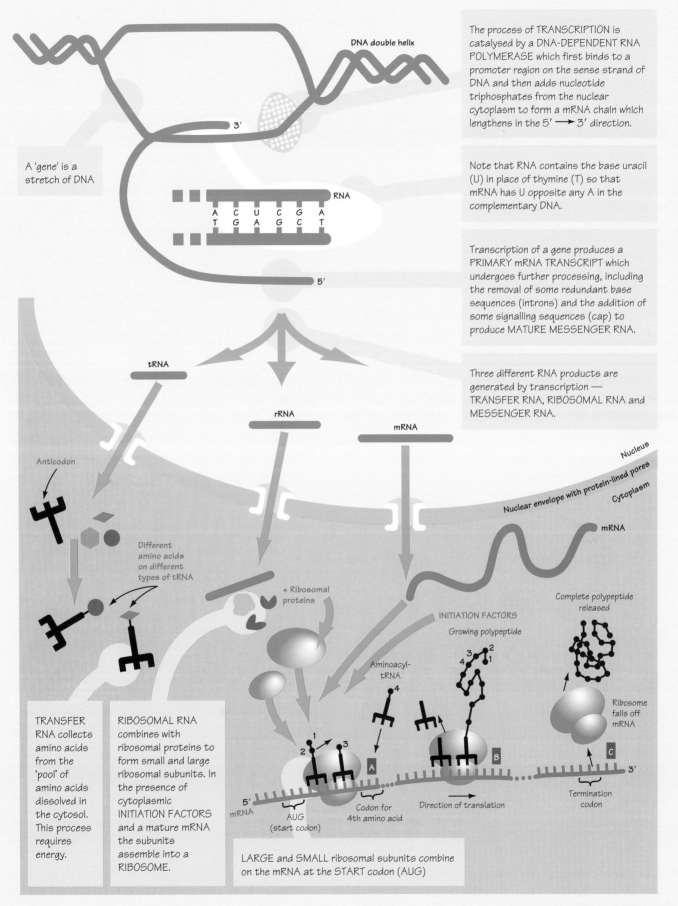

then stay there. These, of course, are the transmembrane or integral proteins. Remember that the starting point for this was saying that proteins do not naturally thread their way through membranes; indeed, part of the function of biological membranes is to separate the various compartments of the cell.

Post-translational modifications

Very few of the proteins made by cells, and especially those exported from the cell, are simply polypeptide chains. Bits of polypeptide are cleaved off, sugar residues are added, cofactors and metal ions are added, and the polypeptides often have to associate with other polypeptides to give multisubunit structures (think of haemoglobin). Many of these transformations, or *post-translational modifications* to the polypeptide, take place within the cavity of the endoplasmic reticulum and the Golgi vesicles, but some of them take place outside the cell. An example of a protein that undergoes extensive post-translational modification is collagen (Figure 6.16).

6.5 Are prokaryotic cells different?

Much of what has been written in this chapter has been about eukaryotic cells; do prokaryotic cells do things in the same way? Remember that they do not have nuclei or an endoplasmic reticulum, for example.

The answer to this is that in general terms they do, although if the processes are looked at in detail a myriad of small differences will be seen. As has been explained, eukaryotes have split genes and mRNA processing; prokaryotes do not. Ribosomes in prokaryotes are different from those in eukaryotes (they are a little smaller) but they are still recognizable as ribosomes and work in essentially the same way. We have to be a little careful here, because both mitochondria and chloroplasts have their own DNA and their own ribosomes (they make some, but not all, of their own proteins) and these turn out to be the prokaryotic type of ribosomes! So there is a common theme, although the prokaryotic and eukaryotic ways of doing things must have evolved a long time ago and have diverged quite a lot since then.

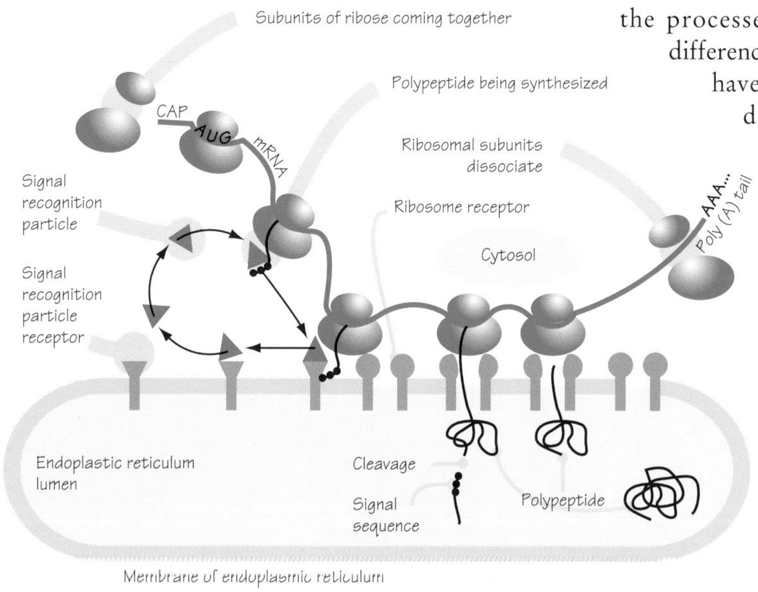

Figure 6.15 The signal mechanism. When newly synthesized proteins are intended for export to other parts of the cell or to the exterior, they have to cross membranes. The mRNA for such proteins contains a sequence which codes for the so-called signal sequence (rather like a postcode) which enables the polypeptide to thread its way from the ribosome, where it is being made, through the membrane of the endoplasmic reticulum. There are receptor proteins in the membrane for the signal recognition particles (also proteins). When these recognize the signal sequence they dock with their receptors. Ribosome receptor proteins, also in the membrane, create a channel through which the newly synthesized polypeptide passes. As soon as the signal sequence has threaded its way through the membrane it is cleaved off by a protease.

6.6 Further reading

Brown, T. (1991) DNA unravelled. Biological Sciences Review **4(2)**, 25–26

Bulleid, N. (1992) How are proteins made? Biological Sciences Review **4(5)**, 2–4

Crick, F. (1989) What mad pursuit. Penguin Books, London

Darnell, J.E. (1985) RNA. Scientific American **253 (Oct)**, 54–64

Grunstein, M. (1992) Histones as regulators of genes. Scientific American **267 (Oct)**, 40–47

Jones, S. (1993) The language of the genes. Flamingo Books, London

Judson, H. (1979) The eighth day of creation. Simon and Schuster, New York

McLennan, S. (1994) The control of gene expression. Biological Sciences Review **6(5)**, 2–4

Naylor, P. (1994) Electrophoresis. Biological Sciences Review **7(1)**, 35–37

Nurse, P. (1992) The new genetics. Biological Sciences Review **5(2)**, 15–18

Radman, M. and Wagner, R. (1988) The high fidelity of DNA duplication. Scientific American **259 (Aug)**, 24–30

Rosenfield, I., Ziff, E. and van Loon, B. (1983) DNA for beginners. Writers and Readers Publishing Inc. U.S.A.

Tjian, R. (1995) Molecular machines that control genes. Scientific American **272 (2 Feb)**, 38–45

Wilcox, D. and Grant, M. (1995) Cracking the genetic code. Biological Sciences Review **8(1)**, 14–18

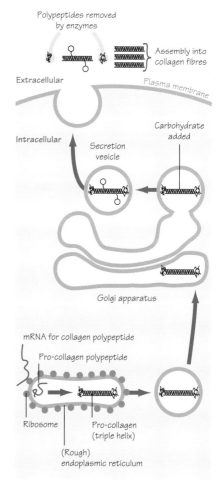

Figure 6.16 Biosynthesis of collagen, an extracellular protein that undergoes a great deal of post-translational modification. The mRNA is translated at the rough endoplasmic reticulum. Inside the endoplasmic reticulum the procollagen chains assemble to form a triple helix. As this passes through the endoplasmic reticulum and then the Golgi, proline and lysine residues are modified, cross-linking takes place and short lengths of carbohydrate are added. Exocytosis brings the collagen outside the cell where enzymes cleave the propeptides at each end to produce insoluble collagen. This assembles to make strong fibres outside the cell.

6.7 Examination questions

1. Read through the following account of protein synthesis, then write on the dotted lines the most appropriate word or words to complete the account.

   During protein synthesis, amino acids are linked by .... to form polypeptides. This process occurs at the ribosomes. The function of the ribosome is to hold the .... in such a way that its .... can be recognized and paired with the complementary .... on the tRNA. The molecules manufactured by ribosomes situated on the .... of the cell may be converted to glycoproteins in the ....

   (6 marks)

   [ULEAC (Biology) June 1995, Module 1, No. 4]

2. The diagram shows the information flow in protein synthesis.

   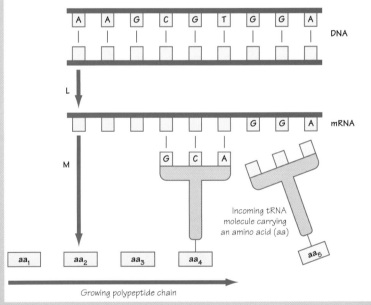

   (a) Write in the boxes in the diagram the missing initial letters of the bases on the (i) DNA strand, and (ii) mRNA.

   (b) (i) Write in the boxes in the diagram the missing initial letters of the three bases on the incoming tRNA molecule. (ii) What name is given to this triplet of bases on the tRNA molecule?

   (c) Name the process shown by (i) arrow L, (ii) arrow M.

   (d) In which organelle does the process shown by arrow M take place?

   (7 marks)

   [Oxford Local Biology 1, June 1986, No. 16]

3. Khorana (1965) artificially synthesized mRNA molecules consisting of a regular alternating sequence of two bases, for example ...AGAGAGAG... etc. He then used these as substitutes for natural mRNA during protein synthesis.

   (a) What bases do the letters A and G represent? (2 marks)

   (b) Which bases would be complementary to (i) A and (ii) G in a DNA molecule? (2 marks)

   (c) (i) One purpose of Khorana's experiment was to test whether or not the theory of the triplet code was correct. Using X and Y to represent different types of amino acids, predict the amino acid sequence in the polypeptides Khorana would have obtained after using artificial mRNA with the sequence ..AGAGAGAGAGAGA-GAGAGAGAGAG...etc., given that the triplet code theory is correct. (1 mark)
   (ii) If the theory had been wrong, and the code was a series of duplets (pairs) or quadruplets (fours) instead of triplets, what would have been the amino acid sequence(s) in his polypeptide(s), given ...AGAGAGAGAGAGAG... etc. as mRNA? (1 mark)

   (d) Khorana went on to synthesize trinucleotide polypeptide sequences such as ...AGCAGCAGCAGC... etc., and found these generated three types of polypeptides, namely ...XXXXX..., ...YYYYY..., and ...ZZZZZ... Explain how this result could be obtained. (2 marks)

   (e) As a result of these experiments Khorana not only confirmed that the genetic code was indeed a series of triplets, as had been suggested some years previously, but also showed that it was a non-overlapping and 'unpunctuated' (no unread bases) code. Explain how the above results support one of these conclusions: either that it was non-overlapping, or unpunctuated. (1 mark)

   (f) Suggest one other piece of evidence/argument which supports the view that the genetic code is based on a series of triplets. (4 marks)

[Oxford Local Biology 1, June 1991, No. 7]

4. The base sequences from two strands of DNA are shown in the diagram below. The two sequences are from the same section of a chromosome: the first is the normal sequence and the second is a mutant form.

Normal DNA	AATCAGGTTA
Mutant DNA	ATACCAGTTA

   (a) Describe and name two point mutations in the mutant DNA. (2 marks)

   (b) After transcription, a strand of mRNA is produced. Write the mRNA sequence which would be produced from the mutant DNA. (1 mark)

   (c) (i) Describe how the polypeptide chain produced from this mRNA sequence may differ from the normal polypeptide. (1 mark)
   (ii) Suggest how this might affect the activity of the polypeptide. (2 marks)

[ULEAC (Biology) January 1995, Module 1, No. 2]

5. An enzyme catalysing the breakdown of certain types of cell walls was found in two different forms, called X and Y. The activity of the two forms of the enzyme was investigated. The graph shows the results of an experiment in which the two forms of the enzyme were added separately to suspensions of the cell wall material. All results were obtained at pH 7.5 and 22°C, with standard volumes and concentrations of substrate and enzyme.

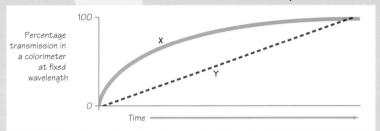

(a) (i) Explain how change in light transmission through the enzyme–substrate mixture allows the course of the reaction to be followed, and (ii) suggest why the cell wall suspensions were shaken thoroughly before being placed in the colorimeter. (3 marks)

(b) Which of the following, A to D, would provide the best control for the investigation? Explain your answer.
A Enzyme and water
B Cell walls and water
C Boiled enzymes and water
D Cell walls and boiled enzymes.
(2 marks)

The amino acid sequences of the two proteins (enzymes X and Y) were determined. Each contained 136 amino acids joined together in a single chain. The only differences in sequence were noticed in a central part of the proteins. In the diagram below, the amino acids concerned are given, together with their numbered individual positions in the proteins. The sequence of 'bases' (nucleotides) corresponding to this central sequence is given for protein X.

Protein X	86	87	88	89	90	91	92	93	
N-terminal end...	Lys	Ser	Pro	Ser	Leu	Asn	Ala	Ala	... C-terminal end
DNA sequence for protein X	TTT	TCA	GGT	AGT	GAA	TTA	CGA	CGA	
mRNA sequence for protein X									
N-terminal end...	Lys	Val	His	His	Leu	Met	Ala	Ala	... C-terminal end
Protein Y	86	87	88	89	90	91	92	93	

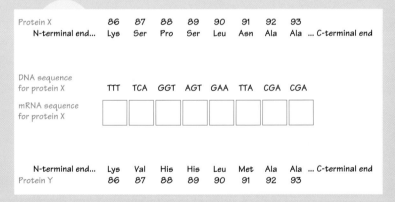

(c) Enter the appropriate mRNA sequence in the grid provided above for this central part of protein X. (2 marks)

Further investigations showed that enzyme Y was the result of two mutations to the DNA coding for enzyme X. In the first mutation a single nucleotide was deleted (lost) from the DNA. In the second mutation a single nucleotide was inserted (added) to the DNA.

Genetic dictionary
(giving only the mRNA codons of the amino acids shown in X and Y).

Amino acid	Codons which can be used
Lys	AAA
Ser	AGU or UCA
Pro	CCA
Leu	CUU or UUA
Asn	AAU
Ala	GCU
Val	GUC
His	CAU or CAC
Met	AUG

(d) Using the genetic dictionary provided above, work out as accurately as you can (i) the sequence of bases in the mRNA coding for protein Y (complete the boxes below), (ii) the identity of the DNA base (nucleotide) deleted in the first mutation, (iii) the identity of the DNA base (nucleotide) inserted in the second mutation. (4 marks)

mRNA sequence for protein Y

| 86 | 87 | 88 | 89 | 90 | 91 | 92 | 93 |

(e) Using the information provided in the question, suggest a reason for the difference in the enzymic activities of X and Y.

[Northern Examination and Assessment Board (formerly JMB) (Biology) June 1991, Paper 1, No. 7]

6. The following pathway shows some of the metabolic conversions involving the amino acids phenylalanine and tyrosine. Normally only very small amounts of phenylalanine are changed to phenylketones.

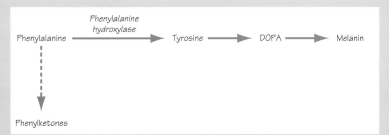

Production of the enzyme phenylalanine hydroxylase, which converts phenylalanine to tyrosine, is controlled by a gene with the alleles *P* and *p*. Enzyme production normally begins in the baby towards the end of pregnancy or within the first few days after birth.

Individuals who are homozygous for the recessive allele *p* lack the enzyme phenylalanine hydroxylase. These individuals suffer from phenylketonuria (PKU), a condition that can lead to brain damage as a result of accumulating phenylketones.

(a) If two heterozygous individuals have a child, what is the chance that this child will have PKU? (1 mark)

(b) Why is it necessary for PKU patients to follow a diet which is low in phenylalanine? (1 mark)

(c) For a normal individual, tyrosine need not be included in the diet. Explain both this observation and whether it is also true for a PKU sufferer. (2 marks)

New-born babies are tested to see if they have PKU. In the test, a mutant strain of the bacterium *Bacillus subtilis* is used. This mutant is unable to make the amino acid phenylalanine which must therefore be provided if the bacteria are to grow.

The *B. subtilis* mutant is spread evenly over the surface of an agar plate. A small blood sample is taken from a 5 to 10 day old baby and absorbed on a filter paper disc. The disc is immediately placed on the surface of the agar plate. Filter paper discs containing known amounts of phenylalanine are also applied to the agar surface. (The agar medium contains all the necessary nutrients for the growth of wild type *B. subtilis* but it does not contain phenylalanine.) The agar plate is incubated at 37°C for 36 hours and then examined for bacterial growth.

The diagram shows an agar plate, before and after incubation, with blood samples from two babies as well as the phenylalanine discs.

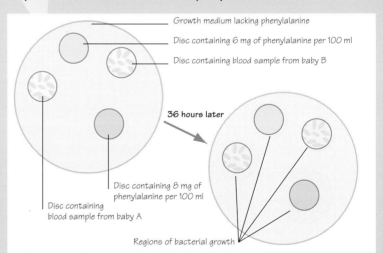

(d) (i) Estimate the amount of phenylalanine in the blood sample from baby A. (1 mark)

(ii) Baby B is a normal individual. Explain how the results suggest that baby A has PKU. (2 marks)

(e) Tyrosinase converts tyrosine to DOPA which can then be converted to melanin. Tyrosinase is inhibited by high levels of phenylalanine. Explain why a PKU patient fed on a normal diet containing tyrosine is likely to have reduced levels of skin and hair pigmentation. (2 marks)

[Northern Examinations and Assessment Board (formerly JMB) (Biology) 1992, Nuffield Biology I, No. 5]

7.1 Introduction

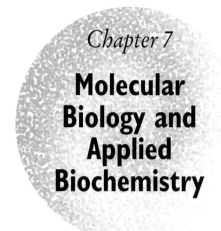

Exploitation is a word that often has a pejorative ring to it — Britain exploited her colonies, the strong exploit the weak, and so on — but one of the reasons for doing science is to use it. Humans have been exploiting their knowledge of Biology for thousands of years by using yeast to make bread and beer, carrying out plant and animal breeding, and in many other ways. Governments and especially medical research charities give billions of pounds every year to support scientific research, the fruits of which they hope will be able to be applied to the problems that beset humankind, or at least to make money.

This chapter is about exploiting biological knowledge to achieve new things, or to do the same things more efficiently or more rapidly. At one extreme it offers new cures and therapies and new food products, and, at the other, it poses severe ethical problems. Should we be 'interfering' with nature in this way: should new genes, for example, be released into the environment in 'engineered organisms'?

This is an enormous area and we shall scarcely scratch the surface of what has been done so far and what is likely to be achieved in the next few years. Biochemistry has only a few basic principles, but these have very wide applications. These principles include the tremendous catalytic power of enzymes, the remarkable specificity with which macromolecules can recognize other molecules, and, not least, the universality of the genetic code and all the details of nucleic acid metabolism and protein synthesis that go with this.

Molecular biology provides the possibility to do things by design — to engineer things — and this is nowhere more apparent than in the field of plant breeding. Ten or twenty years might be a fair time-scale for developing a new strain of plant with new, improved features by traditional plant breeding. With gene transfer techniques this may now be achieved in one generation of plants rather than in many generations. More importantly, however, genes may be transferred across the species barrier — which is not usually possible by traditional methods. Thus the possibility exists of taking the nitrogen-fixing genes from a micro-organism and transferring them to a cereal crop plant, thus reducing the need to apply nitrogenous fertilizers to the soil.

Figure 7.1 Biotechnology and genetic engineering are always in the news these days.

We can only give a flavour of what is actually happening here. It is a vast and exponentially growing area of science/industry. There are fortunes to be made (as well as terrible ecological disasters to be avoided). This means that the sorts of things discussed here are also debated daily in newspapers and news programmes, not only to keep people informed, but also, sometimes, to change people's attitudes. For the student, regular reading of *New Scientist* and *Scientific American*, as well as the reports from science correspondents in the daily newspapers, provide a very rich feast of what is currently happening, and students with even a little biochemistry background will be in a better position to understand these important developments (Figure 7.1).

7.2 The technology

Starting with the double helix of Watson and Crick in 1953, knowledge of DNA has increased exponentially. Key discoveries include the genetic code, how protein synthesis takes place, what a gene is in chemical terms, and how to determine the sequence of DNA (Table 7.1). At all of these stages, biochemical knowledge of enzymes, cofactor requirements and generally of how to do things in the test tube ('*in vitro*') has also increased. This has culminated in the technology by which DNA may be manipulated — sequenced, cut, identified, duplicated and transferred in and out of cells and organisms. Such is the confidence with these new techniques, mostly invented since the mid-1970s, that we can now contemplate replacing defective genes in individuals with genetic diseases with the 'correct' version of the gene: in other words, *gene therapy* has become a real medical possibility. Most of these techniques depend upon the specificity of enzymes and on the ability of macromolecules to recognize other molecules. The techniques themselves are mostly fairly easy to carry out in the laboratory — it does not take a virtuoso performance or years of experience — but what is important is devising the strategy for what you want to do, knowing what can be done and doing things in the right order. This is the essence of practical molecular biology.

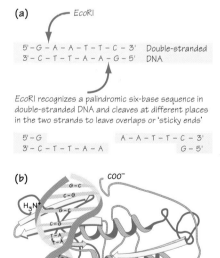

Figure 7.2 Restriction enzymes cleave DNA at specific places. (a) Here the enzyme EcoRI cuts unequally where it finds a specific six-base palindromic sequence (see Table 7.2). (b) The structure of EcoRI is known. The sketch shows how the enzyme protein embraces the DNA double helix. The 'arrows' represent β-strands and the coils α-helix in the protein; the double-stranded DNA is shown vertically.

Table 7.1
Some key dates in Molecular Biology since the Watson and Crick 'double helix' paper of 1953 (Nature **171**, 737)

1953	Watson and Crick publish their paper which proposes a double helical structure for DNA with the bases on the inside of the helix, matching A:T, C:G [Sanger published the first amino acid sequence of a protein, insulin, at about this time.]
1957	DNA polymerase discovered by Kornberg
1962	Restriction enzymes discovered
1966	The genetic code deciphered
1972–3	Cloning in bacteria first carried out
1975	Southern blotting invented
1975	Sanger publishes his method for DNA sequencing
1981	First transgenic animals
1985	Polymerase chain reaction invented

The other key point about the availability of the techniques of molecular biology of DNA manipulation is that experiments in Biology can now be carried out, and questions asked, that could not have been addressed by existing techniques. If you want to know the role played by a particular amino acid residue in a protein or a particular gene in an organism, you can take out that amino acid or gene, or alter it, put it back, and see what happens. The power is awesome!

Cutting DNA

The ability to 'cut' a double-stranded DNA molecule does not sound all that revolutionary, but in fact this possibility using specific enzymes provided the key to both sequencing DNA and manipulating bits of it in all sorts of different ways. The enzymes that do this cutting — and there are hundreds of different types — are microbial enzymes called *restriction enzymes*, or, strictly, restriction *endonucleases*. Their role in bacterial cells seems to be to disable any foreign or extraneous DNA from being active should it get into a given cell. (The DNA of a given bacterial species is protected against itself being disabled in this way.) *Endonuclease* means that the DNA is cleaved at points internally rather than being nibbled away at the ends (which we would call *exonuclease* activity).

The important point is that given restriction enzymes hydrolyse or 'cut' DNA at specific places in the DNA sequence. The enzymes typically recognize a four- or six-base sequence and cut in the middle of it. The sequences recognized are palindromic and the cut is often unequal. This is explained in Figure 7.2. Restriction enzymes are named by the organisms from which they originated: *Alu*I comes from *Arthrobacter luteus*, and *Hin*dIII from *Haemophilus influenzae* serotype d, and so on. (This does not help you very much unless you work with these every day. Most suppliers of restriction enzymes give away charts showing several hundred enzymes and their specificities and the sequences at which they cut; see Table 7.2, page 192.)

The consequence of cutting a long piece of DNA (such as a virus genome or a bacterial chromosome) with a given restriction enzyme is that a whole series of double-stranded DNA fragments is generated. Although there may be thousands of these from a large piece of DNA, the same fragments will always be produced whenever this experiment is carried out with a given piece of DNA with that particular restriction enzyme. Furthermore, although the sequences of the bases in the DNA fragments will not be known, the sequences *at the ends* (i.e. where the cut has taken place) will be known because of the specificity of the restriction enzyme. The different fragments may be sorted by separating them by electrophoresis on an agar gel, which sieves them by size (Figure 7.3).

Now because all the fragments end in the same sequence, and indeed may have little 'overhangs' because the cut was not in the same place in the two DNA strands, if we put such DNA fragments together under the right experimental conditions, the overhangs get together because of base pairing. Sometimes these overhangs are referred to as *sticky ends*. If we took DNA from two different organisms, cut each with the same restriction enzyme, and then put them together, we would find some 'recombination' of DNA from the two species. Addition of an enzyme called DNA ligase would restore the continuous double strand. This shows how we can make *recombinant DNA* (Figure 7.4). This is a key step in sticking together genes or bits of genes.

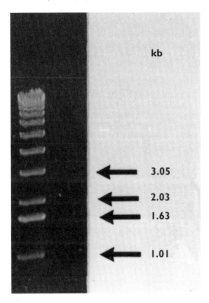

Figure 7.3 Electrophoresis of DNA represents separation by size. All fragments of DNA have the same charge-to-mass ratio as there is one negatively charged phosphate for every base–sugar–phosphate unit. However, when DNA fragments of different sizes are subjected to electrophoresis in an agar gel, the sieving action of the gel retards the larger fragments while allowing the small ones to move rapidly towards the anode. Separation is thus on the basis of size, not charge. The photograph shows agar gel electrophoresis of fragments of the double-stranded DNA from a small virus and the number of kilobase pairs (kb) in each length of DNA.

Figure 7.4 Two pieces of double-stranded DNA may be joined together using DNA ligase. If the pieces of DNA have been produced using the same restriction enzyme, then the overhanging 'sticky ends' ('cohesive ends') match up by complementary base pairing. The ligase then makes the join.

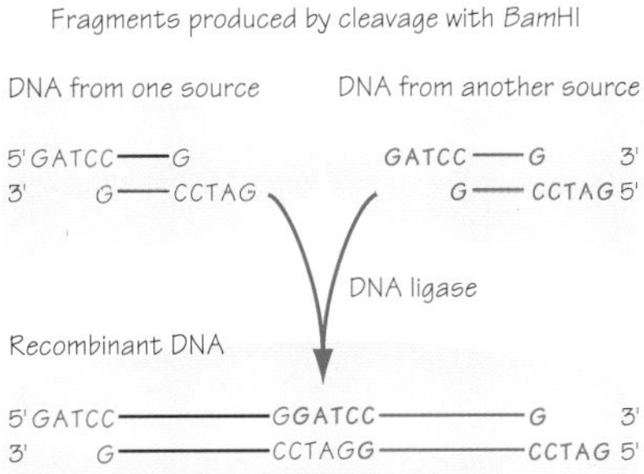

Fragments produced by cleavage with BamHI

```
DNA from one source            DNA from another source

5' GATCC——G                     GATCC——G        3'
3'     G——CCTAG                     G——CCTAG 5'
                    ↓ DNA ligase
Recombinant DNA

5' GATCC————————GGATCC————————G        3'
3'     G————————CCTAGG————————CCTAG 5'
```

Figure 7.5 Southern blotting after DNA electrophoresis. The aim of the experiment is to find out if the original DNA contains a sequence corresponding to that in a DNA probe — a piece of DNA complementary to a certain sequence (e.g. a gene or part of a gene). The DNA is first cut into pieces using a restriction enzyme and the pieces separated by gel electrophoresis. A multitude of bands in the agarose gel are produced — maybe 10,000 — but are not visible at this point. The bands nearer to the positive electrode are the smaller ones. The pattern of bands is blotted on to a nitrocellulose membrane to give a replica. The DNA forming the bands sticks very tightly to the nitrocellulose and heating the membrane makes the DNA single stranded. The fragment(s) that match the DNA probe are then labelled and may be identified by their radioactivity.

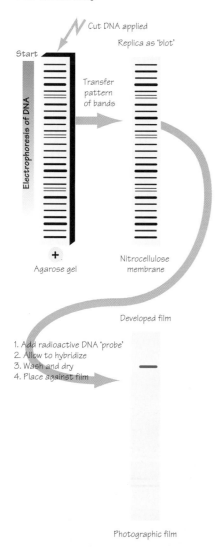

Table 7.2
Restriction enzyme cleavage sites

Example:

*Eco*RI recognizes the following palindromic* sequence in double-stranded DNA:

$$\begin{array}{c} \downarrow \\ 5' - G - A - A - T - T - C - 3' \\ 3' - C - T - T - A - A - G - 5' \\ \uparrow \end{array}$$

In the Table below the recognized sequence (it may be four-base or six-base) of one strand is given, and the cleavage site is indicated by an arrow. When the arrow falls in the middle of the sequence recognized the cut is symmetrical and there are no sticky ends ('blunt ends').

Enzyme	Recognition sequence	From ...
*Eco*RI	G↓AATTC	*Escherichia coli*
*Hin*dIII	A↓AGCTT	*Haemophilus influenzae*
*Xho*I	C↓TCGAG	*Xanthomonas holcicola*
*Bam*HI	G↓GATCC	*Bacillus amyloliquifaciens*
*Alu*I	AG↓CT	*Arthrobacter luteus*
*Taq*I	T↓CGA	*Thermus aquaticus*
*Hae*III	GG↓CC	*Haemophilus aegyptius*

*An example of a palindromic sentence is 'Madam I'm Adam'

Detecting specific bits of DNA

Cutting the DNA from even a bacterial chromosome generates thousands of bits of DNA, and cutting a eukaryotic cell's DNA may yield hundreds of thousands of bits. How can specific bits (genes or parts of genes) be identified? In order to do this we use the power of a specific sequence of single-stranded DNA to recognize and pair with its complementary sequence.

If some double-stranded DNA in solution is heated to nearly 100°C the two strands separate because the hydrogen bonds linking the base pairs together are broken. Rapidly cooling such a solution produces single-stranded DNA because the complementary strands do not have time to find their partners and tend to form random coils as the solution cools. In contrast, if the solution is allowed to cool slowly, the partners do have time to find each other and the double helix reforms itself. This process is called *annealing*.

Suppose now we were to do this same experiment, but, at the point of slow cooling, we added some more DNA, labelled in some way, that had a sequence complementary to a short sequence in the original DNA? Some of it would anneal to this DNA, so identifying a certain specific region of it. In this way we might find out whether the DNA from one organism had a sequence (i.e. a gene) similar to a DNA sequence (gene) in another organism.

The technique of *blotting* enables this to be made more precise and to be used with DNA fragments separated by gel electrophoresis (see Figure 7.3). After the electrophoresis has been carried out, the agarose gel is pressed against a nitrocellulose or nylon membrane. This makes a 'print' or 'replica' of the pattern of bands (people call it a *blot*) on the nitrocellulose or nylon. DNA sticks quite tightly to the membrane material, and baking the membrane in an oven at 80°C for a time makes it stick permanently, while it also becomes single stranded. The pattern of DNA bands at this stage is invisible. If one now came along with a radioactive, isotopically labelled DNA molecule that is complementary to some part of the DNA, added it to the membrane and allowed annealing to take place, any bands originally on the gel that are complementary will take up isotopically labelled DNA as the molecules anneal. Placing the washed and dried blot against photographic film, followed by development, will reveal a pattern showing where the radioactive DNA has annealed (Figure 7.5).

This method of blotting was devised by Professor Ed Southern in 1975 and is still called a 'Southern blot'. The technique may also be performed with RNA on the electrophoresis gel, such that DNA–RNA hybrids are produced. This is called a Northern blot to distinguish it from a Southern blot. Both are extremely valuable techniques.

DNA sequencing

Around this time (1975) it was discovered how to determine the sequence of bases in long DNA molecules. Two methods were devised — one enzymic and the other chemical. Previously people had thought that this would be a daunting and probably impossible task; in fact it turned out to be somewhat easier than sequencing a protein (see Spread 7.1). Sequences of several hundred bases can be determined per week by a single laboratory worker and now there are automated methods available. The race is on to determine the complete sequence of the human genome in the HUGO project (see Spread 7.3, page 204) If one worker can sequence 500 bases in a week and the human genome is 3×10^9 bases long, calculate how many person-years it will take to sequence the whole genome.

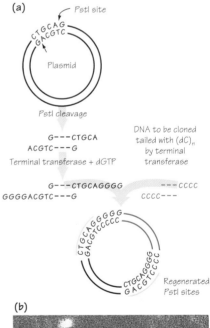

Figure 7.6 (a) Inserting a piece of DNA into a bacterial plasmid vector for cloning. The plasmid is cut with a restriction enzyme (here PstI) to open it. Then an enzyme called terminal transferase is used to add a string of Gs to the end of each strand. Terminal transferase is non-specific about adding nucleotide units: if you give it dGTP, it adds Gs. The DNA it is wished to clone is similarly 'tailed', but now with Cs. If the opened plasmid and the incoming DNA are put together, the Cs and Gs come together and a ligase makes the join good. Hence the DNA is inserted into the 'recombinant' plasmid. This method has two advantages over simply using the overhangs generated by the restriction enzyme (see Figure 7.4). First, the dozen or so Cs and Gs form a stronger union than simply the short 'sticky ends'. Secondly, and more importantly, you will see that the PstI sites are regenerated. Thus after the plasmid has been amplified in bacteria and isolated and purified, the use of PstI again will cut out the inserted piece of DNA from the plasmid. (b) Electron micrograph of a plasmid.

DNA sequencing

objectives
- To recognize the reactions involved in the dideoxy DNA sequencing method
- To explain how DNA sequencing electrophoresis gels are run

The history of sequencing biological macromolecules is interesting. In 1953, Fred Sanger, working in Cambridge, determined for the first time the amino acid sequence of a protein, actually the two chains (21 and 30 residues respectively) of INSULIN. Insulin is quite a small protein and since then literally thousands of proteins have been sequenced, increasingly by automated methods. The work was recognized as a great step forward and Sanger was awarded the Nobel Prize in 1958.

Proteins have up to 20 different amino acids in them, and sequencing, although difficult at first, was seen to be feasible. In contrast, DNA, with only four bases, apparently presented a much greater problem. (The chances of identical sequences coming up are much greater when there are only four monomer units compared with when there are 20.) Nevertheless, it was realized that it was of vital importance to be able to determine DNA sequences, and several groups of research workers set about conquering this problem, and, in the 1980s, succeeded. One of the methods devised — and, indeed, the most commonly used method —utilizes enzymes. This method was first performed by the same Fred Sanger, who was awarded a second Nobel Prize because the work was recognized as being so important. Other workers devised a chemical method for determining sequences.

The workers involved realized that once a rapid and reliable method of DNA sequencing became available it would theoretically be possible to determine the sequence of the whole human genome — all 3×10^9 bases — although this would take some time. Out of this grew the HUMAN GENOME ORGANIZATION (see Spread 7.3), or HUGO, which is aimed to do just that in the early years of the next millenium. There are just so many bases that it was clear that automated DNA sequencing would be vital if sensible progress was to be made.

In this Spread we look at the Sanger method for DNA sequencing: this method is known as the DIDEOXY TECHNIQUE. You will see that it was devised based on a knowledge of how the DNA replicating enzymes work and also of the precise chemical nature of DNA and its links between units.

Photograph of the ABI PRISM™ 377 DNA Sequencer. Courtesy of PE Applied Biosystems.

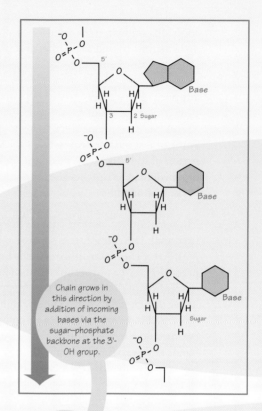

Chain grows in this direction by addition of incoming bases via the sugar–phosphate backbone at the 3'-OH group.

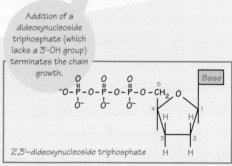

Addition of a dideoxynucleoside triphosphate (which lacks a 3'-OH group) terminates the chain growth.

2',3'-dideoxynucleoside triphosphate

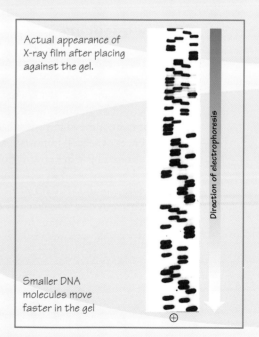

Actual appearance of X-ray film after placing against the gel.

Direction of electrophoresis

Smaller DNA molecules move faster in the gel

194 Life chemistry and molecular biology

7.1 DNA sequencing

Today we know the base sequences of many thousands of genes. This helps us to understand how genes function and to identify precisely how mutations lead to genetic diseases. In the future it will be helpful as a basis for GENE THERAPY (see Spread 7.4).

The structure of DNA

DNA is made up of base–sugar–phosphate units joined together. The joins are through the sugar–phosphate backbone, and the numbers 5' and 3' indicate the position of the –OH on the deoxyribose sugar ring through which the sugar–phosphate bonds are made. Because DNA is deoxyribonucleic acid, there is no –OH on carbon 2 of the sugar ring, and, indeed, nowhere else through which this type of bond can be made.

The bonds are made by the action of the enzyme DNA polymerase which inserts the complementary base (as a nucleotide) to make double-stranded DNA.

The new, incoming, nucleotide units are linked on by joining them to the 3'-OH at the end of a DNA molecule. You will remember that the chain only grows in one direction (see Spread 6.1).

Sanger knew that if DNA polymerase was given the dideoxynucleoside triphosphates (e.g. ddATP, ddCTP), then the enzyme reaction was terminated. What actually happened was that DNA polymerase would continue adding new base–sugar–phosphate units until it added a dideoxy unit. When this happened there is no –OH on which to add the next unit. Consequently if you added just one of the dideoxynucleoside triphosphates, e.g. ddATP, along with the normal dCTP, dGTP and dTTP, DNA polymerase went along adding sugar–base–phosphate units until it had added an A; then it stopped. (This would of course be against a T in the complementary strand.)

Now a little ingenuity was required. If you added both dATP and ddATP plus the other three deoxynucleoside triphosphates (dCTP, dGTP, dTTP), then whenever DNA polymerase came up against a T in the complementary chain it could add either dATP or ddATP. Obviously if it added the latter then chain growth ceased. Thus with this type of reaction mixture, you tended to finish up with a selection of newly made bits of DNA, all of different lengths, but all ending in A. Similarly, if you added both dCTP and ddCTP you would get a set of newly made chains all ending in C, and so on.

Sanger realized that if he could separate these newly made bits of DNA on the basis of size he could start to get a DNA sequence. A method is available for this, namely ELECTROPHORESIS in a gel. All DNA molecules have the same charge/mass ratio (the larger they are the more negative charges they have). In electrophoresis they should all move at the same rate, therefore. However, if the electrophoresis takes place in a gel there is a sieving effect, such that small molecules move more rapidly than large ones. After electrophoresis the bands can be stained to see where they are. More usually, radioactive dideoxynucleoside triphosphates are used so that the newly formed chains all have radioactivity. Putting the dried gel against X-ray film then reveals the position of the bands.

In order to sequence a piece of DNA you need to make it single stranded and then add a short primer from which DNA polymerase can work. You then run four separate reactions: the first one contains dATP, dCTP, dGTP and dTTP along with radioactive ddATP. The second reaction contains dATP, dCTP, dGTP and dTTP along with radioactive ddCTP. [Can you say what the third and fourth reactions contain?]

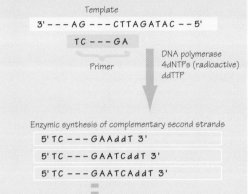

After the enzyme action has taken place, a sample from each of the four reaction tubes is placed on a slab gel and electrophoresis is carried out. After autoradiography a pattern of bands is observed (smallest DNA molecules at the bottom of the gel because they have moved fastest). At this point you can simply 'read off' the DNA sequence:

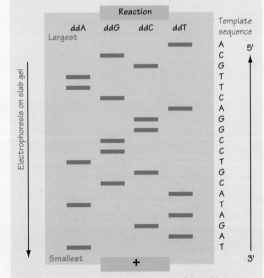

This technique has been very successful and has now been automated. Instead of radioactive markers, the dideoxynucleoside triphosphates are made to have different colours. These colours are read automatically by the machine which then prints out a DNA sequence.

Chapter 7 • Molecular Biology and Applied Biochemistry

Figure 7.7 (a) Selecting clones by Southern blotting. After a piece of DNA has been cloned into a bacterial cell, it is necessary to pick out (select) the resultant bacterial colonies on an agar plate that contain this piece of DNA. The Figure shows the steps required to do this. A replica of the colonies on the plate is taken by blotting lightly on to a nitrocellulose disc. The 'print' of colonies on the nitrocellulose is lysed to release the DNA and this is baked on to the sheet and made single stranded. Immersion in a solution of the radioactively labelled probes for the DNA (an 'oligo') allows this to hybridize with the DNA released from the colonies. If the nitrocellulose is now dried and placed against photographic film in the dark, on development black spots may be seen corresponding to colonies which contain the DNA. We can now go back to the original agar plate from which the blot was made and pick off some of the identified colonies. These may be grown up to produce lots of the DNA that has been cloned. Remember that this DNA is in the plasmids in the bacteria. (b) A developed film indicating the positions of colonies that have picked up probe.

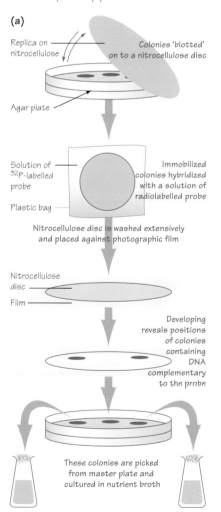

DNA cloning

In Biology, a clone is a set of identical cells or organisms that are derived from a single cell or organism. On a bacteriological Petri dish, each of the small 'colonies' that can be seen each represent millions of bacterial cells derived by binary fission from one original cell and is therefore a clone. *DNA cloning* means producing lots of copies of one piece of DNA, possibly a gene, using bacterial cells to provide the machinery for this replication. Until about 1986 this was the only way to amplify pieces of DNA. Chemical synthesis was not possible and in fact doing it in a bacterial cell you do not need to know what the sequence is anyway. The importance of amplifying was that the amount of starting material available, typically the DNA present in one cell, was usually too small to analyse by any of the methods available.

Cloning had another important function as well — that of selecting bits of DNA of interest. It is possible to start with a fairly large piece of DNA, divide it up into pieces using a restriction enzyme, and then insert the pieces into colonies of bacteria. The starting material might be a million different fragments of DNA, but it is easy to grow up a culture of bacterial cells containing millions of cells. In general it can be organized so that each individual DNA fragment goes into at least one bacterial cell in a culture. Every bacterial cell is, of course, capable of growing up into a clone of identical cells, and hence amplification of the DNA that the cell contains is possible. If the bacterial cell containing the length of DNA of interest can be selected, then we have an excellent system for picking one length of DNA out of millions of pieces and of amplifying that particular piece. This gives enough DNA to analyse, sequence, or use for other purposes. The technology here consists of letting the bacteria do all the hard work of making more of a given piece of DNA.

The method involves placing the desired pieces of DNA into bacterial plasmids — small circles of DNA that occur in bacterial cells in addition to their main genomic DNA molecule, and which replicate when the cell divides (Figure 7.6, page 193).

Selecting the colony of interest (i.e. the one containing the bit of DNA we are interested in) was initially a problem, but one that could be solved in a number of ways. One way was to let the DNA, in other words the DNA inserted into the bacteria, be expressed by the bacteria. This means having the bacteria produce mRNA and translate it into protein. The protein produced by a few thousand bacterial cells could be identified using specific antibodies against the protein, and hence the colony on a Petri dish doing the expression could be identified. Another way of doing it was to have a piece of DNA that is complementary to the DNA of interest. This might be a bit of a gene or it could be a short length of synthetic (i.e. chemically synthesized) DNA. It is now possible to make bits of DNA of known sequence of 20–30 bases in length. Such bits are known as 'oligos', short for oligonucleotides. Making such sequences chemically is quite easy these days with an automated synthesizer. The colonies containing the piece of DNA of interest may be identified by Southern blotting (Figure 7.7).

A great deal of work on DNA sequences and genes was carried out in the early 1980s so that the amount of knowledge increased enormously, providing a basis for future work. In many instances it was quicker and easier to sequence the DNA rather than the protein for which it coded, using the genetic code to write the corresponding amino acid sequence.

In fact, for cloning, the DNA being selected and amplified does not have to be a whole gene or even a gene at all. The bacterial machinery amplifies the stretch of DNA indiscriminately.

The polymerase chain reaction

Cloning is still used a great deal for amplifying DNA lengths, but in 1985 a revolution occurred that gave a different way of selecting and amplifying DNA. In principle it is very simple, and indeed the inventor of the method, Kary Mullis, was somewhat amazed that no one had ever tried it before. He assumed first off that they had, but for some reason the method had not worked. He tried it and found that it did indeed work. The method, called the *polymerase chain reaction* or PCR, has resulted in major changes to a number of procedures in molecular biology. It is also changing the way in which genetic diseases are diagnosed and how protein engineering is performed.

The method requires some knowledge of DNA sequences to be available in order to make a pair of primers (see below), and in 1975 such knowledge would not have been available. By 1985, however, it was available and the method took off exponentially! It is explained in detail in Figure 7.8. Basically, it consists of the chemical construction of a pair of short primers (sequences of DNA) that 'bracket' a region of DNA of interest within a long piece of DNA or even the whole human genome. The following sequence is run as a series of cycles: (i) heating the original DNA to separate the strands, (ii) adding the primers which anneal to the separated strands, and (iii) making new DNA by the addition of the enzyme DNA polymerase. At each cycle the amount of DNA synthesized as a copy of that bracketed by the primers doubles in amount. Consequently exponential amplification results, increasing the amount of DNA by one, two, four, eight, sixteen... and so on. PCR machines these days typically carry out 20–40 cycles over a period of hours, amplifying 1 picogram of DNA into micrograms —sufficient for most analytical methods and for many other uses (Figure 7.9, page 198).

One of the brilliant strokes in devising this method was to use the DNA polymerase from a thermophylic bacterium from hot springs, *Thermus aquaticus*, usually called *Taq* polymerase. This enzyme works well at over 70°C and is not denatured by temperatures of 95°C. Thus the cycle of DNA denaturation (strand separation) by heating to nearly boiling can be carried out all in the same tube without the need for adding new enzyme at each turn of the cycle (Figure 7.9).

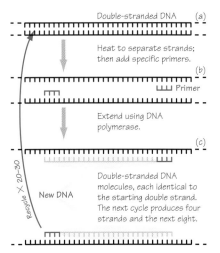

Figure 7.8 The polymerase chain reaction (PCR) is used to amplify specific stretches of DNA. The two strands of the double helix are separated **(a)** and 'primers' are added **(b)** under conditions where they find their complementary sequences and hybridize with them. The primers are short sequences of DNA made synthetically with base sequences 'chosen' or 'designed' to match specific sequences in DNA so that they 'bracket' a certain stretch of DNA. (This may be several thousand base pairs long.) Clearly, it is necessary to know the sequences of DNA near the stretch one wishes to 'bracket'. When the primers have hybridized (or annealed), then the enzyme DNA polymerase is added, which only works in one direction. This results in the synthesis of DNA as shown in **(c)** so that two double strands (of this piece of DNA) now exist where there was only one before. Repeating the whole cycle gives four strands, repeating again gives eight, and so on, doubling each time, so that the amplification factor is enormous. The other key 'trick' is to use DNA polymerase from a thermophilic bacterium that is not denatured by the nearly boiling temperatures needed to separate the two strands of DNA at each cycle. There is no need to add new enzyme at each cycle and the PCR machine (Figure 7.9) can be left running overnight to complete, say, 30 cycles.

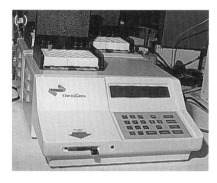

Figure 7.9 (a) Photograph of a 'PCR machine' or thermocycler, consisting of a metal block into which the reaction tubes fit. A program controls the temperature of the block and the time at each of the temperatures, for a chosen number of cycles. The graph (b) shows a typical time and temperature program.

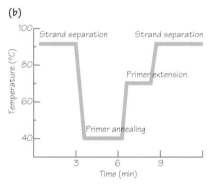

7.3 Transgenic organisms

In the above section we have dealt with a number of techniques for manipulating DNA. One of these was the method for taking a piece of DNA from any source, and inserting it into a bacterial cell — in other words cloning it. The recipient bacteria may or may not express the protein the gene codes for, but the method of getting 'foreign' DNA into a bacterial cell was essentially rather simple. Bacteria with the foreign DNA are said to have been *transformed*: their genome has been changed — new genetic characters have been inserted. Such bacteria may then be said to be *transgenic*, and it soon became the goal to make transgenic animals and plants. This was important as it would allow genes to be transferred across the species barrier in eukaryotes — a given gene from any organism to be put into the cells of any other organism. This works because the genetic code is universal. The next section is about transgenic animals and plants.

We might pause here to remember that prokaryotic and eukaryotic genes differ from one another. Eukaryotic genes come in sections, with introns and exons, but prokaryotic genes do not. Simply putting a eukaryotic gene into a bacterial cell will not work because the bacterial machinery cannot distinguish between introns and exons and tries to transcribe the whole lot — usually unsuccessfully. The answer to this problem was to make DNA starting with mRNA, which, of course, has already been edited (Figure 7.10). This is still a very useful way of getting genes expressed, that is, of producing valuable proteins, using the enormous synthetic capacity of great vats of bacteria containing a foreign gene.

Nevertheless, it was obviously desirable to be able to transfer genes between eukaryotic cells and ways of doing this were soon found. The first transgenic mice were produced in 1981.

Transgenic animals

Transgenic mice, and other animals, are usually produced by the rather tedious (but successful) method of microinjecting the DNA or gene into the eggs of the recipient organisms (Figure 7.11, page 202). It is not a brilliantly efficient method, but it is successful, although skilled operators are needed to carry out the microinjection. Typical figures are as follows (for a skilled technician): out of every 100 eggs that are collected perhaps 85 are suitable for injection. Of those injected about 60 survive the injection procedure, but only six of the injected eggs placed in the host mother result in live births, and only one or two of these will be transgenic animals. Nevertheless, the procedure does work, and selected offspring are bred to produce a transgenic line of animals. There are a number of variables that at present are difficult to control. One is the 'dose' of gene injected: several hundred copies of the gene may be injected, but an uncertain number of these are expressed. Furthermore, the injected DNA may be incorporated into the chromosomes of the host cell at random, may possibly not be stable, or may not be expressed. It may also be incorporated in the middle of one of the hosts' own genes that is vital for life! Since gene therapy is already starting with humans with genetic defects (see below), some of these problems need to be solved.

One of the first successes with transgenic mice was the production of a strain that produced more than the normal amount of growth hormone, giving a much larger than normal mouse (Figure 7.12, page 202).

Producing transgenic animals was a challenge, but why was it so important? First of all, it offered another way of finding out about eukaryotic genes and how they function. Animals — usually mice — can have genes 'added', as described above, and the function of the added gene may be studied. Genes that are only active during embryonic development may be made active at some later stage, so revealing their role in development. Genes can also be 'taken away', in the sense that specific genes can be made faulty or disabled. Mice which have been rendered transgenic in this way are called *knockout mice*. They behave like mutants who lack a functional gene, and this again may reveal the function of that gene. In the past, mutants, especially of micro-organisms which are easy to obtain, have been used extensively to study metabolic pathways. Making mutants of higher organisms to order was not possible, although the investigation of human mutants often gave valuable clues. For example, the appearance of phenylpyruvate and phenyl-lactate in the urine of individuals with phenylketonuria indicated that something was wrong with their phenylalanine metabolism. Soon the pathway of phenylalanine metabolism was worked out and the precise enzyme deficiency identified. This allowed a treatment for the disease to be devised — simply the removal of phenylalanine from the diet. A cure was not, however, possible, but may be in the future, thanks to gene therapy (see below).

Transgenic mice, in addition to providing basic scientific knowledge about how genes function, may be very valuable as a means of studying human diseases. They have already been used in the investigation of cystic fibrosis, for example, through the production of mice containing the faulty human gene.

To summarize, transgenic animals can give us a great deal of information about how genes work and they can provide 'models' of human disease that give clues as to the cause of the disease and might be used to test out potential treatments. They will no doubt find many other uses too, and the technology of transferring genes between organisms will be applicable eventually to the replacement of faulty genes in humans.

Transgenic plants

As well as transgenic animals, many different types of transgenic plants have been produced for a variety of reasons. The principles are the same as those for producing transgenic animals, but the ways in which genes may be transferred are different for plants (Spread 7.2). The fruits from transgenic plants are already appearing in our supermarkets and there is considerable concern about the ethical aspects of releasing transgenic organisms, here plants, into the world.

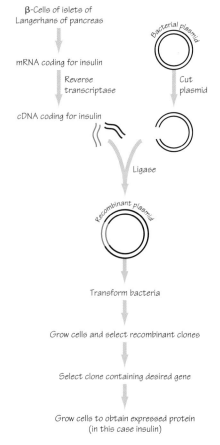

Figure 7.10 Oversimplified scheme for producing human insulin from bacteria. The gene for insulin will not be translated by bacteria because it contains introns which the organism will not recognize. Therefore complementary DNA (cDNA) is first made from the insulin mRNA using reverse transcriptase (see page 211). This is then inserted into a bacterial cell and the desired clones are selected and grown up to express the insulin. Usually bacterial promoters are included with the cDNA sequence to ensure expression. Note that insulin has two polypeptides, but in fact is synthesized as one polypeptide that is subsequently cut. Bacteria are unable to do this and the task is carried out instead by an enzyme which is added after the protein has been purified from the bacteria.

Transgenic plants

objectives
- To outline how transgenic plants may be produced by transferring 'foreign' genes into them using *Agrobacterium* plasmids
- To discuss the potential dangers in making such gene transfers
- To recognize that the technology for making gene transfers into monocotyledons, which include a great number of important crop plants, is not as well developed as that for dicotyledons

TRANSGENIC means containing a (functional) gene from some other organism, deliberately transferred in order to add new (presumably desirable) characteristics to a species. This has been done by plant and animal breeding for centuries, although there are severe problems associated with making crosses between species, which limits what can be done. However, because the genetic code is practically universal, genes (i.e. DNA) will work in any cell; the only problem is how to transfer this DNA to achieve its orderly integration into the recipient genome. Plant genetic engineering to produce transgenic plants gets over the species barrier; conventional plant breeding is not able to do this. Monocot genes may be expressed in dicots and indeed there is no reason why human genes may not be expressed in green plants.

Are transgenic plants safe to release?

Plants with modified genomes have already been produced that have increased resistance to disease and to herbicides, as well as having other changed characteristics. There is an ongoing debate about whether such 'unnatural' life forms should be released into the environment. Could they become uncontrollable and could they cross-breed with existing flora (it is impossible to stop pollen drifting in the wind or being carried by insects!) with effects detrimental to the environment?

Ten years ago there was considerable and vigorous opposition to all transgenic releases. Indeed, at the present time all laboratory work on transgenic plants is carried on in closely confining conditions. Although there was little scientific data on which to base conclusions it was thought to be better safe than sorry. Various organizations came into being to monitor possible environmental effects, such as PROSAMO, Planned Release of Selected and Modified Organisms, and ACRE, the Advisory Committee on the Release to the Environment. These organizations are concerned with risks to human health and the environment from growing the crop, but not from eating it. [The safety of food produced from transgenic plants is the responsibility of the Ministry of Agriculture (MAFF).] The sort of worry was whether potatoes with modified genes could transfer these genes to related *Solanaceae* such as black nightshade and woody nightshade by cross-pollination. Evidence is now starting to accumulate that allows rational decision-making. For example, when transgenic and non-transgenic plants were touching (in the case of potatoes) there was a cross-pollination rate of 24%; for plants 3 m apart the cross-pollination rate fell to 2% and at 10 m apart it was 0.017%. This was between potatoes; there was no evidence of cross-pollination with other *Solanaceae*.

Since the first cautious approach, scientists are now more relaxed about the release of genetically modified plants. In general, it is now believed that any risk is rather small; but the monitoring process will continue, with each case being considered on its merits and the risks being assessed as far as possible in the light of scientific evidence. About 10 experimental releases have been made in the U.K. since the first in 1987.

It is to be expected that more and more transgenic plants will be released as our experience increases. Release will be based on experimental evidence and the work will proceed with caution.

PRODUCING TRANSGENIC PLANTS

Plants cells often have a close association with certain bacteria, e.g. *Agrobacterium tumefaciens*. This bacterium readily infects dicots (tobacco, carrot) and when it contains a plasmid (the so-called Ti plasmid) it will cause the formation of a plant tumour called a crown gall.

A new gene can be inserted into the Ti plasmid and, via this route, incorporated into the plant cell. The region of the Ti plasmid responsible for tumour formation (the T region) is usually opened up and the new DNA inserted. The result is that the Agrobacterium still has infective properties; but galls are not formed. Hence the new gene is transferred into the plant cells.

The vir, for virulent, region of the Ti plasmid is responsible for integrating the new DNA into the plant's genome.

Usually the bacteria containing the Ti plasmid with the new gene are incubated with a leaf disc from the plant chosen. Eventually this leaf disc grows to produce a new plant.

TOTIPOTENCY

Plant cells, unlike most animal cells, are totipotent. This means that plant cells, under the appropriate conditions, can dedifferentiate into cells with the potential for producing a whole new plant.

A transgenic plant laboratory

Plant biotechnology: is it all about producing tasteless tomatoes with a longer shelf life?

7.2 Transgenic plants

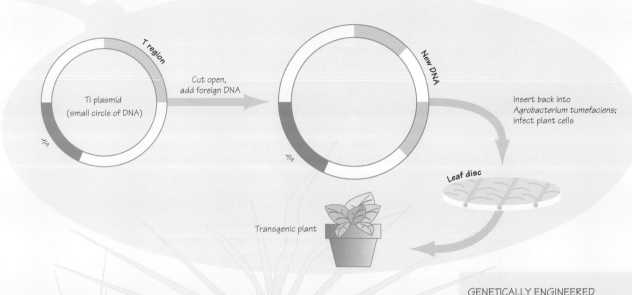

MONOCOTYLEDONS

Reliable techniques for introducing foreign genes into monocots are still being developed; these plants are not susceptible to Agrobacterium infection. This group of plants contains some extremely important crops, such as rice, corn, wheat and barley, sugar cane, etc. Consequently there is great interest in genetically modifying them. A goal, for example, would be to place the genes for nitrogen fixation in rice plants.

Direct introduction of DNA into rice has been achieved by the technique of electroporation, in which the cells are subjected to an electric field. This creates transient, microscopic holes in the plasma membrane, thereby allowing the entry of DNA. Microscopic laser beams have also been used. However, the subsequent incorporation of the foreign DNA into the genome of the cell is haphazard and of low efficiency.

Rice has 24 chromosomes and its genome is about one-tenth the size of the human genome. Other cereals, such as wheat and maize, have much larger genomes than that of rice. However, much of the genome of cereals is highly conserved, giving rise to the notion that 'the grasses' can be considered a single genetic system. This means that information and discoveries about their genomes are probably applicable to all types.

GENETICALLY ENGINEERED PLANTS IN THE FIELD

Several trials have taken place with genetically modified plants; including:

- incorporation of an insecticidal protein into tobacco plants; this prevents insects from feeding.

- production of pest-resistant seeds: for example, a team of US and Australian researchers developed a weevil-resistant variety of garden peas in 1994, offering the potential for similar varieties to be produced in chickpea, mung beans and kidney beans. (Brazilian farmers lose 20–40 per cent of their beans to storage pests.) The gene inserted is that coding for the inhibitor of the enzyme α-amylase, which prevents the weevil larvae from digesting starches in the seeds. The gene is only expressed in the pea seeds.

- insertion of antiherbicidal genes into crop plants: when a field is sprayed the weeds are susceptible, but the crop plants are resistant.

- introduction of genes into potatoes to make them resistant to potato leaf roll virus (PLRV), an important viral disease of potatoes.

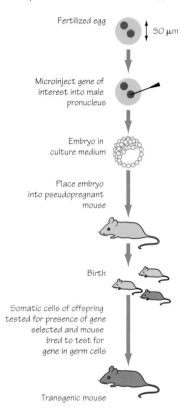

Figure 7.11 Basic procedure for making transgenic mice by micro-injecting DNA into the male pronucleus of the fertilized egg.

Figure 7.12 The transgenic mouse on the left carries an extra gene for rat growth hormone. (Courtesy of R. Brinster, University of Pennsylvania, U.S.A.)

7.4 Genetic diseases

At the beginning of the present century physicians coped with the ravages of infectious diseases caused by bacteria and viruses — tuberculosis, smallpox, diphtheria, polio, whooping cough, tetanus, and so on. They had no life-saving antibiotics and the use of immunization of whole populations was in its infancy. Treatment consisted of trying to keep patients alive for a sufficient time for their own immune systems to cope with the disease. Hopefully the survivors would have their own active immunity the next time there was an epidemic, but many died from these microbial diseases. In the developed countries we now have widespread immunization programmes and these once-dread diseases take up little of the doctor's time.

At the end of the twentieth century doctors are dealing with cancer and the diseases of old age, because people are living longer, at least in the developed countries. Doctors are also spending much more time on diseases of genetic origin, of which perhaps 5000 are recognized and many more still to be discovered. We have already considered the examples of sickle cell disease and phenylketonuria, but there are many more — some simple point mutations and some much more complicated multilocus conditions. Patients are being kept alive by modern medical treatment and we are finding out more and more about the diseases at the gene level. In the very near future the Human Genome Project (HUGO; Spread 7.3) will reveal the complete sequence of the human genome, and we shall start to look into individuals' genomes and see which faulty genes they carry. You may not want to know all of this information; for example, that you carry the gene for Alzheimer's disease that will develop when you are in middle age!

In this section we deal with two aspects of genetic diseases: one is their diagnosis using molecular biology methods, and the other is the possibility of gene therapy for individuals with genetic disease.

Diagnosis

When doctors diagnose disease they collect together information from various sources, and then, based on their knowledge and experience, reach a decision. The information sources include what they can see and feel, what the patient tells them including his or her history, and the results of laboratory tests. For example, a high glucose concentration in the blood or urine can mean that the patient is suffering from diabetes. The diagnosis will decide the treatment, and for diseases with similar symptoms, a careful differential diagnosis may be required to rule out some possibilities and confirm others.

The tests that are done are usually carried out on samples of blood or urine from the patient, and it is important that as much information as possible is obtained from such samples, and that, whenever possible, the samples are obtained non-invasively. Obtaining a urine specimen is obviously non-invasive, taking a blood sample involves a syringe prick, and obtaining a sample of cerebrospinal fluid is quite an uncomfortable procedure. It is not only the comfort of a sick patient that has to be considered here. Carrying out invasive tests puts the patient at further risk, including from infection, and may put the medical practitioner at risk from legal proceedings if things go wrong.

Hospitals run clinical chemistry laboratories that are organized to carry out a battery of tests on urine and blood samples from patients. Many of these tests are done by automatic 'analysers', and typically the whole range of analyses is carried out whether or not the physician specifically asks for them (Table 7.3). The analyses are usually for substances (glucose, urea, haemoglobin, various ions) and, in blood, certain enzymes that give clues about disease conditions. More rare conditions may require special analyses to be carried out manually which tend to be much more expensive to do. They will only be performed after other possibilities have been ruled out or if the initial diagnosis clearly warrants it.

Table 7.3
Substances tested for by clinical chemistry laboratories on blood samples

Ions	Na^+, K^+, Cl^-, HCO_3^-, Ca^{2+}
Organic constituents	Glucose, lactate, citrate, bilirubin
Nitrogenous constituents	NH_4^+, urea, creatinine, uric acid, glutamine, phenylalanine
Lipids	Triacylglycerol, cholesterol, total fatty acids
Proteins	Total, albumin, globulins, lipoprotein
Clotting factors	Fibrinogen, plasminogen, prothrombin
Blood cells	Haemoglobin, red cell count, packed cell volume, white cell count, platelets
Enzymes	Acid phosphatase, alkaline phosphatase, creatine kinase, lactate dehydrogenase, amylase, aspartate transaminase (SGOT), alanine transaminase (SGPT), serum lipase, α-galactosidase, α-mannosidase, α-glucosidase, N-acetyl-β-glucosaminidase, γ-glutamyltransferase

It should be emphasized that uncomfortable and potentially dangerous invasive methods such as taking biopsies (small pieces of tissue) from muscle or liver, or bone marrow sampling, will only be carried out if all other ways of deciding on the diagnosis have been exhausted.

The traditional clinical chemistry laboratory, as we have said, provides the bedside physician with information about substances and enzymes, and also about blood parameters (do the patient's red cells show sickling; is there anaemia?). Some of these obviously relate to genetic disease. Sickle cell anaemia is one example, phenylpyruvic acid in the urine is another. Can the diagnosis of genetic disease be confirmed by examining the patient's DNA and is there any value in doing this? (It may, for example, be perfectly obvious that a patient has sickle cell disease from his appearance and symptoms and from an examination of the red cells.)

The answer to this question is "it depends". It might help to diagnose the heterozygous condition, it might help to track the genetic origin of

The Human Genome Project

objectives

- To comprehend the size of the task involved in sequencing the human genome
- To appreciate that, in addition to the scientific interest, there are financial and ethical issues too

The human genome — the total length of DNA in a diploid human cell nucleus — is about 3×10^9 base pairs. This corresponds to about 2 metres of DNA per nucleus. This is not the largest — some amphibians and some plants have even longer genomes (see chart).

In general, the more complex the organism, the longer the DNA, although, as the chart shows, this is not a fixed rule. Thus human cells contain about 700 times more DNA than *Escherichia coli*, a typical bacterial cell. There is a question of whether all this DNA codes for proteins or has some other function of which we are unaware. Calculations show that 3×10^9 base pairs is enough to code for about 3×10^6 average sized proteins, but that the actual number of protein-coding genes in a human cell is likely to be much less than this, probably about 60,000.

On the basis that the genetic message in the genome specifies all the inherited characteristics of an organism, as soon as it became possible to sequence DNA (Spread 7.1), biologists were keen to know the complete sequences of genomic DNA molecules. Obviously, for a small viral genome only encoding a few proteins, this represented a feasible task that could be completed with existing technology and in a time span of months or years. Very soon complete sequences of viral genomes started to appear. Polyoma virus contains a single molecule of DNA of length 1.7 μm, corresponding to 5100 base pairs; this is sufficient to code of about 1700 amino acid residues. Assuming a typical protein has about 250 amino acid residues, then six or seven proteins are encoded. Sequencing the 3×10^9 base pairs of the human genome is a problem of a completely different order. To put this in perspective, the background diagram on the right shows the actual sequence of bases of the DNA of a small bacteriophage ($\varnothing \times 174$) that attacks *E. coli*, containing 5057 base pairs. On the same scale as this, the genome of *E. coli* would occupy 500 times this area and the human genome about 250,000 times this area!

It is now technically possible to determine the entire base sequence of the human genome and there seems no doubt that this will indeed be achieved in the next 5–10 years. However, it is not just a matter of saying "let's get on with it"; there are a number of other important factors to consider, such as who will pay for it, how will the information be stored, and, not least, how will the information be used?

Comparison of genome sizes in a variety of organisms from the smallest bacteria (mycoplasmas) through to flowering plants (note that this is on a logarithmic scale!).

	10^6	10^7	10^8	10^9	10^{10}	10^{11}
Flowering plants						
Birds						
Mammals						
Reptiles						
Amphibians						
Bony fish						
Cartilaginous fish						
Echinoderms						
Crustaceans						
Insects						
Moulds						
Algae						
Fungi						
Gram-positive bacteria						
Gram-negative bacteria						
Mycoplasma						

Base pairs

THE YEAST GENOME

Wednesday 24 April 1996 marked the end of a huge project involving scientists in 37 laboratories to sequence the entire yeast genome. The sequence contains about 14 million nucleotides encoding perhaps 6000 genes. This is the first complete genome for a *eukaryotic organism* to be sequenced. The sequence itself makes rather boring reading. It may be accessed on the World Wide Web at the following address: http//genome-www.stanford.edu/Saccharomyces/
Figure shows yeast cell growing on an acrylic plate as seen under the scanning electron microscope. Magnification × 1600. (Courtesy of J. Verran, the Manchester Metropolitan University.)

THE COST

At the current estimate the cost (labour and chemicals) of sequencing DNA is about $1 a base; thus to determine the entire sequence will cost $1 billion at least. There are other costs too, such as that of sorting out all the bits of DNA (you cannot just start at one end and carry on sequencing) and dealing with the computing problems. On the other hand automated technology will probably make sequencing considerably cheaper per base in the next few years.

Another way of looking at the cost is in terms of number of bases and time. Some people estimate that the project will cost $60 million a year for 15 years, again giving a cost of nearly $1 billion. If the rate of sequencing can be increased to one million base pairs per day by automation, the project will still take nearly 10 years to complete. A similar project, at least in terms of finance, was that undertaken in the 1960s to put a man on the moon. Some people thought this immoral when millions of people in the world were dying of hunger, and the same arguments have been set against proceeding with the Human Genome Project. Such arguments are unlikely to stop the project, however.

7.3 The Human Genome Project

DATA STORAGE

DNA base sequences do not make very exciting reading for humans, and storage of this information is something that can only be done using a computer. Computers of course are happy with this sort of linear data, and systems are being developed to store such sequences. It is important to realize, however, that it is not just storage. For the data to be of any use, it must be possible to search it to find sequences, for example to look at homologies and to find mutations. Here, again, computer algorithms can be designed to do this.

ETHICAL ISSUES

Assuming that a complete human DNA sequence becomes available, then questions start to arise as to who should be able to have access to and use the information:

Who owns genetic information?

How should genetic information be used?

Who decides who should be screened?

Are there limits to prenatal testing?

Who should be employed?

Who should be insured?

Such issues need to be widely discussed, and interested parties include parents of children with genetic diseases, science educators, religious groups, people from ethnic minorities as well as the medical profession. Obviously insurance companies and lawyers have an interest too.

In contrast, new knowledge about new genes could bring untold medical benefits for individuals with, or likely to have, genetic disease, especially in an era when gene therapy is just beginning (Spread 7.4).

In the double helix of DNA, 10 base pairs occupy 3.4 nm of a helix. If the human genome contains 3×10^9 base pairs, how long is this DNA?

OTHER ISSUES

Sequencing the human genome is not straightforward, especially because of the belief that, of the base sequence in the DNA, only perhaps 5% codes for proteins or other nucleic acids (e.g. ribosomal RNA) or represents control regions. The rest may be junk; it certainly consists of introns, repeated elements and other non-functional (as far as we know) sequences. Some workers say that the money should be concentrated on sequencing targeted function segments of DNA. This would speed up things enormously in terms of finding mutations responsible for specific diseases and providing means for their diagnosis and treatment. Nevertheless, the Project is going ahead and is co-ordinated worldwide by HUGO, the Human Gene Organization.

OTHER GENOMES

Other, more modest, projects are continuing in order to find the sequences of smaller genomes or chromosomes. The sequence of *E. coli*'s genome is eagerly sought, but, in fact, the sequence of another bacterial genome, that of *Haemophilus influenzae*, was reported in July 1995. This bacterium can cause ear infections and meningitis. Its genome sequence is 1.8×10^6 base pairs long, comprising 1743 genes. As is typical for bacteria there is little non-coding DNA. There were two interesting things about this particular project. One was that more than 40% of the genes identified seemed to have no counterparts of known function among the genes that have been sequenced from other organisms. These therefore represent 'new' genes to be studied. The other thing is that the experimental approach used was rather revolutionary and included dropping the sequencing accuracy demanded from the original target of 99.99% to 99.90%, making the process much quicker.

Other genomes that have been partially sequenced include those of bakers' yeast (yeast has 17 chromosomes; see adjacent page), and a continuous segment of length 1.1×10^6 base pairs of chromosome III of the nematode worm, *Caenorhabditis elegans* (out of a total of about 100×10^6 base pairs).

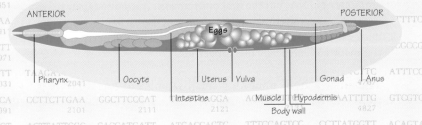

Body plan of *Caenorhabditis elegans*. The whole animal is just over 1 mm long: it consists of exactly 959 somatic cells plus 1000–2000 germ cells.

Chapter 7 • Molecular Biology and Applied Biochemistry

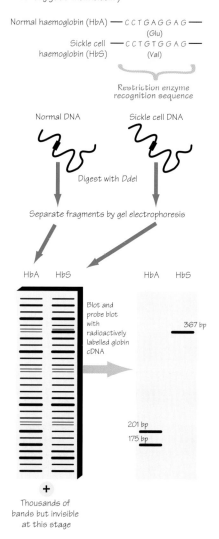

Figure 7.13 Prenatal diagnosis of sickle cell disease. A sample of DNA is obtained from the foetus (see Figure 7.14) and digested with the restriction enzyme *Dde*I. The normal HbA gene is cleaved by this but, because of the A to T change, the HbS gene is not. When blotted with the cDNA for globin, hybridization takes place and the radioactivity is detected when the probe has found its partner. For normal haemoglobin (HbA) there are two bands because the gene has been cut by the restriction enzyme. HbS only shows one band corresponding to a longer piece of DNA (bp = number of base pairs). (What result would be obtained with a heterozygous individual?)

the disease in the family, and in the future it will be vital before gene therapy is attempted. It is not particularly difficult to do, although not all laboratories have the facilities. The tests normally involve using a piece of DNA and matching it up by hybridization with the patient's DNA, to see if it has an abnormal base sequence or not. It does not matter where the DNA comes from: all the patient's cells (except the gametes) have the same DNA, so a few white cells from the blood or some epidermal cells from inside the cheek will do. Figure 7.13 illustrates one method of showing that a patient has sickle cell disease or sickle cell trait by using specific restriction enzymes. Similar methods are used for other diseases. The PCR has also improved methodology enormously: a pair of DNA primers may be used which will amplify the normal gene but not the mutated one. This method can give sufficient amplified DNA even from a few cells for a band to be seen on an electrophoresis gel, which is visible to the naked eye after staining. This is a great advantage because it means that radioactive isotopes do not need to be used; these are both expensive and potentially dangerous to laboratory staff.

Prenatal diagnosis

One of the most important and controversial areas of diagnosing genetic disease is the determination of whether an unborn baby is likely to be born with a serious genetic disease. This may be important information for the way the pregnancy is managed by the doctor and for what precautions are taken in preparation for the birth itself. Most controversially, however, it gives the parents the opportunity of deciding whether the best way forward, taking all the factors into consideration, is to have an abortion. In the case of Down's syndrome infants, an analysis of the chromosomes, of course, gives this information. The availability of genetic testing for an ever increasing number of genetic diseases by the methods described above increases the information available as well as increasing the possibility for potential parents to have to make heart-rending decisions.

It is now a fairly routine procedure to obtain a few cells from a foetus for chromosome and/or DNA analysis. The original method was *amniocentesis*, where a long needle is placed through the mother's abdominal wall to take some of the fluid containing cells from the amniotic fluid surrounding the foetus (Figure 7.14). This method, although invasive, is very safe and is used extensively, but had two disadvantages. One was that it could not be carried out until the foetus was a certain size, usually after the end of the third month of pregnancy. The second was that only a very few cells were obtained in the amniotic fluid and these had to be grown up in cell culture for a week or two in order to obtain sufficient material to analyse. The second disadvantage has become less important since the advent of the PCR to amplify tiny amounts of DNA, but the first one remains. It is a disadvantage because, if parents are contemplating an abortion, the earlier they can make the decision the better so that the procedure will be, relatively speaking, less traumatic and medically less dangerous.

The other way of obtaining small samples of foetal DNA is by a technique called *chorionic villus sampling*. In this procedure a small amount of tissue from the placenta is taken (remember that the placenta is foetal tissue; Figure 7.14). This procedure may be carried out very early on in the pregnancy, and, consequently, decisions about abortion taken, typically well before the end of the third month. The procedure does not seem to affect the foetus significantly, although it has to be borne in mind that any invasive procedure has risks attached to it.

In the future, increasing knowledge about DNA as well as the HUGO project will mean that it will be possible to test for more and more genetic diseases, and the ethical and personal problems arising from this can only increase, too.

Gene therapy

Modern medicine treats many diseases by means of a *transplant*. Kidney and heart transplants are commonplace, as are bone-marrow transplants, and livers and lungs are also transplanted. Such measures are reasonably successful, one of the main problems being to try to get a good degree of tissue compatibility so that the grafted organ is not rejected by the recipient's immune system. This is difficult in other than identical twins, but quite good matches can be made. It is often a case of a potential recipient awaiting a donor organ of a suitable tissue type. If an organ is transplanted there is a difficult balance to be struck between suppressing the recipient's immune system with drugs such as cyclosporin A to prevent rejection of the graft, but giving the patient sufficient antibiotic cover to prevent infection after major surgery with the immune system suppressed.

In the case of genetic diseases, treatment is often possible by diet or drugs such that the patient can live a reasonable life, but it is always a treatment rather than a cure. If a gene is lacking or is mutated so that it does not produce a functional gene product (i.e. a protein), then it may be asked why not transplant a functional gene rather than an organ such as a kidney? The big advantage to this would be that genes or the genetic code know no species barriers, and such a transplantation should not, therefore, affect the immune system. The big problem is how to deliver the gene to the right cells and get it incorporated into the chromosomes such that it is expressed at the appropriate time. In some cases the mutation causes lack of an enzyme protein, while in others, such as cystic fibrosis, it is a membrane protein that acts as an ion channel that is missing. When there is a faulty gene causing havoc, as is the case in sickle cell disease, it may be necessary to inactivate the existing mutant gene before replacing it with a good one.

Many things need to be discovered and many new techniques remain to be developed, but medical scientists are already aware of the possibilities and the first clinical trials are already taking place. The simplest case to treat is one in which a soluble blood protein is absent, such as in the condition of haemophilia. Implanting a small group of cells capable of making the missing protein and secreting it into the blood may be sufficient to produce a cure. Some examples of how gene therapy might be carried out are given in Spread 7.4.

Figure 7.14 Prenatal diagnosis of genetic disease. **(a)** Amniocentesis allows 10–20 cm^3 of amniotic fluid to be extracted from around the foetus. This contains cells derived from foetal skin. However, it may take up to 2 weeks to culture the cells and carry out chromosome analysis, and longer if sufficient cells are to be grown for enzyme analysis. **(b)** Chorionic villus sampling may be carried out much earlier in the pregnancy. Chromosome and enzyme analysis may be performed directly on the tissue obtained. Furthermore, DNA may be extracted for analysis.

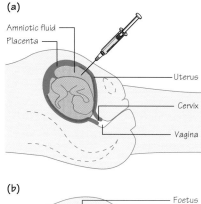

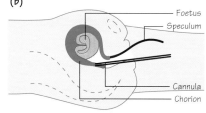

Gene therapy

objectives

- To describe how genetic diseases caused by mutations may potentially be cured by gene therapy (human genetic engineering)
- To explain that genes may be transferred as naked DNA, albeit rather inefficiently, or more efficiently by using disabled viruses to effect the transfer

GENE THERAPY is genetic engineering with humans, that is to say, attempting to put new genes into human cells. Such genes may be lacking in an individual or inactive as a result of a mutation. An example of gene therapy might be to attempt to replace the gene for sickle cell haemoglobin by the 'correct' gene for normal adult haemoglobin. In fact, this would be rather a difficult thing to do with present-day techniques, namely, stopping the action of one gene and inserting another. An easier proposition would be to simply replace a gene that was lacking. There are good prospects for achieving this in individuals with haemophilia who lack one of the clotting factors in their blood due to a mutation.

GERMLINE or SOMATIC THERAPY?

The GERMLINE cells of the body are those of the ovary and testis that give rise to the eggs and sperm that will eventually form the next generation of individuals. The SOMATIC cells are all the other, diploid, cells of the body, which do not pass their genetic material on to the next generation.

In principle either germline or somatic cells could be subjected to gene therapy. At present, however, germline therapy is held to be unacceptable, mainly because we know so little about what we are doing (rather little is known about the control and integration of genes) and of course it would affect all future generations. Somatic cell therapy, however, is generally accepted as a possible way of treating otherwise incurable genetic diseases. It is still in its infancy; there is still a great deal we do not know, but there is no doubt that it will become an accepted treatment within the next 5–50 years, depending upon how the techniques of gene transfer improve.

Possible routes and cells for introducing new genes into somatic cells

CENTRAL NERVOUS SYSTEM (*ex vivo*): fibroblasts, myoblasts, glial cells

LUNG (*in vivo*): DNA–liposome complexes; aerosol

VASCULAR (*in vivo*): DNA–protein complexes

SKIN (*ex vivo*): fibroblasts, keratinocytes

BONE MARROW (*ex vivo*): stem cells

SKELETAL MUSCLE (*in vivo*): naked DNA, DNA–liposome complexes (*ex vivo*): myoblasts

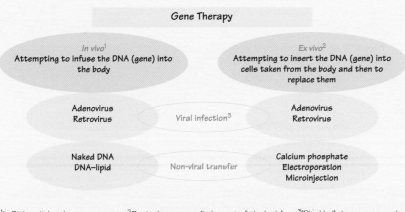

Gene Therapy

In vivo[1]
Attempting to infuse the DNA (gene) into the body

- Adenovirus / Retrovirus
- Naked DNA / DNA–lipid

Viral infection[3]

Non-viral transfer

Ex vivo[2]
Attempting to insert the DNA (gene) into cells taken from the body and then to replace them

- Adenovirus / Retrovirus
- Calcium phosphate / Electroporation / Microinjection

[1] In Biology it has been customary to talk about *in vivo* (things that are carried out 'in life', i.e. in the body), compared with those performed *in vitro* ('in glass', i.e in the test tube).

[2] Ex vivo here means 'taken out of the body' for manipulation. (Some molecular biologists now talk about doing experiments *in silico* — changing DNA base sequences in the computer and trying to predict what the effects of these 'theoretical mutations' might be!)

[3] 'Disabled' viruses are used. Their replicating machinery has been eliminated; however, they can still infect cells and can carry pieces of foreign DNA (genes) within their genomes.

Life chemistry and molecular biology

7.4 Gene therapy

WILL THE AIDS VIRUS TURN OUT TO BE USEFUL FOR DELIVERING GENES TO CELLS FOR GENE THERAPY?

The biggest problem with gene therapy using viruses as the vectors to deliver genes to human cells is that they do not work very well. There is now hope that a harmless form of HIV (see Spread 7.5), constructed by a group of researchers at the Salk Institute in San Diego, may provide a good delivery system. The group deleted all the genetic elements that the virus needs to copy itself. They also swapped its surface protein with that from a different virus, to give it the ability to enter a range of cells and to make it more stable. This work is still a long way from the clinic, but it is another step towards the production of a reliable vector. HIV has produced the most feared 'plague' this century, but may ultimately provide salvation to sufferers from genetic disease. [See report in Science (1996), **272**, 195.]

INTRODUCTION OF NEW GENES INTO HUMAN SOMATIC CELLS CAN BE PERFORMED USING BOTH VIRAL AND NON-VIRAL MECHANISMS

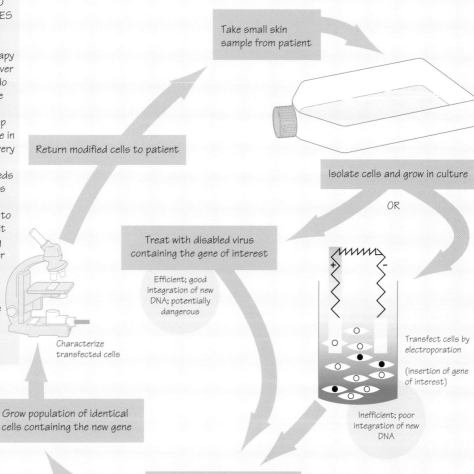

Take small skin sample from patient

Isolate cells and grow in culture

OR

Treat with disabled virus containing the gene of interest

Efficient; good integration of new DNA; potentially dangerous

Transfect cells by electroporation (insertion of gene of interest)

Inefficient; poor integration of new DNA

Characterize transfected cells

Return modified cells to patient

Grow population of identical cells containing the new gene

Select transfected cells

LIPOSOMES

Liposomes are spherical configurations of cationic lipids. As the electron micrograph shows, they are multilayered structures. They are capable of delivering their contents (e.g. DNA) to cells by fusing with the plasma membrane. They are currently being tested as a possible means of delivering the gene for cystic fibrosis into the human lung.

Liposomes seen in the electron microscope by freeze fracture.

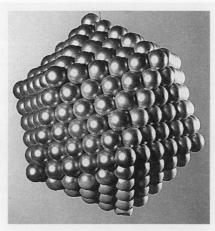

Model of an icosahedron similar to that in adenovirus, made with ping-pong balls.

ADENOSINE DEAMINASE DEFICIENCY (ADA) and SCID

Individuals lacking the enzyme adenosine deaminase because of a mutation accumulate its substrate, deoxyadenosine, to toxic levels. This leads to destruction of the T cells of the immune system and the infant is left without defence against invading organisms — a condition known as 'SCID' or severe combined immuno-deficiency. Affected children usually die by the age of 2 years, although efforts may be made to protect them against environmental organisms by having them live in a sterile 'bubble'.

The first attempts to treat such children by gene therapy were made in 1990. They were only partially successful. Some of the treated children are now being kept alive by (very expensive) injections of the pure enzyme, ADA, rather than by gene therapy.

Chapter 7 • **Molecular Biology and Applied Biochemistry** 209

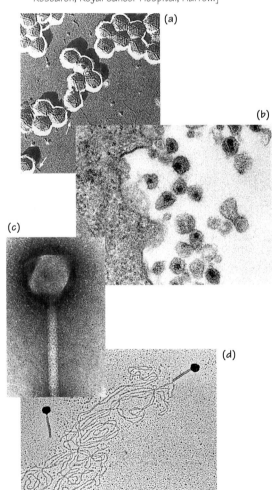

Figure 7.15 Some viruses seen under the electron microscope. (a) Adenovirus particles, shadowed with heavy metal. The arrows show the 'spikes' (see Figure 7.18). Magnification × 50,000. (b) HIV particles leaving a human white blood cell; each virus is surrounded by a piece of membrane from the cell from which it 'budded'. Magnification × 22,000.
(c) A bacteriophage showing the 'head' which contains the DNA (top) and the 'tail plate' which attaches to a bacterial cell during infection. Magnification × 1,700,000.
(d) When phage particles are subjected to osmotic shock their DNA is released. Magnification × 340,000. [Photographs courtesy of: (a) G.E. Blair, (c) J.H. Parish, (d) V. Virrankoski-Castrodeza, all Department of Biochemistry and Molecular Biology, University of Leeds;
(b) D. Robertson, Institute for Cancer Research, Royal Cancer Hospital, Harrow.]

Viruses

Viruses are particles consisting mainly of some nucleic acid surrounded by a protein coat (Figure 7.15). They have no independent existence but when they infect a cell — and viruses are known that infect bacteria, plants and animals — they typically take over the cellular machinery so that more virus particles are produced. Viruses are extremely small and may only be seen under the electron microscope. Nevertheless, they cause extremely serious diseases in humans (smallpox, polio, AIDS), many irritating but not usually fatal diseases (influenza, measles, colds) (Table 7.4), as well as being capable of devastating plant crops. Viruses represent biological information at its simplest, in the virus nucleic acid, whose only aim is to reproduce itself by parasitizing other cells. The amount of information carried by a given virus varies enormously, but ranges from very small to very tiny! 'Very tiny' here means information to code for a few enzymes and a coat protein, for example; everything else required for the virus' replication is supplied by the host cell.

Table 7.4
Viruses that cause disease in humans

Class	Name	Disease (example)
DNA viruses	Adenovirus	Respiratory infections in children
	Vaccinia virus	Smallpox, cow pox
	Hepadna virus	Hepatitis B
	Herpes virus	Herpes simplex
RNA viruses	Picorna virus	Polio
	Toga virus	Rubella
	Paramyxovirus	Measles
	Rhabdovirus	Rabies
	Orthomyxovirus	Influenza

Some viruses have a genome that is made up of double-stranded DNA but others have single-stranded DNA. Yet others have instead RNA, sometimes double stranded and sometimes single stranded (Table 7.5).

Where viruses have only single-stranded nucleic acid, it can be shown that they always have a stage in their life cycle inside the host cell when there is a double-stranded form. Thus they replicate their DNA by the semi-conservative method just like other organisms do. An example here is the small single-stranded DNA virus ØX174 which infects the bacterium *Escherichia coli*. The virus itself is an icosahedron 25 nm in diameter and its genome is 5500 base pairs long, equivalent to about five genes.

Table 7.5
Viruses may have DNA or RNA as their genome, and it may be single stranded (ss) or double stranded (ds)

Name	Diameter (nm)	Nucleic acid	Approximate number of genes
Bacteriophage ⌀ X174	25	ss DNA	5
Smallpox (Vaccinia)	200	ds DNA	240
Polio virus	27	ss RNA	8
Tobacco mosaic virus	18 × 300	ss RNA	6
Reoviruses	60	ds RNA	22

Many viruses have single-stranded RNA as their genome and some have double-stranded RNA. Double-stranded RNA is unusual in nature, but it comprises the genome of reoviruses. Included in this class are the rotaviruses that cause vomiting, diarrhoea and fever in babies. Where viruses have single-stranded RNA, as is the case with tobacco mosaic virus (TMV), a second complementary strand of RNA is synthesized when the virus enters the plant cell. This then serves as the template for the synthesis of a large number of new RNA strands complementary to itself, and these become packaged in newly synthesized coat protein as new virus particles. However, this is not the only mode of replication of single-stranded RNA viruses, and the machinery of the retrovirus has turned out to be extremely important recently because of the AIDS virus, but also because some retroviruses can cause tumours. The retroviruses replicate by producing a complementary DNA copy of the RNA genome once inside the host cell. The enzyme that catalyses this is called *reverse transcriptase* and has proved to be an extremely valuable tool in molecular biology.

For the purposes of delivering genes to a person's cells a modified retrovirus would provide an ideal vehicle. As will be explained below, at a certain stage of the virus's life cycle, it produces double-stranded DNA by copying its RNA, and this DNA becomes incorporated into the host cell DNA. If this were organized properly, by inserting the gene we wish to transplant into the viral RNA, it would succeed in delivering the gene. It would also be vital to disable the virus, that is to say, cut down its genome (and this can be done by molecular biology methods) to the absolute minimum that allows it to enter a cell, make DNA from the RNA, and incorporate this into the host cell DNA. Any genes responsible for other activities, such as making new virus particles or causing toxic effects, would be removed from the viral genome.

It is also important to realize that animal viruses only attack cells that have specific receptors for them on their surfaces. These cell surface receptors are probably glycoproteins that are involved in cell–cell recognition under normal circumstances. When viruses link on to them, changes may then take place so that the virus is actually taken up into the cell. This is a case of the virus exploiting the cellular machinery. The consequence is that most animal viruses are highly specific for the host cells they can attack. This too offers a way in which they can be controlled.

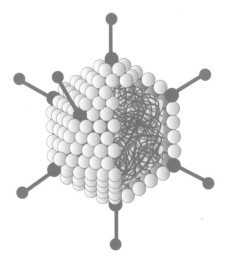

Figure 7.16 Structure of adenovirus (see electron micrograph in Figure 7.15a). The virus coat, made up of protein subunits, is an icosahedron, with fibres extending from the 12 vertices. The cutaway reveals the DNA in the centre of the particle (although it is not known how this is packaged).

A number of laboratories are working on producing disabled viruses (usually adenovirus; Figure 7.16) as a way of delivering genes to human cells. Adenoviruses derive their name from the tissue from which they were first isolated; namely, human adenoids. They can cause diseases of the respiratory tract and the eye. Most children become infected with adenoviruses in early life (exhibiting symptoms such as stuffy nose and cough), but the majority of people rapidly become immune to them.

Retroviruses

A problem with gene therapy is that of inserting a correct gene into millions of body cells. One of the ways in which a gene — a piece of DNA — might be delivered to a large number of cells is by the use of *retroviruses*. Viruses such as human immunodeficiency virus (HIV) can cause very severe consequences when they invade and inactivate white blood cells of the immune system, causing acquired immunodeficiency syndrome (AIDS). Nevertheless, a knowledge of the biology of retroviruses may equally provide a means of carrying out gene therapy — another instance of exploiting biological knowledge.

RNA tumour viruses

Viruses were discovered in around 1910 as infectious particles that were so small they could pass through filters that held back bacteria. At about that time an agent was found that caused malignant tumours called *sarcomas* in chickens. Cell-free filtrates of the tumours, injected into healthy chickens, caused the development of tumours. Much later on, other viruses were discovered that could cause tumours in other animals such as mice. The viruses were classified as oncornaviruses from *onco*, a lump or a tumour, and RNA. Subsequently it was discovered that a number of human cancers were highly likely to be caused by viruses. These include various tumours caused by the Epstein–Barr virus and carcinoma of the liver by hepatitis B virus. It is also possible that carcinoma of the cervix may be caused by the herpes virus.

[We should be very clear here that not all cancers, and probably not the majority of cancers, are caused by viruses. There are many causes, including so-called carcinogenic compounds, of which many are known and which may be present in, for example, cigarette smoke, as well as a failure of the mechanisms that normally control cell division. A tumour is the result of uncontrolled cell division, and tumour cells can migrate and initiate tumour formation in other parts of the body — a process known as *metastasis*. This again is failure of the normal methods of control. Liver or skin cells do not normally wander off to other parts of the body; certain chemical signals keep them in their proper place.]

The important finding was that RNA tumour viruses could make DNA from RNA using the enzyme reverse transcriptase (RTase). The viral particle gains access to a cell, the viral RNA is uncoated in the cytosol and RTase synthesizes DNA. Meanwhile the RNA is destroyed by digestion and the RTase continues to synthesize a second, complementary, strand of DNA, thus making double-stranded DNA. Other retroviruses, such as the AIDS virus, work in a similar fashion, although they do not cause cancer (Spread 7.5).

An obligatory step in the continuation of the life cycle of retroviruses is the integration of the viral DNA into the host cell's DNA. This integration can occur nearly anywhere in the host genome. Subsequently, replication of the virus can take place by a rather complex and incompletely understood mechanism.

Oncogenes

Studies of the avian sarcoma virus many years after its discovery as a tumour-causing virus revealed that just one viral gene (out of 20 or so), called *src*, is required to induce cancer in chickens. Surprisingly, it was found that normal chicken cells uninfected by the virus had a very similar gene to *src*. The one from the virus was called v-*src* and that from the cells, c-*src*. Both genes code for a particular enzyme (see Table 7.6), but in the viral version the 19 carboxy-terminal residues are replaced by 12 unrelated residues. It just happens that these 19 residues in the cellular protein represent a control region of the protein, and in the viral version this control is absent. Consequently the virally encoded enzyme is active all the time, rather than being switched on and off at appropriate times as is the cellular gene, and this is what causes the tumour. The viral version of the gene, v-*src*, is called an *oncogene*.

Since then many other oncogenes have been discovered. They are all characterized by being similar to certain cellular proteins but lacking small regions that control their activities, and hence they tend to be active all the time, causing cancer. The cellular gene is sometimes called a *proto-oncogene* (e.g. c-*src* is a proto-oncogene). What seems likely to have happened is that, in the course of viral genes being incorporated into the host genome during the virus's life cycle, the proto-oncogene has been picked up accidentally by the virus from the genome of the host cell but at some stage has undergone mutation. Some examples of oncogenes are given in Table 7.6.

Table 7.6
Oncogenes from retroviruses and the proteins or enzymes for which they code

Oncogene	Retrovirus	Protein/enzyme
src	Avian sarcoma virus	Tyrosine kinase
erb B	Avian erythroblastosis virus	Growth factor receptor
ras	Mouse sarcoma virus	GTP-binding protein
fos	Osteosarcoma virus	Nuclear proteins

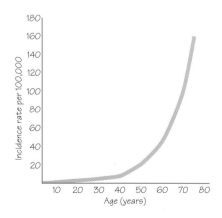

Figure 7.17 The incidence of cancer in a population increases with age. (The figures here are for colon cancer in England and Wales for 1 year.) The incidence rises steeply as a function of age. If only a single mutation were required to trigger cancer and this mutation had an equal chance of occurring at any time, the incidence would be independent of age. These data suggest that about six independent rare events must occur and their effects accumulate before a cancer occurs. (Redrawn with permission from the U.S. Department of Health, Education and Welfare: Vital Statistics of the United States, vol. II, Mortality, Washington DC, U.S. Government Printing Office.)

Viruses, HIV and AIDS

objectives
- To explain the structure of viruses that infect human cells
- To outline the progression of HIV in destroying the immune system
- To describe the class of viruses called retroviruses

Viruses are acellular and parasitic. A typical virus is a simple structure composed of a nucleic acid core completely enclosed within a protein coat (see diagram opposite). Some also have an outer lipid coat or envelope.

Viruses are classified on the border between living and non-living. They are considered living since they contain genetic material and are capable of reproduction, but non-living since they have no cellular organization. They certainly show no signs of life outside the cells of their host.

RETROVIRUSES are RNA viruses that use the enzyme REVERSE TRANSCRIPTASE to synthesize DNA from their RNA template. The best known retrovirus is human immunodeficiency virus (HIV), the agent thought to be responsible for acquired immunodeficiency syndrome (AIDS).

Viral diseases are often difficult to treat since antibiotics cannot be used (there are few metabolic processes distinct from those of the host cell for them to inhibit). VACCINES may be ineffective since the viruses may mutate, thereby changing the antigenic factors on their capsids. CHEMOTHERAPY may damage host cells as well as inhibiting viral replication. The most common forms of control are vaccination and removal of the source of infection. ACYCLOVIR, a thymidine analogue, is a drug specific to the treatment of *Herpes simplex*, since it is activated only in the presence of *Herpes simplex* virus (HSV) thymidine kinase. If it is applied topically it prevents multiplication of HSV but does not eradicate it from the body so that infection may flare up again on another occasion.

AMANTADINE has a different effect — it prevents the influenza virus from entering the cells and is therefore most effective when given before the infection has spread widely.

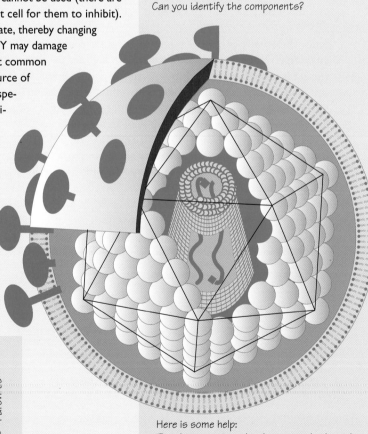

A TYPICAL VIRUS

THE AIDS VIRUS (HIV)
Can you identify the components?

Here is some help:
The glycoprotein molecules protrude through the lipid membrane. An icosahedral shell underlies the membrane and itself encloses a vase-shaped structure. The diploid RNA is enclosed in the 'vase'.

Relative sizes of some viruses compared with a typical Bacterium

Poxvirus, Rhabdovirus, Herpes virus, Adenovirus, Parovirus

Escherichia coli

2 μm

214 Life chemistry and molecular biology

7.5 Viruses, HIV and AIDS

The CORE is the region within the capsid. It is not cytoplasm, and contains no organelles.

The GENETIC MATERIAL may be either DNA or RNA, but never both. The number of genes is very small; these genes contain only the information necessary for replication of viral subunits and their assembly into a complete virus or VIRION, once within the host cell.

The PROTEIN COAT is composed of a large number of identical units called CAPSOMERES which self-assemble into a highly symmetrical CAPSID. The form of the assembled capsid may be used in the classification of viruses.

SURFACE EXTENSIONS of the capsid (glycoproteins) are responsible for the antigenic properties of the virus.

A LIPOPROTEIN ENVELOPE surrounds the capsid in some viruses, usually the larger ones. The envelope is frequently derived from the host cell and encloses the virion as it is dispersed from the host. This envelope may be important in the viruses' ability to penetrate the host cell's defences. The human immunodeficiency virus is surrounded by such a lipoprotein envelope.

MODES OF TRANSMISSION OF AIDS

1. Sexual intercourse (unprotected).
2. Intravenous drug use (sharing needles).
3. From mother to infant (across placenta, during delivery, in milk).
4. From contaminated blood products (transfusion, clotting factors).

CLINICAL COURSE (may take 10 years)

1. Mild, influenza-like illness for 1 month; then slowly over years ......
2. Swollen lymph nodes, fever, night sweats, weight loss, and ......
3. 'Opportunistic' infection with organisms we are usually resistant to [Pneumocystis, Toxoplasma, Herpes, Candida (thrush)].
4. Death (no cure, little effective treatment).

A SHORT HISTORY OF RETROVIRUSES AND AIDS

1910   Rous discovers avian sarcoma virus which causes tumours in chickens.

1930   Other tumour-causing or 'oncogenic' (Greek oncos; a lump) viruses discovered.

1979   Gallo isolates a retrovirus from lymphocytes of a leukaemia patient.

1981   A new disease reported in male homosexuals — a profound immunodeficiency — called acquired immunodeficiency syndrome, or AIDS.

1983   A virus — human immunodeficiency virus or HIV — isolated. Does not cause tumours but infection leads to the destruction of cells of the immune system (T-helper cells). This leads eventually to death from infections.

LIFE CYCLE OF A RETROVIRUS

The best known retrovirus, HIV, is thought to be responsible for AIDS. The provirus may remain inactive once incorporated into the host DNA (one reason why HIV-positive individuals might not develop full-blown AIDS immediately), and some have been shown to act as molecular 'switches', turning on the activity of genes called ONCOGENES, the products of which may be responsible for some forms of cancer.

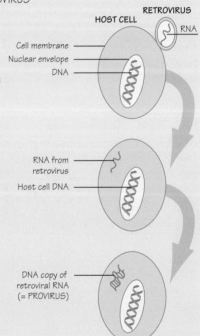

Electron micrograph of lymphocyte showing HIV particles budding from the cell membrane. Magnification × 12,000.

Chapter 7 • **Molecular Biology and Applied Biochemistry**

Figure 7.18 Progression of cancer. (a) The normal tissue, here uterus, has a layer of epithelium, the cells of which are constantly dividing in the basal layer only at a rate that matches their loss from the surface. (b) In dysplasia the differentiation of the cells is incomplete and cells several layers above the basal layer are still dividing. (c) In malignancy there is little differentiation and cells are crossing the basal lamina and invading the underlying connective tissue. Several years may elapse between dysplasia and the fully malignant stage. [Modified with permission from B.A. Alberts et al. (1994), Molecular Biology of the Cell, 3rd edn., Garland, New York (1994).]

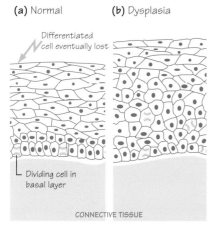

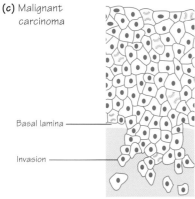

What does biochemistry say about cancer?

From what was said previously it is clear that some cancers are caused by viruses, and we have a fairly good idea of how this can happen at the molecular level. Cancer can also be caused by carcinogens in the environment by different mechanisms. Cancer is a group of diseases: it does not have a single cause. A knowledge of biochemistry and molecular biology can help us to understand the mechanisms by which tumours originate and such knowledge may eventually lead to cures. It is likely that most tumours originate from a change in a single cell, and that mutations — several of them — result in a change in the processes that control cell division.

We can make a rough calculation as follows. In the course of a lifetime probably 10^{16} cell divisions take place in a human body, and, even in an environment that is free of mutagens, mutations occur spontaneously with a frequency of about 10^{-6} mutations per gene per cell division. Consequently, in a lifetime, every single gene is likely to have undergone mutation on about 10^{10} separate occasions. Amongst the resulting mutant cells, one might expect that there would be quite a lot that disturb the genes involved in the regulation of cell division. On this basis, it is puzzling why cancer occurs so infrequently, relatively speaking. The answer must be that for a cancerous cell to be produced requires several independent rare events to occur together in one cell. Data for populations show that cancer incidence increases as a function of age and this must mean that a cell must accumulate several disruptive mutations before it turns into a tumour cell (Figure 7.17, page 213). Increased knowledge of how the proteins that control the cell cycle function will give us more indication of how cancer cells develop and how they may be controlled.

Finally, we may reflect that the normal multicellular organism is a rather peculiar society: self sacrifice rather than competition is the rule. This is in contrast to free-living organisms such as bacteria which constantly compete with one another. A small advantage will win the competition. In the multicellular organism any mutation that gives rise to selfish behaviour will jeopardize the whole organism's survival. Therefore the characteristic feature of cells in a multicellular organism is *control*. Cells must not divide uncontrollably and they must stay within their tissues rather than wander off to colonize new locations in the body (Figure 7.18). The mechanisms by which cell division and migration are controlled are gradually becoming understood at the molecular level; and, of course, this is the aim of biochemistry — to understand how life works at the molecular level.

7.5 Further reading

Aldridge, S. (1996) The thread of life. The story of genes and genetic engineering. Cambridge University Press.

Brown, T. (1992) Cloning genes. Biological Sciences Review **5(1)**, 17–20

Capecchi, M.R. (1994) Targeted gene replacement. Scientific American **270 (Mar)**, 34–41

Chilton, M.D. (1983) A vector for introducing new genes into plants. Scientific American **248 (June)**, 50–59

Darby, S. and Roman, E. (1995) Leukaemia and ionizing radiation. Biological Sciences Review **7(3)**, 11–13

Denning, D.W. (1995) HIV and AIDS: an update for 1995. Biological Sciences Review **7(4)**, 30–33

Evan, L. (1993) The Human Genome Project. Biological Sciences Review **5 (5)**, 2–4

Greene, W.C. (1993) AIDS and the immune system. Scientific American **269 (Sept)**, 66–73

Gull, D. (1989) Gene cloning for beginners. Biological Sciences Review **2(5)**, 25–28

Jain, R.K. (1994) Barriers to drug delivery in solid tumours. Scientific American **271 (July)**, 42–49

Kumar, A. and Forster, B. (1995) Biotechnology of the potato. Biological Sciences Review **7(5)**, 31–33

Lomonossoff, G. (1995) Vaccine production in plants. Biological Sciences Review **8(1)**, 31–33

MacDonald, C. (1995) Immortal cells and cancer. Biological Sciences Review **7(4)**, 6–7

Martin, P. (1995) Genetic profiling in forensic science. Biological Sciences Review **7(5)**, 38–40

Pickering, C. and Beringer, J. (1995) Modern genetics. Biological Sciences Review **6(5)**, 34–37

Prosser, J. (1994) Tracking genetically engineered organisms. Biological Sciences Review **6(4)**, 30–32

Read, A. (1993) Gene therapy. Biological Sciences Review **6(1)**, 22–24

Sykes, B. (1992) Cloning in a test tube — the polymerase chain reaction. Biological Sciences Review **4(3)**, 23–24

Whitelaw, B. (1995) Pharmaceuticals from transgenic sheep. Biological Sciences Review **7(3)**, 25–26

7.6 Examination questions

1. Briefly explain what is meant by the following terms:
   (a) recombinant DNA;
   (b) reverse transcriptase;
   (c) an endonuclease enzyme;
   (d) genetic fingerprinting. (8 marks)

   [Oxford Local Biology June 1991, Paper 2, No. 9]

2. Read the passage below and answer the questions which follow.
   "Micropropagation is a rapid method for cloning desirable plants. Tissue explants from the parent are sterilized and transferred to a multiplication medium. After a callus has formed it is divided into individual segments which are transferred to a shoot elongation medium, illuminated and allowed to grow. These are further subdivided and transferred to a rooting medium where they remain until roots have developed."
   (a) Explain the meaning of the following terms: cloning; callus. (4 marks)
   (b) Name one part of the parent plant that could be used to provide the explant. (1 mark)
   (c) Give the name of a solution which could be used to sterilize the explants before their transfer to growth medium. (1 mark)
   (d) Which plant growth substances would be added to the following media in the highest concentration: multiplication medium; rooting medium? (2 marks)

   [O & C (Biology) 1991, Paper 2.8, No. 1]

3. Give an account of the production and uses of genetically modified organisms. (10 marks)

   [ULEAC (Biology) June 1995, Module 1, No. 8]

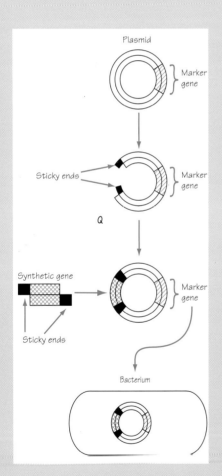

4. In some genetic engineering processes a synthetic gene is inserted into a bacterial host. This is shown in the diagram opposite.
   (a) (i) What term is used to describe the function of the plasmid in this process? (1 mark)
       (ii) Name the type of enzyme used at Q to cleave (cut) the DNA. (1 mark)
       (iii) Explain the role of the 'sticky ends' on the synthetic gene and the plasmid. (3 marks)
   (b) In this technique, a synthetic gene has been used. Suggest one method of making a synthetic gene. (2 marks)
   (c) Genetically engineered human insulin is now widely used in the treatment of diabetes. What are the advantages of using genetic engineering techniques in the production of human insulin? (3 marks)

   [ULEAC (Biology) January 1993, Paper 4A, No. 1]

Index

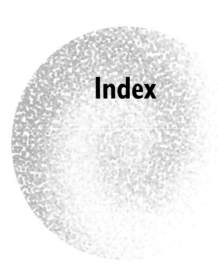

Page references to *Tables are in italic* and those to **Figures are in bold**.

Absorption, **103**
Accessory pigment, 103
Acetyl-CoA, 116, 117, 120, 139
Acetyl-coenzyme A, see Acetyl-CoA
Aconitase, 121
Actin, 30, 31, *37*, **149**, 153
Action spectrum, 103
Activation energy, 76
Active site, see also Enzyme, 72, 80–83
Active transport, 22, 23, 154, 156
Acquired immunodeficiency syndrome, see AIDS
Acyclovir, 214
ADA, 208, 209
Adenine, 56, 57, **61**
Adenosine deaminase, see ADA
Adenosine diphosphate, see ADP
Adenosine triphosphate, see ATP
Adenovirus, 208, **211**
ADP, 99, 101
Affinity constant (K_a), 156
Agrobacterium tumefaciens, 200
AIDS, 209, 210, 212, 214, 215
AIDS virus, see HIV
Alcohol dehydrogenase, 123
Aldolase, 125
Aldose sugar, 47
Allosteric enzyme, see also Enzyme, 91
Allosteric inhibitor, **91**
Amino acid, 37, 38, *39*
Ammonia, 13, 127, 130
Ammonium, 146, 147
Ammonotelic organism, *127*
Amniocentesis, 206, 207
Amoeba, 21, 149
Amphipathic molecule, 19
Amylopectin, 52, 53
Anabolism, 6, 102, 136
Animal cell, 30, 31
Annealing, 192
Anticodon, 180, 181
Antifreeze, 92
Antimycin A, 113
Antiparallel DNA strand, **56**, 162
Antiport system, 22, 23
Antiporter, 156
Apoptosis, 171
Applied biochemistry, 189
Archaebacteria, 26
Arginine, **130**
Argininosuccinate, 130
Aspartate, 130

ATP, 99, 100, **101**
  as energy currency, **136**
  generation, 101
  in heterotrophs, 110, 114, 115
  production in mitochondria, 114
  production during photosynthesis, 105
ATPase, 109, 114
ATP synthetase, **119**
Autoradiography, **195**
Autotroph, **10**, 11
Autotrophic nutrition, 136, 138, 139

Bacteria, 28
Bacterial photosynthesis, 109, 110
Bacteriochlorophyll, 28, 110
Bacteriophage, **170**
Base pair, **56**, **57**, 165
Beaumont, 71
Beri-Beri, 121
Berzelius, 71
Bilayer, 66
Biochemistry, 2
Biogeochemical cycle, 11
Bioluminescence, 101
Biomass, **9**
Biosphere, 1, 14
Biosynthesis, 6, 136, 137
  of fats, 150, 151
  of polymers, 148
  of polysaccharides, 150, 151
Biotechnology, 189
Blue-green algae, see Cyanobacteria
Bonding in biological systems, 4,5
Brush border, 30, 31
Buffer solution, 25

C3 pathway, 141
C4 pathway, 141
Caenorhabditis elegans, 205
Calcereous rock, 12
Calvin cycle, 138, 139, 140, 142, 143
Calvin's 'lollipop' experiment, **141**
Cancer, **213**, 216
Capsid, 215
Capsomere, 215
Carbohydrate, 47–49, 52, 53, 117, 122
Carbomyl phosphate, 130
Carbon, 3
  cycle, **12**
  fixation, 11
Carbon dioxide fixation, 141, 142, 143, **144**
Carboxypeptidase, **82–84**
Cardiac muscle, **149**

Carotene, **103**
Carotenoid, **105**
Catabolism, 6, 102
Catalysis, 76
Catalyst, 72
Cech, 72
Cell, 18
    structure, 18
    wall, 28–31
Cell adhesion protein, 22, 23
Cell biology, 2
Cellular building block, 6
Cellulose, 53, **54**
Centrifugation, 25
Centriole, 30, 31
Chemical diversity, 86
Chemiosmotic theory, see Mitchell's chemiosmotic theory
Chitin, 54
Chlorophyll, 103
Chloroplast, 19, **32**
    envelope, 106, 107
    structure, 106, 107
Cholera toxin, 37, 157
Cholesterol, 22, 23
Chorionic villus sampling, 206
Chromatin, 168, 169
Chromosome, 163, **169**
Chymotrypsin, 81
Cilium, 26, **145**, 148
Citrate, 117
Citrate synthase, 121
Citric acid cycle, see Krebs tricarboxylic acid cycle
Citrulline, **130**
Clathrin, 22
Cloning, **196**
Coated pit, 22
Codon, 167, 171, 176, 180, 181
Coenzyme, 72, 85, *86*
    coenzyme A, 116
    coenzyme Q, 112, 118
    in oxidative phosphorylation, *112*
    in photosynthesis, *112*
    reduced, 118
Cofactor, 42, 82, 83, 85
Cohesion, 16, 17
Collagen, *37*, 41, 182
Colloid, 16, 17
Competitive inhibition, 90, 92, 93
Complementary DNA, 162
Condensation reaction, 36, 150, 151
Condensing enzyme, see Citrate synthase
Coupled reaction, 102
Covalent bond, 3, 4, 5

Crassulacean acid mechanism (CAM), 141, **144**
Creatine phosphate, **155**
Cristae, 27
Cyanide, 113
Cyanobacteria, 21, 26
Cyclic electron flow, 109
Cytochrome, 43, 112
    cytochrome b/f, *112*, **107**
    cytochrome c, **46**
    cytochrome c oxidase, *112*, 119
Cytoplasmic streaming, 149
Cytosine, 56, 57, **61**
Cytoskeleton, 30
Cytosol, 32

Dark phase of photosynthesis, 137
dATP, 163
Deamination, 127
Decomposer, 9, 11, 13
Degradative metabolism, 6
Dehydrogenation, 128
Denaturation, **42**
Denitrification, 13
Deoxyadenosine triphosphate, see dATP
Deoxyribonucleic acid, see DNA
Diagnosis, 202
Dicotyledon, 200, 201
Dictyosome, see Golgi apparatus
Dideoxy technique, 194, 195
Dideoxynucleoside triphosphate, 195
Diffusion, 154
2,4-Dinitrophenol, 113
Disaccharide, 49, 51–53, 150, 151
Disulphide link, 4, **42**
DNA, 55–57, **59**
    annealing, 192
    cloning, 196
    detection, 192
    mitochondrial, 115
    packaging, 168, 169
    sequencing, 193–**195**
    structure, **163**, 166, 195
    replication, 161–165
    template, 177
DNA helicase, 164
DNA ligase, 164, 165
DNA polymerase, 163–165, 194, 195
Double helix, 58, **161**
Dynamic recognition, 83
Dynein, **145**

EC number, 73, 74
Ecosystem, 1, 8, 9
Electrical discharge apparatus, **1**

Electron
    carrier, **49**
    transport chain, 118, 119, 128, 129
    transport system, 111, 114, 115
Electron microscopy, 21, **24**
Electrophoresis, 174, 175, **191**, 194, 195
Electrostatic bond, 4, 5
Endocytic vesicle, 30, 31
Endocytosis, 20, **21**
Endonuclease, 191
Endoplasmic reticulum, 19, **26**, 27
Energy
    from carbohydrates, 122
    currency, 136
    from fats, 126
    from proteins, 127
Energy flow, **11**
    in biosphere, **14**
    through ecosystem, 8, 9
Enolase, 124, 125
Enzyme activity, see Enzyme
Enzyme specificity, see Enzyme
Enzyme, 71
    active site of, 80–83
    activity, 78, 79
    allosteric, 91
    catalysis, 71
    classification, 73–75
    effect of pH, 86, 87, **88**, 89
    effect of temperature, 86, 87, 88, **89**
    inhibition, 87
    kinetics, 90
    and rate enhancement, 77
    specificity, 84, 85
    substrate binding to, 82, 83
Equilibrium, 73
Erythrocyte, see Red blood cell
Essential amino acid, 136
Essential fatty acid, 136
Ethanol, 123
Ethical issues, 205
Ethylene glycol, 92
Euchromatin, 168
Eukaryote, 21, 26
Excretion, 10
Exocytosis, 20, **21**
Exon, 178

Facilitated diffusion, 155
FAD, 86, 92, *112*
$FADH_2$, 112, 114, 118, 129
Fat, 62, 64, 65
    biosynthesis, 150, 151
    obtaining energy from, 117, 126
Fat body, 138, 139

Fatty acid, 63, 128, 129, 150, 151
    melting point, 66
    oxidation of, **128**, **129**
Fatty acid synthase, 151
Feedback inhibition, 91
Fermentation, 71, 123
Ferredoxin, 13, 105, 107, 109, *112*
Fibrous protein, 41
Fimbria, see Pilus
Fischer projection formula, **51**
Flagellum, 26, 28, 29, 137, **140, 145**, 148, *149*
Flavin adenine dinucleotide, see FAD
Flavin adenine dinucleotide, reduced form, see $FADH_2$
Fluid-mosaic model, 20
Fluorescence, 103
Fluorocitrate, 121
Fossil fuel, 12
Frame-shift mutation, 176
Free ribosome, 30, 31
Fructose, 48, 49
Fructose 1,6-bisphosphate, 125
Fructose 6-phosphate, 125
Fumarate, 121, 130, 139
Furanose ring, 52, 53

β-**G**alactosidase, 172
Gated channel, 22
Gene, 170
    expression, 172, 198, 199
    knockout, 199
    therapy, 190, 207–209
Genetic code, 166, 167, **176**
Genetic disease, 202
Genetic engineering, 171, **201**
Germline cell, 208, 209
Globular protein, 82
Glucose, 48, 49, 82, 83, 122, 124, 125
    energy from, 102
    structure of, **52**, 53
Glucose 6-phosphate, 125
Glutamate, 130, 147
Glutamate cycle, 145
Glutamate dehydrogenase, 130, 147
Glyceraldehyde 3-phosphate, 138
Glycerate 3-phosphate, 138
Glycerol, **64**, 128, 129
Glycocalyx, 20, 22, 23, 48, 49
Glycogen, 49, 51, 53, 124, 125
    biosynthesis, 150, 151
Glycolipid, 48, 49
Glycolysis, 122, **123**, 124, **125**, 126
    relationship to TCA cycle, **117**
Glycoprotein, 48, 49
Glycosaminoglycan, 48, 49, 55

Glycosidic bond, **52**, **53**
Glyoxylate cycle, 138, 139
Glyoxysome, 138, **139**
Golgi apparatus, 19, **27**, 30, 31
Gram-positive, 28, 29
Gram-negative, 28, 29
Griffith's experiment, **167**
Guanine, 56, 57, **61**

Haemoglobin, 37, **39**, **43**
    oxygen binding curve of, **177**
    in sickle cell disease, 174
Haworth projection, **51**
α-Helix, 41, **43**
Heparin, **55**
Herpes simplex, 214
Herrick, 174
Hershey and Chase experiment, 166, **170**
Heterochromatin, 168
Heterotroph, **10**, 11, 110
Heterotrophic nutrition, 136
Hexokinase, 37, 124, 125
Hexose, 49
Hexose phosphate, 138
Histone, 168
HIV, 208, 209, 212, **214**, 215
Hormone, 64, 65
HUGO, see Human genome project
Human genome project, 193–195, 204, 205
Human immunodeficiency virus, see HIV, see also AIDS
Hydrocarbon chain, 65
Hydrogen, 3
Hydrogen bond, 4, 5, 14, 17
Hydrolase, 74, 75
Hydrolysis, 36
Hydrophilic, 15–17
Hydrophilic membrane, 155
Hydrophobic, 4, 5, 15–17
Hydrophobic membrane, 155

Immunoglobulin, *37*
Induced fit hypothesis, 82, 83, 92, 93
Induced fit model, 81
Induction, 172, 173
Ingram, 174
Inhibitor,
    of enzyme activity, 87, 92, 93
    of oxidative phosphorylation, 113
    of TCA cycle, 120, 121
Initial rate of reaction, 77, 78, 79
Initiation codon, 170
Initiation factor, 181
Inner mitochondrial membrane, 114, 115

Inorganic chemistry, 2
Inorganic phosphate (P_i), 102
Insecticide, 93
Insulin, *37*, 194, 199
Intron, 178
Ionic bond, 4, 5
Iron–sulphur protein, 112
Irreversible inhibitor, 92, 93
Irritability, 10
Isocitrate, 121, 139
Isocitrate dehydrogenase, 121
Isoelectric point, 44, 45
Isomerase, 43, 74, 75
Isomerism, 50

Jacob, 172

Katal, 77
Keratin, *37*, 41
α-Ketoglutarate, see α-Oxoglutarate
K_m, see Michaelis constant
Kornberg, 190
Krebs tricarboxylic acid cycle, 113, 117, 120, 121
    in plant cells, 138, 139
Kuhne, 71

*L*ac repressor, **172**
Lactate dehydrogenase, 73, 82, 83, 123
Lactose, 48, 49, 172, 173
Lactose transport system, **157**
Leader sequence, 179
Leucocyte, 30, 31
Ligase, 74, 75
Light reaction of photosynthesis, 103, 104
Light-dependent reaction, 106, 107
Lignin, 139
Lineweaver–Burk equation, 78, 79
Lipase, 129
Lipid, 62, **64**, **65**
Lipid bilayer, 20
Liposome, 208, 209
Lock and key hypothesis, 81
Locomotion, 10, 148
Long-chain fatty acid, **63**, **65**
Lyase, 74, 75
Lymphocyte, 215
Lysosome, 19, 30, 31
Lysozyme, 28

Macromolecule, 3, 35
Malaria, 174, 175
Malonate, 120
Maltose, 48, 49

Index 221

Maximum velocity (V_{max}), 77–79
Membrane, 18, **19, 20**, 21–23
    asymmetry, 20
    potential, 156
    protein, **155**
    transport, 154
Meselson and Stahl experiment, **162**
Mesosome, 28, 29
Messenger RNA, 60–62, 167, 177, 180, 181
Metabolic energy, 136
Metabolic pathway, 90
Metabolism, 90
Metabolite, 67
Metachronal rhythm, 148
Metastasis, 212
Micelle, **63**
Michaelis constant (K_m), 78, 79
Michaelis–Menten equation, 78, 79
Microfilament, 30, 31
Microtubule, 30, 31, **145**
Microvillus, 30, 31
Mitchell's chemiosmotic theory, 112, 113, 118, 119
Mitochondria, 19, **27**, 110–112, 114, 115
Mitochondrial DNA, 115
Molecular biology, 2, 58, 189
Molecular motor, 152
Mollicute, 26
Monocotyledon, 200, 201
Monod, 171
Monomer, 35, **36**
Monosaccharide, 49, 50, 52, 53
Motif, 42, 178
mRNA, see Messenger RNA
Mullis, 197
Muscle, 149, 152–154
    ATP store, 153
    cardiac, **149**, *152*
    contraction, **152**
    fibre, 149
    smooth, **149**, *152*
    striated, **149**, *152*
    structure, **152**
Mutation, 40, 163, 171, 174, 175, 178
Myelin, 64
Myofibril, 149
Myosin, 152, **153**

NAD, 85, 112
NADH, 85, 112, 129
NADPH, 104, **107**
Negative feedback, 91
Nerve gas, 93
Nexin, **145**

Niacin, 82
Nicotinamide–adenine dinucleotide, see NAD
Nicotinamide–adenine dinucleotide phosphate, see NADPH
Nitrate, 13, 147
Nitrate reductase, 13, 147
Nitrification, 13, 146
Nitrite reductase, 13, 147
Nitrogen, 3, *127, 131*
Nitrogen cycle, **12**, 146, **147**
Nitrogen fixation, 12, 137, 144, 146, 147
Nitrogen-fixing bacteria, 13, 147
Nitrogenase, 147
Non-competitive inhibition, 90, 92, 93
Non-cyclic electron flow, 109
Northern blot, 193
Nucleic acid, 55–57
Nucleoside, 101
Nucleosome, 168, 169
Nucleus, 19, **26**, 27, 30, 31
Nutrition, 10

Oil, **15**, 62, 64, 65
Okazaki fragment, 165
Oligosaccharide, 48, 49
Oncogene, 213, 215
Operator gene, 172, 173
Optical activity, 50
Optical isomer, **50**
Organelle, 19, 26, 30, 31
Organic chemistry, 2
Organic molecule, **3**
Origin of life, 1, 2
Ornithine, **130**
Oxaloacetate, 117, 120, 130, 139
Oxidation of fatty acids, 128, 129
β-Oxidation pathway, 117, 128, 129
Oxidative decarboxylation, 123
Oxidative phosphorylation, 114, 115
    fatty acids, 128, 129
    inhibitors and uncouplers, *113*
    mitochondrial, 111
    TCA cycle, 120, 121
Oxidoreductase, 74, 75
α-Oxoglutarate, 117, 130, 139
α-Oxoglutarate dehydrogenase, 121
Oxygen, 3
    binding curve of haemoglobin, **177**
    debt, 123

Packaging DNA, 168, 169
Palmitate, **129**
Paper chromatography, 174, 175
Paramecium, 148

Passive diffusion, 154
Pauling, 41, 174
PCR, **197**, 206
Penicillin, 28
Pentose, 49
Pentose phosphate pathway, 108
Pepsin, *37*, 84
Peptide bond, 4, 37, **38**
Permease, 155
Peroxisome, 30, 31
pH, 16, 17
    effect on enzymes, 86, 87, **88**, 89
Phosphatide, 64, 65
3′,5′-Phosphodiester bridge, 4, 57
Phosphoenolpyruvate, 125
Phosphofructokinase, 125
Phosphoglucose-isomerase, 125
Phosphoglyceraldehyde, 117, 125
Phosphoglyceraldehyde dehydrogenase, 125
Phosphoglycerate, 125
Phosphoglycerokinase, 125
Phosphoglyceromutase, 125
Phospholipid, 19, 64–66, 150, 151
Phospholipid bilayer, 22, 23
Photosynthesis, 8, 102
    ATP production during, 105
    bacterial, 109
    dark phase, 137
    light reaction, 103, 104
Photosystem I, 104, 106, 107
Photosystem II, 104, 106, 107
Phycocyanobilin, **103**
Pilus, 28, 29
Plant cell, **18**
    metabolism, 138, 139
    structure, 30, 31
Plasmalemma, see Plasma membrane
Plasma membrane, 28, 30, **31**
Plasmid, 28, 29, **193**
Plasmodesmata, 30
Plastoquinone, 109, 112
β-Pleated sheet, 41, **43**
Pneumococcus, 166
Point mutation, 176
Polarity, 36
Polymer, 148
Polymerase chain reaction, see PCR
Polynucleotide, 56, 57
Polypeptide, 38
Polyribosome, 60, 61, **180, 181**
Polysaccharide, 49, 51–53, 150, 151
Polysome, 60, 61, 180, 181
Pore protein, 22
Post-translational modification, 182, **183**
Prenatal diagnosis, 206, 207

Primary consumer, 8
Primary producer, 8
Primary protein structure, 38, 44, **45**
Primase, 165
Primer, 197
Producer, 11
Progress curve, 77–**79**
Prokaryote, 21, 26, **28**, **29**
Promoter, 170, 172, 173
Prosthetic group, 42, 82, 83, 85
Protease, *84*
Protein, 36, *37*, 44, 45
  binding specificity, 46
  biosynthesis, 177
  conformation, 45
  energy from, 127
  engineering, 177
  motif, 42, 178
  structure, 40
  synthesis, 180, 181
  TCA cycle, 117
Proteoglycan, 55
Proton gradient, 118, 119, 105
Proto-oncogene, 213
Protoplasmic streaming, 149
Purine, 56, 57
Putrefaction, 13, 146
Pyranose ring, 52, 53
Pyrimidine, 56, 57, 61
Pyrophosphate, 100, 101
Pyruvate, 117, 121, 123, 125, 139
Pyruvate kinase, 125

Q$_{10}$, 88
Quaternary structure, 42, 44, 45
Quinone, *112*

Reaction, initial rate of, 78, 79
Reaction centre, 104
Recombinant DNA, 191
Red blood cell, **19**, 176
Reduced nicotinamide–adenine
    dinucleotide, *see* NADH
Reducing sugar, **52**
Regulation of gene activity, 172–173
Relative molecular mass (M_r), 35
Replication of DNA, 62, 161, **164**, **165**
Repressed protein, 170
Repressor protein, **172**, **173**
Reproduction, 10
Respiration, 10, 120, 121
Restriction enzyme, 190–*192*
Retinol, *see* Vitamin A
Retrovirus, 212, 214, 215
Reverse transcriptase, 211, 212, 214, 215

Rhodopsin, 23
Ribonucleic acid, *see* RNA
Ribosomal RNA, 59–61, 179, 180, 181
Ribosome, 27, 60, 61, 179, 180, 181
  eukaryotic, 61
  prokaryotic, 28, 29, 61
Ribozyme, 72
Ribulose 1,5-bisphosphate, **49**, 106, 142, 143
Ribulose 5-phosphate, 138
RNA, 55, 59–61, 161
  polymerase, 177, 180, 181
  processing, 178
  tumour virus, 212
Root nodule, 146
Rotenone, 113
Rough endoplasmic reticulum, 30, 31
rRNA, *see* Ribosomal RNA
Rubisco, 142, 143

Sanger, 190, 194
Saponification, 63
Sarcomere, 149, 153
Sarcoplasmic reticulum, 152
Schleiden, 18
Schwann, 18
Schwann cell, 64
SCID, 208, 209
Secondary consumer, 9
Secondary protein structure, 41, 44, **45**
Semi-conservative, 162
Sequencing, 194, 195
Severe combined immunodeficiency,
    see SCID
Sickle cell anaemia, *see* Sickle cell disease
Sickle cell disease, 40, 171, 174, 176, 203, 206
Signal mechanism, **182**
Signal recognition particle, **182**
Silk fibroin, *37*, 41
Site-directed mutagenesis, 177
Skeletal muscle, *see* Muscle
Sliding filament model, **154**
Smooth endoplasmic reticulum, 30, 31
Smooth muscle, *see* Muscle
Sodium–potassium pump, **157**
Solar flux, 8
Solubility, 44
Soluble protein, 41
Somatic cell, 208, 209
Southern blot, **192**, **193**, **196**
Sphingolipid, **65**
Splicing, **179**
Starch, **49**, 51–53, 106, 150, 151
START codon, 176, 181
Stearic acid, **65**
Stereospecificity, 82, 83

Steroid, **64**, **65**
Sticky end, 191
STOP codon, 176
Striated (skeletal) muscle, *see* Muscle
Stroma, 104, 106
Subcellular fractionation, 21, **25**
Substrate, 72
Substrate-level phosphorylation, 120,
    124, 125
Succinate, 121, 139
Succinate dehydrogenase, 119, 120
Succinyl-CoA, 121
Sucrose, 48, 49
Sucrose density gradient, **25**
Sucrose synthesis, 151
Sulphanilamide, 92
Sumner, 71
Supercoiling, 168, 169
Superoxide dismutase, **75**
Symport system, 22, 23
Symporter, 156
Syncitium, 149, 152

Taq polymerase, 197
Targeting, 179
TCA cycle, *see* Krebs tricarboxylic
    acid cycle
Temperature, effect on enzymes, 86–88, **89**
Tertiary protein structure, 42, 44, **45**
Thermal denaturation, 88, 89
Thermolysin, **75**
Thermophile, 86, 197
Thiamin pyrophosphate, 121
Thylakoid membrane, 106, **107**
Thylakoid network, 104
Thymine, 56, 57
Tonoplast, 30
Totipotency, 200
Transamination, 139, 147
Transcription, 60–62, 77, 180, 181
Transfer RNA, 60–62, 179–181
Transferase, 74, 75
Transformation of *Pneumococcus*, 166, **167**
Transgenic
  mouse, 198, **202**
  organism, 198
  plant, 199–201
Transition state, 76
Translation, 59–62, 167, 178
Transmembrane protein, 22, 23
Transmission electron microscope, **24**
Transport across membrane, 154
Triacyglycerol, 63–65, 138, 139, 150, 151
Triglyceride, *see* Triacylglycerol

Index 223

Trifluorocarbonylcyanide phenylhydrazone, 113
Triose-phosphate, 43
Triplet code, 167
tRNA, see Transfer RNA
Trypsin, **72**, **80**, 81
Tumour, 212
Turnover number, 77, *80*

Ubiquinone, see Coenzyme Q
Ultrastructure, 21
Uncoupler of oxidative phosphorylation, 113
Uniport system, 22, 23
Unity of life, 6, 7
Unsaturation, 66
Uracil, **61**
Urea, 2, 127, **130**

Urea cycle, **130**
Urease, 71
Ureotelic organism, *127*
Uric acid, 127
Uricotelic organism, *127*

Vacuole, **30**
Van der Waals interaction, 4, 5
Vector, **193**
Virchow, 18
Virus, 210, **214**, 215
Vital force, 2
Vitamin, 64, 65
   A, **64**
   B group, 82, *86*
V_{max}, see Maximum velocity

Water
   biological importance, 14
   properties, 16, 17
Watson and Crick, 58, 59, 167, *190*
Wax, **65**
Wöhler, 2

X-ray crystallography, 38, 94, 153
X-ray diffraction pattern, **40**

Yeast, 204, 205

Z scheme, 105
Zinc finger, **172**
Zwitterion, 38, 39